Contents

Fundamentals of Physics

THE OPEN YALE COURSES SERIES is designed to bring the depth and breadth of a Yale education to a wide variety of readers. Based on Yale's Open Yale Courses program (http://oyc.yale.edu), these books bring outstanding lectures by Yale faculty to the curious reader, whether student or adult. Covering a wide variety of topics across disciplines in the social sciences, physical sciences, and humanities, Open Yale Courses books offer accessible introductions at affordable prices.

The production of Open Yale Courses for the Internet was made possible by a grant from the William and Flora Hewlett Foundation.

RECENT TITLES

Paul H. Fry, *Theory of Literature*
Christine Hayes, *Introduction to the Bible*
Shelly Kagan, *Death*
Dale B. Martin, *New Testament History and Literature*
Giuseppe Mazzotta, *Reading Dante*
R. Shankar, *Fundamentals of Physics*
Ian Shapiro, *The Moral Foundations of Politics*
Steven B. Smith, *Political Philosophy*

Fundamentals of Physics

Mechanics, Relativity, and Thermodynamics

R. SHANKAR

Yale

UNIVERSITY PRESS

New Haven and London

Published with assistance from the foundation established in memory of
Amasa Stone Mather of the Class of 1907, Yale College.

Yale University Press books may be purchased in quantity for educational,
business, or promotional use. For information, please e-mail sales.press@
yale.edu (U.S. office) or sales@yaleup.co.uk (U.K. office).

Set in Minion type by Newgen North America.
Printed in the United States of America.

ISBN: 978-0-300-19220-9
Library of Congress Control Number: 2013947491

A catalogue record for this book is available from the British Library.

This paper meets the requirements of ANSI/NISO Z39.48-1992
(Permanence of Paper).

10 9 8 7 6 5 4 3 2

To my students
for their friendship and inspiration

Deep and original, but also humble and generous, the physicist
Josiah Willard Gibbs spent much of his life at Yale University. His father was
a professor of sacred languages at Yale, and Gibbs received his bachelor's
and doctorate degrees from the university before teaching there until his death in
1903. The sculptor Lee Lawrie created the memorial bronze tablet pictured
above, which was installed in Yale's Sloane Physics Laboratory in 1912. It now
resides in the entrance to the J. W. Gibbs Laboratories, Yale University.

Preface

Given that the size of textbooks has nearly tripled during my own career, without a corresponding increase in the cranial dimensions of my students, I have always found it necessary, like my colleagues elsewhere, to cull the essentials into a manageable size. I did that in the course Fundamentals of Physics I taught at Yale, and this book preserves that feature. It covers the fundamental ideas of Newtonian mechanics, relativity, fluids, waves, oscillations, and thermodynamics without compromise. It requires only the basic notions of differentiation and integration, which I often review as part of the lectures. It is aimed at college students in physics, chemistry, and engineering as well as advanced high school students and independent self-taught learners at various stages in life, in various careers.

The chapters in the book more or less follow my 24 lectures, with a few minor modifications. The style preserves the classroom atmosphere. Often I introduce the questions asked by the students or the answers they give when I believe they will be of value to the reader. The simple figures serve to communicate the point without driving up the price. The equations have been typeset and are a lot easier to read than in the videos. The problem sets and exams, without which one cannot learn or be sure one has learned the physics, may be found along with their solutions at the Yale website, http://oyc.yale.edu/physics, free and open to all. The lectures may also be found at venues such as YouTube, iTunes (https://itunes.apple.com/us/itunes-u/physics-video/id341651848?mt=10), and Academic Earth, to name a few.

The book, along with the material available at the Yale website, may be used as a stand-alone resource for a course or self-study, though some instructors may prescribe it as a supplement to another one adapted for the class, so as to provide a wider choice of problems or more worked examples.

To my online viewers I say, "You have seen the movie; now read the book!" The advantage of having the printed version is that you can read it during take-off and landing.

In the lectures I sometimes refer to my *Basic Training in Mathematics*, published by Springer and intended for anyone who wants to master the undergraduate mathematics needed for the physical sciences.

This book owes its existence to many people. It all began when Peter Salovey, now President, then Dean of Yale College, asked me if I minded having cameras in my Physics 200 lectures to make them part of the first batch of Open Yale Courses, funded by the Hewlett Foundation. Since my answer was that I had yet to meet a camera I did not like, the taping began. The key person hereafter was Diana E. E. Kleiner, Dunham Professor, History of Art and Classics, who encouraged and guided me in many ways. She was also the one who persuaded me to write this book. Initially reluctant, I soon found myself thoroughly enjoying proselytizing my favorite subject in this new format. At Yale Universtity Press, Joe Calamia was my friend, philosopher, and guide. Liz Casey did some very skilled editing. Besides correcting errors in style (such as a long sentence that began in first person past tense and ended in third person future tense) and matters of grammar and punctuation (which I sprinkle pretty much randomly), she also made sure my intent was clear in every sentence.

Barry Bradlyn and Alexey Shkarin were two graduate students and Qiwei Claire Xue and Dennis Mou were two undergraduates who proofread earlier versions.

My family, from my wife, Uma, down to little Stella, have encouraged me in various ways.

I take this opportunity to acknowledge my debt to the students at Yale who, over nearly four decades, have been the reason I jump out of bed on two or three days a week. I am grateful for their friendship and curiosity. In recent years, they were often non-majors, willing to be persuaded that physics was a fascinating subject. This I never got tired of doing, thanks to the nature of the subject and the students.

CHAPTER 1

The Structure of Mechanics

1.1 Introduction and some useful tips

This book is based on the first half of a year-long course that introduces you to all the major ideas in physics, starting from Galileo and Newton, right up to the big revolutions of the twentieth century: relativity and quantum mechanics. The target audience for this course and book is really very broad. In fact, I have always been surprised by the breadth of interests of my students. I don't know what you are going to do later in life, so I have picked the topics that all of us in physics find fascinating. Some may not be useful, but you just don't know. Some of you are probably going to be doctors, and you don't know why I'm going to cover special relativity or quantum mechanics. Well, if you're a doctor and you have a patient who's running away from you at the speed of light, you'll know what to do. Or, if you're a pediatrician, you will understand why your patient will not sit still: the laws of quantum mechanics don't allow a very small object to have a definite position and momentum. Whether or not you become a physicist, you should certainly learn about these great strides in the human attempt to understand the physical world.

Most textbooks are about 1,200 pages long, but when I learned physics they were around 400 pages long. When I look around, I don't see any student whose head is three times as big as mine, so I know that you cannot digest everything the books have. I take what I think are the really essential parts and cover them in these lectures. So you need the lectures to find out what's in the syllabus and what's not. If you don't do that, there's a danger you will learn something you don't have to, and we don't want that, right?

To learn physics well, you have to do the problems. If you watch me online doing things on the blackboard or working through derivations in the book, it all looks very reasonable. It looks like you can do it yourself and that you understand what is going on, but the only way you're going to find out is by actually doing problems. A fair number are available, with their solutions, at http://oyc.yale.edu/physics/phys-200. You don't have to do them by yourself. That's not how physics is done. I am now writing a paper with two other people. My experimental colleagues write papers with four hundred or even a thousand other people when engaged in the big collider experiments like the ones in Geneva or Fermilab. It's perfectly okay to be part of a collaboration, but you have to make sure that you're pulling your weight, that everybody makes contributions to finding the solution and understands it.

This calculus-based course assumes you know the rudiments of differential and integral calculus, such as functions, derivatives, derivatives of elementary functions, elementary integrals, changing variables in integrals, and so on. Sometime later, I will deal with functions of more than one variable, which I will briefly introduce to you, because that is not a prerequisite. You have to know your trigonometry, to know what's a sine and what's a cosine and some simple identities. You cannot say, "I will look it up." Your birthday and social security number are things you look up; trigonometric functions and identities are what you know all the time.

1.2   Kinematics and dynamics

We are going to be studying Newtonian mechanics. Standing on the shoulders of his predecessors, notably Galileo, Isaac Newton placed us on the road to understanding all the mechanical phenomena for centuries until the laws of electromagnetism were discovered, culminating in Maxwell's equations. Our concern here is mechanics, which is the motion of billiard balls and trucks and marbles and whatnot. You will find out that the laws of physics for this entire semester can be written down on the back of an envelope. A central purpose of this course is to show you repeatedly that starting with those few laws, you can deduce everything. I would encourage you to think the way physicists do, even if you don't plan to be a physicist. The easiest way to master this subject is to follow the reasoning I give you. That way, you don't have to store too many things in your head. Early on, when there are four or five formulas, you can memorize all of them and you can try every one of them until something works, but, after a couple of weeks, you will have hundreds of formulas, and you cannot memorize all of them. You cannot resort to trial and error. You have to know the logic.

The goal of physics is to predict the future given the present. We will pick some part of the universe that we want to study and call it "the system," and we will ask, "What information do we need to know about that system at the initial time, like right now, in order to be able to predict its future evolution?" If I throw a piece of candy at you and you catch it, that's an example of Newtonian mechanics at work. What did I do? I threw a piece of candy from my hand, and the initial conditions are where I released it and with what velocity. That's what you see with your eyes. You know it's going to go up, it's going to follow some kind of parabola, and your hands get to the right place at the right time to receive it. That is an example of Newtonian mechanics at work, and your brain performed the necessary calculations effortlessly.

You only have to know the candy's initial location and the initial velocity. The fact that it was blue or red is not relevant. If I threw a gorilla at you, its color and mood would not matter. These are things that do not affect the physics. If a guy jumps off a tall building, we want to know when, and with what speed, he will land. We don't ask why this guy is ending it all today; that is a question for the psych department. So we don't answer everything. We ask very limited questions about inanimate objects, and we brag about how accurately we can predict the future.

The Newtonian procedure for predicting the future, given the present, has two parts, *kinematics* and *dynamics*. Kinematics is a complete description of the present. It's a list of what you have to know about a system right now. For example, if you're talking about a piece of chalk, you will want to know where it is and how fast it's moving. Dynamics then tells you why the chalk goes up, why it goes down, and so on. It comes down due to the force of gravity. In kinematics, you don't ask for the reason behind anything. You simply want to describe things the way they are, and then dynamics tells you how and why that description changes with time.

I'm going to illustrate the idea of kinematics by following my preferred approach: starting with the simplest possible example and slowly adding bells and whistles to make it more and more complicated. In the initial stages, some of you might say, "Well, I have seen this before, so maybe there is nothing new here." That may well be true. I don't know how much you have seen, but it is likely that the way you learned physics in high school is different from the way professional physicists think about it. Our priorities, and the things that we get excited about, are often different; and the problems will be more difficult.

1.3   Average and instantaneous quantities

We are going to study an object that is a mathematical point. It has no size. If you rotate it, it will look the same, unlike a potato, which will look different upon rotation. It is not enough to just say where the potato is; you have to say which way its nose is pointing. The study of such extended bodies comes later. Right now, we want to study an entity that has no spatial extent, a dot. It can move around all over space. We're going to simplify that too. We're going to take an entity that moves only along the x-axis. So you can imagine a bead with a straight wire going through it, which allows it to only slide back and forth. This is about the simplest thing. I cannot reduce the number of dimensions. I cannot make the object simpler than a mathematical point.

To describe what the point is doing, we pick an origin, call it $x = 0$, and put some markers along the x-axis to measure distance. Then we will say this guy is sitting at $x = 5$. Now, of course, we have to have units and the unit for length is going to be the meter. The unit for time will be a second. Sometimes I might not write the units, but I have earned the right to do that and you haven't. Everything has got to be in the right units. If you don't have the units, and if you say the answer is 42, then we don't know if you are right or wrong.

Back to the object. At a given instant, it's got a location. We would like to describe the object's motion by plotting a graph of space versus time. A typical graph would be something like Figure 1.1. Even though the plot is going up and down, the object is moving horizontally, back and forth along the spatial x-axis. When it is at A, it's crossing the origin from the left and going to the right. Later, at B, it is crossing back to the left. In the language of calculus, x is a function of time, $x = x(t)$, and the graph corresponds to some generic function that doesn't have a name. We will also encounter functions that do have a name, like $x(t) = t$, $x(t) = t^2$, $x(t) = \sin t$, $\cos t$, and so on.

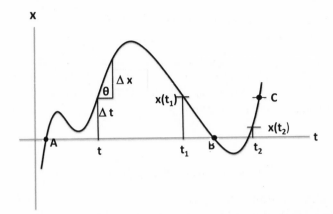

Figure 1.1   Trajectory of a particle. The position, $x(t)$, is measured vertically and the time, t, is measured horizontally.

Consider $\bar{v}$, the *average velocity* of an object, given by

$$\bar{v} = \frac{x(t_2) - x(t_1)}{t_2 - t_1} \tag{1.1}$$

where $t_2 > t_1$ are two times between which we have chosen to average the velocity. In the example in Figure 1.1, $\bar{v} < 0$ for the indicated choice of t_1 and t_2 since the final $x(t_2)$ is less than the initial $x(t_1)$.

The average velocity may not tell you the whole story. For example, if you started at $x(t_1)$ and at time t_1 ended up at point C with the same coordinate, the average velocity would be zero, which is the average you would get if the particle had never moved!

The *average acceleration*, $\bar{a}$, involves a similar difference of velocities:

$$\bar{a} = \frac{v(t_2) - v(t_1)}{t_2 - t_1}. \tag{1.2}$$

Now for an important concept, the *velocity at a given time* or *instantaneous velocity*, $v(t)$. Figure 1.1 shows some particle moving a distance Δx between times t and $t + \Delta t$. The average velocity in that interval is $\frac{\Delta x}{\Delta t}$. What you want is the velocity *at* time t. We all have an intuitive notion of velocity right now. When you're driving your car, if the needle says 60 miles per hour, that's your velocity at that instant. Though velocity seems to involve two different times in its very definition—the initial time and the final time—we want to talk about the velocity right now. That is obtained by examining the position now and the position slightly later, and taking the ratio of the change in position to the time elapsed between the two events, while bringing the two points closer and closer in time. We see in the figure that when we do this, both $\Delta x \to 0$ and $\Delta t \to 0$, but their ratio becomes the tangent of the angle θ, shown in Figure 1.1. Thus the velocity at the instant t is:

$$v(t) = \lim_{\Delta t \to 0} \frac{\Delta x}{\Delta t} = \frac{dx}{dt}. \tag{1.3}$$

Once you take one derivative, you can take any number of derivatives. The derivative of the velocity is the *acceleration*, and we write it as the second derivative of position:

$$a(t) = \frac{dv}{dt} = \frac{d^2 x}{dt^2}. \tag{1.4}$$

You are supposed to know the derivatives of simple functions like $x(t) = t^n$ ($\frac{dx}{dt} = nt^{n-1}$), as well as derivatives of sines, cosines, logarithms, and exponentials. If you don't know them, you should fix that weakness before proceeding.

1.4   Motion at constant acceleration

We are now going to focus on problems in which the acceleration $a(t)$ is just a constant denoted by a, with no time argument. This is not the most general motion, but a very relevant one. When things fall near the surface of the earth, they all have the same acceleration, $a = -9.8 \ ms^{-2} = -g$. If I tell you that a particle has a constant acceleration a, can you tell me what the position $x(t)$ is? Your job is to guess a function $x(t)$ whose second derivative is a. This is called integration, which is the opposite of differentiation. Integration is not an algorithmic process like differentiation, though it is governed by many rules that allow us to map a given problem into others with a known solution. If I give you a function, you know how to take the derivative: change the independent variable, find the change in the function, divide by the change in the independent variable, take the ratio as all changes approach zero. The opposite has to be done here. The way we do that is we guess, and such guessing has been going on for three hundred years, and we have become very good at it. The successful guesses are published as Table of Integrals. I have a copy of such a table at home, at work, and even in my car in case there is a breakdown.

So, let me guess aloud. I want to find a function that reduces to the number a when I take two derivatives. I know that each time I take a derivative, I lose a power of t. In the end, I don't want any powers of t. It's clear I have to start with a function that looks like t^2. Well, unfortunately, we know t^2 is not the right answer, because the second derivative is 2, while I want to get a. So I multiply the original guess by $\frac{1}{2}a$ and I know $x(t) = \frac{1}{2}at^2$ will have a second derivative a.

This certainly describes a particle with an acceleration a. But is this the most general answer? You all know that it is just one of many: for example, I can add to this answer some number, say 96, and the answer will still have the property that if you take two derivatives, you get the same acceleration. Now 96 is a typical constant, so I'm going to give the name c to that constant. We know from basic calculus that in finding a function with a given derivative, you can always add a constant to any one answer to get another answer. But if you only fix the *second* derivative, you can also add anything with one power of t in it, because the extra part will get wiped out when you take two derivatives. If you fixed only the third derivative of the function, you can also add something quadratic in t without changing the outcome.

So the most general expression for the position of a particle with constant acceleration a is

$$x(t) = \frac{1}{2}at^2 + bt + c \tag{1.5}$$

where b, like c, is a constant that can be anything.

Remember that $x(t)$ in the figure describes a particle going from side to side. I can also describe a particle going up and down. If I do that, I would like to call the vertical coordinate $y(t)$. You have to realize that in calculus, the symbols that you call x and y are arbitrary. If you know the second derivative of y to be a, then the answer is

$$y(t) = \frac{1}{2}at^2 + bt + c. \tag{1.6}$$

Let me go back now to Eqn. 1.5. It is true, mathematically, you can add $bt + c$ as we did, but you have to ask yourself, "What am I doing as a physicist when I add these two terms?" What am I supposed to do with b and c? What value should I pick? Simply knowing that the particle has an acceleration a is not enough to tell you where the particle will be. Take the case of a particle falling under gravity with acceleration $-g$. Then

$$y(t) = -\frac{1}{2}gt^2 + bt + c. \qquad (1.7)$$

The formula describes every object falling under gravity, and each has its own history. What's different between one object and another object is the initial height, $y(0) \equiv y_0$, and the initial velocity $v(0) \equiv v_0$. That's what these numbers b and c are going to tell us. To find c in Eqn. 1.7 put time $t = 0$ on the right and the initial height of y_0 on the left:

$$y_0 = 0 + 0 + c \qquad (1.8)$$

which tells us c is just the *initial coordinate*. Feeding this into Eqn. 1.7 we obtain

$$y(t) = -\frac{1}{2}gt^2 + bt + y_0. \qquad (1.9)$$

To use the information on the initial velocity, let us first find the velocity associated with this trajectory:

$$v(t) = \frac{dy}{dt} = -gt + b \qquad (1.10)$$

and compare both sides at $t = 0$

$$v_0 = b. \qquad (1.11)$$

Thus b is the *initial velocity*. Trading b and c for v_0 and y_0, which makes their physical significance more transparent, we now write

$$y(t) = -\frac{1}{2}gt^2 + v_0 t + y_0. \qquad (1.12)$$

Likewise for the trajectory $x(t)$ when the acceleration is some constant a, the answer with specific initial position x_0 and initial velocity v_0 is

$$x(t) = \frac{1}{2}at^2 + v_0 t + x_0. \qquad (1.13)$$

In every situation where the body has an acceleration a, the location has to have this form. So when I throw a candy and you catch it, you are mentally estimating the initial position and velocity and computing the trajectory and intercepting it with your hands. (The candy moves in three spatial dimensions, but the idea is the same.)

Now, there is one other celebrated formula that relates $v(t)$, the final velocity at some time, to the initial velocity v_0 and the distance traveled, with no reference to time. The trick is to eliminate time from Eqn. 1.13. Let us rewrite it as

$$x(t) - x_0 = \frac{1}{2}at^2 + v_0 t. \tag{1.14}$$

Upon taking the time-derivative of both sides we get

$$v(t) = at + v_0 \tag{1.15}$$

which may be solved for t:

$$t = \frac{v(t) - v_0}{a}. \tag{1.16}$$

Feeding this into Eqn. 1.14 we find

$$x(t) - x_0 = \frac{1}{2}a\left[\frac{v(t) - v_0}{a}\right]^2 + v_0\left[\frac{v(t) - v_0}{a}\right] \tag{1.17}$$

$$= \frac{v^2(t) - v_0^2}{2a} \tag{1.18}$$

which is usually written as

$$v^2 - v_0^2 = 2a(x - x_0) \tag{1.19}$$

where v and x are assumed to be the values at some common generic time t.

The Structure of Mechanics

1.5 Sample problem

We will work through one standard problem to convince ourselves that we know how to apply these formulas and predict the future given the present. Figure 1.2 shows a building of height $y_0 = 15m$.

   I am going to throw a rock with an initial velocity $v_0 = 10m/s$ from the top. Notice I am measuring y from the ground. The rock is going to go up to point T and come down as shown in Figure 1.2. You can ask me any question you want about this rock, and I can give the answer. You can ask me where it will be 9 seconds from now, how fast will it be moving 8 seconds from now, and so on. All I need are the two initial conditions y_0 and v_0 that are given. To make life simple, I will use $a = -g = -10ms^{-2}$. The position $y(t)$ is known for all future times:

$$y = 15 + 10t - 5t^2. \tag{1.20}$$

Of course, you must be a little careful when you use this result. Say you put t equal to 10,000 years. What are you going to get? You're going to find y is some huge negative number. That reasoning is flawed because you cannot use the formula once the rock hits the ground and the fundamental premise that $a = -10ms^{-2}$ becomes invalid. Now, if you had dug a hole of

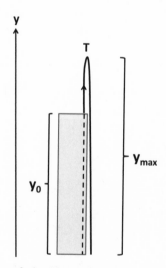

Figure 1.2   From the top of a building of height $y_0 = 15m$, I throw a rock with an initial upward velocity of $v_0 = 10m/s$. The dotted line represents the trajectory continued back to earlier times.

depth d where the rock was going to land, y could go down to $-d$. The moral is that when applying a formula, you must bear in mind the terms under which it was derived.

If you want to know the velocity at any time t, just take the derivative of Eqn. 1.20:

$$v(t) = 10 - 10t. \qquad (1.21)$$

Let me pick a few more trivial questions. What is the height y_{max} of the turning point T in the figure? Eqn. 1.20 tells you y *if you know t*, but we don't know the time t^* when it turns around. So you have to put in something else that you know, which is that the highest point occurs when it's neither going up nor coming down. So at the highest point $v(t^*) = 0$. From Eqn. 1.21

$$0 = 10 - 10t^* \quad \text{which means} \quad t^* = 1s. \qquad (1.22)$$

So we know that it will go up for one second and then turn around and come back. Now we can find y_{max}:

$$y_{max} = y(t^*) = y(1) = 15 + 10 - 5 = 20m. \qquad (1.23)$$

When does it hit the ground? That is the same as asking when $y = 0$, which is our origin. When $y = 0$,

$$0 = 15 + 10t - 5t^2. \qquad (1.24)$$

The solutions to this quadratic equation are

$$t = 3s \quad \text{or} \quad t = -1s. \qquad (1.25)$$

Why is it giving me a second solution? Can t be negative? First of all, negative times should not bother anybody; $t = 0$ is when I set the clock to zero, and I measured time forward, but yesterday would be $t = -1$ day, right? So we don't have any trouble with negative time; it is like the year 300 BC. The point is that this equation does not know that I went to a building and launched a rock or anything. What does it know? It knows that this particle had a height of $y = 15\ m$ and velocity $v = 10\ m/s$ at time $t = 0$, and it is falling under gravity with an acceleration of $-10\ ms^{-2}$. That's all it knows. If that's all it knows, then in that scenario there is no building or anything else; it continues a trajectory both forward in time and backward in time, and it says that one second before I set my clock to 0, this particle would have been on the ground. What it means is that if you had released a rock at $y = 0$ one second before I did with a certain speed that we can calculate ($v(-1) = 20m/s$ from Eqn. 1.21), your rock would have ended up at the top of the building when I began my experiment, with the same height $y = 15m$, and velocity $v_0 = 10 \ m/s$. So sometimes the extra solution is very interesting, and you should always listen to the mathematics when you get extra solutions.

When Paul Dirac was looking for the energy of a particle in relativistic quantum mechanics, he found the energy E was connected to its momentum p, mass m, and velocity of light, c, by

$$E^2 = p^2 c^2 + m^2 c^4, \tag{1.26}$$

in accord with a relation we will encounter in relativity. Now, this quadratic equation has two solutions:

$$E = \pm \sqrt{c^2 p^2 + m^2 c^4}. \tag{1.27}$$

You may be tempted to keep the plus sign because you know energy is not going to be negative. The particle's moving, it's got some energy and that's it. This is correct in classical mechanics, but in quantum mechanics the mathematicians told Dirac, "You cannot ignore the negative energy solution in quantum theory; the mathematics tells you it is there." It turns out the second solution, with negative energy, was telling us that if there are particles, then there must be anti-particles, and the negative energy particles, when properly interpreted, describe anti-particles of positive energy.

So the equations are very smart. When you find some laws in mathematical form, you have to follow the mathematical consequences; you have no choice. Here was Dirac, who was not looking for anti-particles. He was trying to describe electrons, but the theory said there are two roots to the quadratic equation and the second root is mathematically as significant as the first one. In trying to accommodate and interpret it, Dirac was led to the positron, the electron's anti-particle.

Returning to our problem, if you were only asking for the maximum height y_{max}, and not the time t^* when it got there, there is a shortcut using

$$v^2 = v_0^2 + 2a(y - y_0). \tag{1.28}$$

Using $v = 0$, $v_0 = 10m/s$ and $a = -10ms^{-2}$ we find

$$y_{max} - y_0 = 5m \tag{1.29}$$

—that is, the rock reached a maximum height of $20m$ from the ground. You can find the speed when it hits the ground ($y = 0$) using

$$v^2 = 10^2 + 2 \cdot (-10)(0 - 15) = 400 \quad \text{which means} \quad v = \pm 20m/s. \tag{1.30}$$

The root we should take for when it hits the ground is of course $-20m/s$. As mentioned earlier, the other root $+20m/s$ is the speed with which it should have been launched upward, from $y = 0$ at $t = -1$, to follow the dotted trajectory in the figure.

1.6 Deriving $v^2 = v_0^2 + 2a(x - x_0)$ using calculus

I want to derive Eqn. 1.19, $v^2 = v_0^2 + 2a(x - x_0)$ in another way that illustrates the judicious use of calculus.

Start with

$$\frac{dv}{dt} = a \tag{1.31}$$

and multiply both sides by v and write $v = \frac{dx}{dt}$ in the right-hand side:

$$v\frac{dv}{dt} = a\frac{dx}{dt}. \tag{1.32}$$

Now I'm going to do something that is viewed with suspicion, which is just to cancel the dt on both sides. Although I agree that you're not supposed to cancel that d in $\frac{dy}{dx}$, canceling the dt on both sides gives valid results if interpreted carefully. Doing so here gives us

$$vdv = adx. \tag{1.33}$$

This equation tells us that in an infinitesimal time interval $[t, t+dt]$, the variables v and x change by dv and dx, and these changes are related as above *in the limit* $dx, dv, dt \to 0$. Now the limit of $dx \to 0$ or $dv \to 0$ (as compared to their ratio) is of course trivial, and Eqn. 1.33 reduces to $0 = 0$. However, the way we interpret and use Eqn. 1.33 is as follows. Suppose in the finite time interval $[t_1, t_2]$, the variable v changes from v_1 to v_2, and x changes from x_1 to x_2. Let us divide the interval $[t_1, t_2]$ into a very large number N of equal sub-intervals of width dt, and let dx and dv be the changes in x and v in the interval $[t, t+dt]$. The relation between these changes is given in Eqn. 1.33. If we sum up the N changes on both sides of Eqn. 1.33 as $N \to \infty$, the sums converge to nontrivial limits, namely the corresponding integrals:

$$\int_{v_1}^{v_2} vdv = a\int_{x_1}^{x_2} dx \tag{1.34}$$

$$\frac{1}{2}v_2^2 - \frac{1}{2}v_1^2 = a(x_2 - x_1). \tag{1.35}$$

Thus it must be understood that the two sides of a relation like Eqn. 1.33 are to be ultimately integrated between some limits to obtain a useful equality.

Eqn. 1.19 follows upon setting

$$v_2 = v, v_1 = v_0, x_2 = x, x_1 = x_0. \tag{1.36}$$

CHAPTER 2

Motion in Higher Dimensions

2.1 Review

In the last chapter we took the simplest case, of a point particle moving along the x-axis with a constant acceleration a. What is the fate of this particle? The answer is that at any time t, the location of the particle is given by

$$x(t) = x_0 + v_0 t + \frac{1}{2}at^2, \tag{2.1}$$

where x_0 and v_0 are its initial position and velocity. If you took the derivative of this, you would get

$$v(t) = v_0 + at. \tag{2.2}$$

You can easily check, by taking one more derivative, that this particle does indeed have a constant acceleration a. This equation, which gives the velocity of the object at time t, in terms of its initial velocity and acceleration can be inverted to give t in terms of v:

$$t = \frac{v - v_0}{a}. \tag{2.3}$$

Feeding this into Eqn. 2.1 we obtain the result that makes no reference to time:

$$v^2 = v_0^2 + 2a(x - x_0). \tag{2.4}$$

It is understood v and x correspond to some common time.

I showed you in the end how we can use calculus to derive this result. It is important to brush up on your calculus. When a student says, "I know calculus," sometimes that means the student knows it, and sometimes that means he or she once met someone who did. One solution for that is to get a copy of a textbook I wrote called *Basic Training in Mathematics*. This is a little awkward: I don't want to foist my book on you. On the other hand, I don't want to withhold relevant information. If you're going into any science that uses mathematics—chemistry, engineering, or even economics—you should find the contents of that book useful. Don't wait for the movie: it is not coming.

2.2 Vectors in $d = 2$

The next difficult thing is to consider motion in higher dimensions. Everything moves around in $d = 3$. However, I'm going to use only two dimensions for most of the time. Whereas the difference between one dimension and two is very great, that between two and higher dimensions is not. Later we will encounter a few concepts that make sense in $d = 3$ but not $d < 3$. String theorists will tell you that actually we need 9 spatial dimensions plus time to describe superstrings, which will be discussed in depth in Chapter 3,498 of this book.

Picture some particle that's traveling in the $x - y$ plane as shown in Figure 2.1. This is not an x versus t plot or a y versus t plot. It's the actual path the particle traces out on the $x - y$ plane. You might say "Where is time?" One way to mark time is to imagine the particle carries a clock with it, and put markers every second. Four representative markers at $t = 1, 2, 33$, and 34 are shown. It obviously is going much slower between 33 and 34 than between 1 and 2.

The kinematics of this particle requires a pair of numbers x and y. It's more convenient to lump these into a single entity, called a *vector*. The simplest context in which one can motivate a vector and the rules for dealing with vectors is to look at movements in the plane. Let's imagine that when I went camping I walked for 5 *km* from the base camp on the first

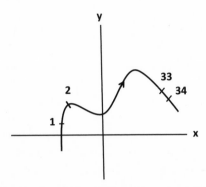

Figure 2.1   Path of a particle in $d = 2$. Equal intervals in time are indicated by markers on the path numbered $1, 2, \ldots, 33$, and 34.

day and another 5 *km* on the second day. How far am I from the base camp? You cannot answer that, even if I promised to move only along the *x*-axis. It's not enough to say I went 5 *km*. I have to tell you whether I went to the right or to the left. So I could be 10 *km*, 0 *km* or -10 *km* from base. If I say not just that I walked 5 *km*, but specified whether it was ± 5 *km*, that takes care of all ambiguity in one dimension.

But in $d = 2$ the options are not just left and right, but an infinity of possible directions. For example, on the first day I could leave the base camp at the origin and move along the arrow labeled **A** to arrive at the point labeled 1 in Figure 2.2(a). The second hike is described by the arrow

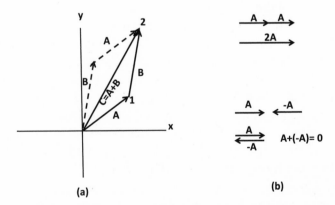

Figure 2.2   Adding vectors. Part (a) shows how to add vectors and that $\mathbf{A} + \mathbf{B} = \mathbf{B} + \mathbf{A}$. Part (b) illustrates the meaning of multiplying a vector by a number (2 in this example) and the null vector **0**.

B, which starts where **A** ended and brings me to 2. These two arrows are examples of *vectors* and I use them here for describing *displacement*, or changes in position. Vectors can be used to describe many other physical quantities, as we will see.

A vector is an arrow that has got a beginning and an end. This is why one says a vector has a magnitude and a direction. The magnitude is how long it is, and direction is its angle relative to some fixed direction, usually the x-axis. When you refer to a vector **A** in your notes, you're supposed to put a little arrow on top like this: $\vec{A}$. In textbooks, vectors are in boldface: **A**. If you don't put an arrow on top or do not use boldface, you're talking about just a number A. When applied to a vector **A**, A stands for its length.

From Figure 2.2(a), we see that there is a very natural quantity that you can call **A** + **B**. One day I moved by **A** and on the next by **B**. If I want to do it all in one shot, what is the equivalent step I should take from the start? It's obvious that the bottom line of my two-day trip is this object **C**. We will call that **A** + **B**. It does represent the sum, in the same sense that if I gave you 4 bucks and then I gave you 5 bucks, you have the equivalent of a single payment of 9 bucks. Here, we are not talking about a single number, but a displacement in the plane, and **C** indeed represents an effective displacement due to **A** and **B**.

So here is the rule for adding two vectors that comes from a study of displacements: you draw the first one and at the end of that first one, you begin the second one, and their sum starts at the beginning of the first and ends at the end of the second.

You can verify, as illustrated in the figure, that **A** + **B** is the same as **B** + **A** where you first draw **B** and from where **B** ends you draw **A**. You will end up with the same point, 2, as shown by the sum of the dotted arrows.

The next thing I want to do is to define the vector that plays the role of the number 0, which has the property that when you add it to any *number*, it gives the same number. The vector **0** that I want to call the zero or *null vector* should have the property that when I add it to any *vector*, I should get the same vector. So you can guess who it is: a vector of no length. I cannot show you the **0**. If you can see it, I'm doing something wrong.

Look at part (*b*) of Figure 2.2. What if I draw **A**, then I add to it another **A** to get **A** + **A**. You have to agree that if there's any vector that deserves to be called 2**A**, it is this guy, **A**, stretched to twice its length. Now we have discovered a notion of multiplying a vector by a number. If you multiply it by 2, you get a vector twice as long and in the same direction. Then we're able to generalize that and say, if you multiply it by 2.6, you get a vector 2.6 times as long. So multiplying a vector by a positive number means to stretch it (or shrink it) by that factor.

Let us keep going. I want to think of a vector that I can call $-\mathbf{A}$. What do I expect of $-\mathbf{A}$? I expect that if I add $-\mathbf{A}$ to $\mathbf{A}$, I should get $\mathbf{0}$, which plays the role of 0 among vectors. What should I add to $\mathbf{A}$ so I get the null vector? It's clear that you want to add a vector that looks like $-\mathbf{A}$ in part (b) of Figure 2.2, because, if you go from the start of $\mathbf{A}$ to the finish of $-\mathbf{A}$, you end up where you started and you get this invisible $\mathbf{0}$ vector. So the minus vector is the same vector flipped over, pointing the opposite way. That's like -1 times a vector. Once you have got that, you can do -7.3 times a vector: just take the vector, rescale it by 7.3 and flip it over. Multiplying a vector by a number is called *scalar multiplication*, and ordinary numbers are called *scalars*. You can do more complicated things. You can take one vector, multiply it by one scalar, take another vector, multiply that by another scalar, and add the two of them. We know what all those operations mean now. You don't have to memorize the rules for all this. The only rule is: "Do what comes naturally." Do what you normally do with ordinary numbers.

2.3   Unit vectors

Let us go back to the same $x - y$ plane. I'm going to introduce two very special vectors. They are the *unit vectors*: $\mathbf{i}$ and $\mathbf{j}$, pointing along the x and y axes and of unit length, as shown in Figure 2.3. If I had a third axis perpendicular to the page, I would draw a $\mathbf{k}$, but we don't need that yet. I claim I can write any vector you give me as a number A_x times $\mathbf{i}$, plus a number A_y times $\mathbf{j}$. There's nothing you can throw at me that lies in the plane that I cannot describe as some multiple of $\mathbf{i}$ plus some multiple of $\mathbf{j}$. It's intuitively clear, but I will just prove it beyond any doubt. Here is some vector $\mathbf{A}$. It is clear from the figure that it is the sum of the dotted horizontal vector and the dotted vertical vector, by the rules of vector addition. The horizontal part, parallel to $\mathbf{i}$, has to be a multiple of $\mathbf{i}$. We know that because we can stretch $\mathbf{i}$ by whatever factor we like. Call that factor A_x, which happens to be positive in this example. It is called the *x-component of* $\mathbf{A}$ or the *projection* of $\mathbf{A}$ along $\mathbf{i}$ or along the x-axis. The vertical part is likewise $\mathbf{j}A_y$ where A_y is the *y-component of* $\mathbf{A}$, *or the projection*

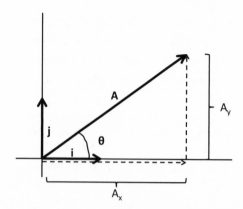

Figure 2.3   The unit vectors **i** and **j** and an arbitrary vector $\mathbf{A} = \mathbf{i}A_x + \mathbf{j}A_y$ built out of them.

of **A** *along* **j** *or along the y-axis.* Therefore I have managed to write **A** as

$$\mathbf{A} = \mathbf{i}A_x + \mathbf{j}A_y. \tag{2.5}$$

We refer to the pair **i** and **j**, in terms of which any vector can be expressed, as *basis vectors* or as the *basis.*

If you gave me a particular vector **A** as an arrow of some length A and orientation θ relative to the x-axis, what do I use for A_x and A_y? You can see from trigonometry that

$$A_x = A \cos\theta \tag{2.6}$$

$$A_y = A \sin\theta. \tag{2.7}$$

Conversely, given the components, the length and angle are

$$A = \sqrt{A_x^2 + A_y^2} \tag{2.8}$$

$$\theta = \tan^{-1}\frac{A_y}{A_x}. \tag{2.9}$$

Eqns. 2.6 to Eqn. 2.9 will be invoked often. So please commit them to memory.

If you give me a pair of numbers, (A_x, A_y), that's as good as giving me this arrow, because I can find the length of the arrow by Pythagoras' theorem and I can find the orientation from $\tan\theta = \frac{A_y}{A_x}$. You have the option of either working with the two components of **A** or with the arrow. In practice, most of the time we work with these two numbers, (A_x, A_y). In particular, if we are describing a particle whose location is the *position vector* **r**, then we write it in terms of its components as

$$\mathbf{r} = \mathbf{i}x + \mathbf{j}y. \tag{2.10}$$

The *changes* in **r** are the *displacement vectors* and examples are **A** and **B** in Figure 2.2 that described the two hikes.

I have not given you any other example of vectors besides the displacement vector, but at the moment, we'll define a vector to be any object that looks like some multiple of **i** plus some multiple of **j**. If I tell you to add two vectors **A** and **B**, you have got two options. You can draw the arrow corresponding to **A** and attach to its end an arrow corresponding to **B**, and then add them, as in Figure 2.2. But you can also do the bookkeeping without drawing any pictures as follows:

$$\mathbf{A} + \mathbf{B} = \mathbf{i}A_x + \mathbf{j}A_y + \mathbf{i}B_x + \mathbf{j}B_y \tag{2.11}$$

$$= \mathbf{i}(A_x + B_x) + \mathbf{j}(A_y + B_y) \tag{2.12}$$

so that the sum **C** is the vector with components $(A_x + B_x, A_y + B_y)$.

In the above, I have used the fact that vectors can be added in any order. So I grouped the things involving just **i** and likewise **j**. Then I argued that since $\mathbf{i}A_x$ and $\mathbf{i}B_x$ are vectors along **i**, their sum is a vector of length $A_x + B_x$ also along **i**. I did the same for **j**.

In summary if

$$\mathbf{A} + \mathbf{B} = \mathbf{C} \tag{2.13}$$

then

$$C_x = A_x + B_x \tag{2.14}$$

$$C_y = A_y + B_y \tag{2.15}$$

which can be summarized as follows:

To add two vectors, add their respective components.

An important result is that $\mathbf{A} = \mathbf{B}$ is possible only if $A_x = B_x$ and $A_y = B_y$. You cannot have two vectors equal without having exactly the same x component and exactly the same y component. If two arrows are equal, one cannot be longer in the x direction and correspondingly shorter in the y direction. Everything has to match completely. The vector equation $\mathbf{A} = \mathbf{B}$ is actually a shorthand for two equations: $A_x = B_x$ and $A_y = B_y$.

2.4 Choice of axes and basis vectors

I have in mind a vector whose components are 3 and 5. Can you draw the vector for me? If you immediately said, "It is $3\mathbf{i} + 5\mathbf{j}$," you're making the assumption that I am writing the vector in terms of $\mathbf{i}$ and $\mathbf{j}$. I agree $\mathbf{i}$ and $\mathbf{j}$ point along two natural directions. For most of us, given that the blackboard or notebook is oriented this way, it is very natural to line up our axes with it. But there is no reason why somebody else couldn't come along and say, "I want to use a different set of axes. The x and y axes or $\mathbf{i}$ and $\mathbf{j}$ are not nailed in absolute space. They are human constructs and we're not wedded to any of them."

Quite often, it's natural to pick the axes in a certain way to suit the problem. If you are studying a cannon ball launched from the earth, it makes sense to pick the horizontal as the x axis and the vertical as the y axis, but, mathematically, you don't have to. Another set of rotated but mutually perpendicular unit vectors $\mathbf{i}'$ and $\mathbf{j}'$ that form another basis can also be rescaled and added to form any given vector $\mathbf{A}$ in the plane. For example, when we study objects sliding down an inclined plane, we will choose our axes parallel and perpendicular to the incline.

If I draw an arrow $\mathbf{A}$ on a blank sheet of paper, it has life of its own without reference to any axes. The *same* vector $\mathbf{A}$ can be written either in terms of $\mathbf{i}$ and $\mathbf{j}$, which is the old basis, or in terms of $\mathbf{i}'$ and $\mathbf{j}'$, the new basis. How do the components (A_x', A_y') in the new basis relate to the components of the old basis? It's a simple problem, but I just want to do it so you get used to working with vectors.

For this we need the very busy Figure 2.4. It shows the old x and y axes and the x' and y' axes obtained by rotating the $x - y$ axes counterclockwise by an angle ϕ. The unit vectors $\mathbf{i}'$ and $\mathbf{j}'$ are likewise rotated versions of $\mathbf{i}$ and $\mathbf{j}$. The components in the two bases are shown by dotted lines and are simply the projections of $\mathbf{A}$ along the various axes. We want to relate (A_x', A_y') to (A_x, A_y).

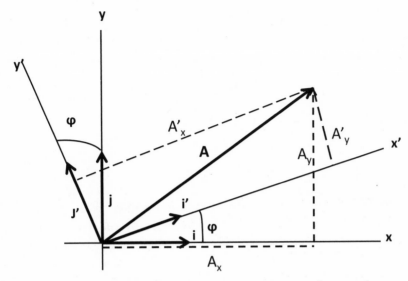

Figure 2.4 The same vector **A** is written as $\mathbf{i}A_x + \mathbf{j}A_y$ in one frame and as $\mathbf{i}'A'_x + \mathbf{j}'A'_y$ in the other. The dotted lines indicate the components in the two frames.

First we express $\mathbf{i}'$ and $\mathbf{j}'$ in terms of $\mathbf{i}$ and $\mathbf{j}$ using the figure:

$$\mathbf{i}' = \mathbf{i}\cos\phi + \mathbf{j}\sin\phi \qquad\qquad (2.16)$$

$$\mathbf{j}' = \mathbf{j}\cos\phi - \mathbf{i}\sin\phi. \qquad\qquad (2.17)$$

Here are the details. The vector $\mathbf{i}'$ has got a horizontal part, which is its length, namely, 1, times $\cos\phi$, and a vertical part that is 1 times $\sin\phi$. How about $\mathbf{j}'$? It is at an angle ϕ relative to $\mathbf{j}$. So its y-component is $\cos\phi$. Finally, its x or horizontal component is $(-\sin\phi)$, where the minus sign comes because it is pointing to the left, along the negative x-axis. All that remains now is to eliminate $\mathbf{i}'$ and $\mathbf{j}'$ in favor of $\mathbf{i}$ and $\mathbf{j}$ in $\mathbf{A} = \mathbf{i}'A'_x + \mathbf{j}'A'_y$ and equate it to **A** written in terms of $\mathbf{i}$ and $\mathbf{j}$:

$$\mathbf{A} = \mathbf{i}'A'_x + \mathbf{j}'A'_y \qquad\qquad (2.18)$$

$$= (\mathbf{i}\cos\phi + \mathbf{j}\sin\phi)A'_x + (\mathbf{j}\cos\phi - \mathbf{i}\sin\phi)A'_y \qquad\qquad (2.19)$$

$$= \mathbf{i}(A'_x\cos\phi - A'_y\sin\phi) + \mathbf{j}(A'_x\sin\phi + A'_y\cos\phi) \qquad\qquad (2.20)$$

$$= \mathbf{i}A_x + \mathbf{j}A_y. \qquad\qquad (2.21)$$

When we equate the coefficients of $\mathbf{i}$ and $\mathbf{j}$ on the right-hand sides of Eqns. 2.20 and 2.21, we obtain the desired expression for A_x and A_y in terms of A'_x and A'_y:

$$A_x = A'_x \cos\phi - A'_y \sin\phi \tag{2.22}$$

$$A_y = A'_x \sin\phi + A'_y \cos\phi. \tag{2.23}$$

So, you can pick your basis vectors any way you like and so can I. Your basis is obtained from mine by a counterclockwise rotation by an angle ϕ. The same entity $\mathbf{A}$, the same arrow which has an existence of its own, independent of axes, can be described by you and by me using different components. Your components with primes on them are related to mine by Eqns. 2.22 and 2.23. This is called the *transformation law* for the vector components under rotation of basis vectors.

Now, you can ask the opposite question. How do I get A'_x and A'_y in terms of A_x and A_y? The quickest way is to replace ϕ by $-\phi$: if we go from the unprimed to the primed system by a rotation ϕ, then rotation by $-\phi$ is the way to go from the primed to the unprimed basis. The result, using $\cos(-\phi) = \cos\phi$ and $\sin(-\phi) = -\sin\phi$, is

$$A'_x = A_x \cos\phi + A_y \sin\phi \tag{2.24}$$

$$A'_y = -A_x \sin\phi + A_y \cos\phi. \tag{2.25}$$

That turns out to be the correct answer. But I want you to think about another way to show this, which often seems to bother some students. If I told you

$$3x + 5y = 21 \tag{2.26}$$

$$4x + 6y = 26, \tag{2.27}$$

you certainly know how to solve for x and y, right? You have got to juggle the two equations, multiply the first by 6, the second by 5, and subtract to isolate x and so on. Why is it when some of you see Eqns. 2.22 and 2.23, you don't realize it's the same kind of problem, where you can multiply Eqn. 2.22 by $\cos\phi$, Eqn. 2.23 by $\sin\phi$ and add to isolate A'_x, for example? For any particular value of ϕ, $\sin\phi$ and $\cos\phi$ are just some numbers. For example, if I pick $\phi = \frac{\pi}{3} = 60°$, $\cos \phi = \frac{1}{2}$ and $\sin \phi = \frac{\sqrt{3}}{2}$. The equations become (for this angle)

$$A_x = \frac{1}{2}A'_x - \frac{\sqrt{3}}{2}A'_y \tag{2.28}$$

$$A_y = \frac{\sqrt{3}}{2}A'_x + \frac{1}{2}A'_y. \tag{2.29}$$

If you multiply the second by $\sqrt{3}$ and add it to the first you obtain

$$A_x + \sqrt{3}A_y = 2A'_x \quad \text{which means} \tag{2.30}$$

$$A'_x = \frac{1}{2}A_x + \frac{\sqrt{3}}{2}A_y \tag{2.31}$$

$$= A_x \cos\frac{\pi}{3} + A_y \sin\frac{\pi}{3} \tag{2.32}$$

in accordance with Eqn. 2.24. So go forth and treat $\sin\phi$ and $\cos\phi$ as plain numbers and juggle Eqns. 2.22 and 2.23 to derive Eqns. 2.24 and 2.25. Along the way of course you will have to use identities like $\sin^2\phi + \cos^2\phi = 1$.

The components of the vector depend on who is looking at the vector. However, there's one quantity that's going to come out the same, no matter who is looking at the vector. It is the length of the vector. It is unaffected by the rotation of axes. It is an *invariant* under rotations. You may verify from Eqns. 2.24 and 2.25 that

$$(A'_x)^2 + (A'_y)^2 = (A_x \cos\phi + A_y \sin\phi)^2 + (-A_x \sin\phi + A_y \cos\phi)^2 \tag{2.33}$$

$$= A_x^2(\cos^2\phi + \sin^2\phi) + A_y^2(\sin^2\phi + \cos^2\phi) \tag{2.34}$$

$$= A_x^2 + A_y^2. \tag{2.35}$$

The $A_x A_y$ term is gone since its coefficient is $2(\cos\phi \sin\phi - \sin\phi \cos\phi)$.

I want to conclude with one important point. We learned that a vector is a quantity that has a magnitude and a direction. A more advanced view of vectors is that they are a pair of numbers (in $d = 2$) which, under rotation of axes, transform as per Eqns. 2.24 and 2.25. Anything that transforms this way is called a vector. We already know about the position vector **r** and the changes in it, the displacement vectors (used in describing the hike). How about more vectors? There turns out to be a very nice way to produce vectors, given one vector like the position vector. And that's the following.

2.5 Derivatives of the position vector **r**

Let's take a particle in the $x - y$ plane that moves from **r** at time t to $\mathbf{r} + \Delta\mathbf{r}$ at time $t + \Delta t$ as in Figure 2.5. At time t its location is

$$\mathbf{r} = \mathbf{i}\, x(t) + \mathbf{j}\, y(t) \tag{2.36}$$

and at $t + \Delta t$ it is

$$\mathbf{r} + \Delta\mathbf{r} = \mathbf{i}(x(t) + \Delta x) + \mathbf{j}(y(t) + \Delta y) \quad \text{so that,} \tag{2.37}$$

$$\Delta\mathbf{r} = \mathbf{i}\Delta x + \mathbf{j}\Delta y \quad \text{and by the usual limiting process,} \tag{2.38}$$

$$\mathbf{v} = \lim_{\Delta t \to 0} \frac{\Delta\mathbf{r}}{\Delta t} = \frac{d\mathbf{r}}{dt} = \mathbf{i}\frac{dx}{dt} + \mathbf{j}\frac{dy}{dt}. \tag{2.39}$$

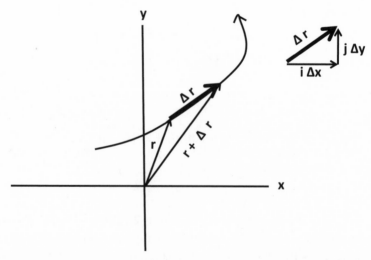

Figure 2.5   The particle moving along some curved path goes from **r** at time t to $\mathbf{r} + \Delta\mathbf{r}$ at time $t + \Delta t$. The velocity **v** is the limit of the ratio $\frac{\Delta\mathbf{r}}{\Delta t}$ as $\Delta t \to 0$, and thus parallel to $\Delta\mathbf{r}$, which eventually becomes tangent to the curve.

When you move just along the x-axis, you wait a small time Δt and you move by an amount Δx, and their ratio gives the velocity in the appropriate limit. When you move in the plane, your position and its change are both vectors.

Can you see why the derivative of a vector is also a vector? Because $\Delta \mathbf{r}$, the difference in the vector between two times, is itself a vector. Dividing it by Δt is like multiplying by $1/\Delta t$, but we know that when we multiply a vector by a number, we simply rescale the vector. So the limit will be some arrow that we call the *instantaneous velocity vector*. It will be tangential to the curve $\mathbf{r}(t)$ and point toward the instantaneous direction of travel.

If I gave you the location of a particle as a function of time, you can find its velocity by taking derivatives. For example, if I say a particle's location is

$$\mathbf{r} = t^2\, \mathbf{i} + 9t^3\, \mathbf{j} \tag{2.40}$$

then its velocity at time t is

$$\mathbf{v} = 2t\, \mathbf{i} + 27t^2\, \mathbf{j}. \tag{2.41}$$

You can take a derivative of the velocity or the second derivative of the position to get the acceleration vector

$$\mathbf{a}(t) = \frac{d\mathbf{v}}{dt} = \frac{d^2\mathbf{r}}{dt^2} = 2\,\mathbf{i} + 54t\,\mathbf{j} \quad \text{(in our example).} \tag{2.42}$$

You can then also multiply $\mathbf{a}$ by the mass m, which is a *scalar* unaffected by rotations, to get a vector $m\mathbf{a}$, which Newton's law equates to another vector, the force $\mathbf{F}$.

Even though we started with one example of a vector $\mathbf{r}$, we're now finding out that its derivative has to be a vector and the derivative of the derivative is also a vector. When you learn relativity, you will find out there's again one vector that's staring at you, the analog of the position vector, but with four components. But more vectors can be manufactured by multiplying vectors by scalars (like mass) or taking derivatives with respect to a parameter that plays the role analogous to time.

Here is an illustration of vector addition and differentiation. Imagine an airplane in flight, as depicted in Figure 2.6. Let $\mathbf{r}_{pg}$ be the location of a

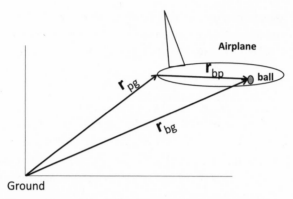

Figure 2.6   The position of the ball relative to (some origin on) the ground $\mathbf{r}_{bg}$ is the vector sum of the position of the ball relative to the (tail of the) plane, $\mathbf{r}_{bp}$, and the position of (the tail of) the plane $\mathbf{r}_{pg}$ relative to the ground.

fixed point in the airplane, say the tail, with respect to a fixed point on the ground. Imagine that in the airplane there is a ball located at $\mathbf{r}_{bp}$ as measured from this fixed point in the airplane. By vector addition the location of the ball with respect to the ground is

$$\mathbf{r}_{bg} = \mathbf{r}_{bp} + \mathbf{r}_{pg}. \qquad (2.43)$$

Upon taking a time derivative and in the same notation, the law of composition of velocities follows:

$$\mathbf{v}_{bg} = \mathbf{v}_{bp} + \mathbf{v}_{pg}, \qquad (2.44)$$

which says the velocity of the ball as seen by a person on the ground is the velocity of the ball relative to the airplane plus the velocity of the airplane relative to the ground. Taking yet another derivative we may relate the accelerations:

$$\mathbf{a}_{bg} = \mathbf{a}_{bp} + \mathbf{a}_{pg}. \qquad (2.45)$$

In the special case of a airplane moving at constant velocity, $\mathbf{a}_{pg} = 0$. Then we find

$$\mathbf{a}_{bg} = \mathbf{a}_{pg}, \qquad (2.46)$$

which means, in this case, the acceleration of the ball is the same as measured by an observer on the ground and an observer on the airplane. These results will be recalled in our study of relativity.

2.6 Application to circular motion

Now we'll take a concrete problem where you will see how to take derivatives to obtain very useful results. I'm going to write a particular case of $\mathbf{r}(t)$:

$$\mathbf{r}(t) = R(\mathbf{i}\cos\omega t + \mathbf{j}\sin\omega t) \tag{2.47}$$

where R and ω are constants. What is going on as a function of time? What's this particle doing? Look at the length squared of this vector:

$$r_x^2 + r_y^2 = R^2(\cos^2\omega t + \sin^2\omega t) = R^2. \tag{2.48}$$

That means the particle is going around in a circle of radius R as shown in Figure 2.7. The x component is $R\cos\omega t$ and the y component is $R\sin\omega t$

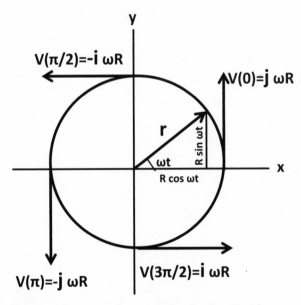

Figure 2.7   The particle moves along a circle of radius R with an angular velocity ω.

where ω is a fixed number. As t increases, this angle ωt increases and the particle goes round and round. Let's get a feeling for ω. As time increases, the angle increases and we can ask how long it will take the particle to come back to the starting point. Suppose the starting point was on the x-axis. As t increases, ωt increases, and the particle will come back at a time T such that

$$\omega T = 2\pi. \tag{2.49}$$

Thus ω is related to the time period T by

$$\omega = \frac{2\pi}{T} = 2\pi f \tag{2.50}$$

where $f = \frac{1}{T}$ is the frequency or number of cycles per second. It is measured in *Hz*, which stands for Hertz. Since in every cycle the particle rotates by 2π, and it completes f revolutions per second, $\omega = 2\pi f$ is called the *angular velocity* and measures the radians swept out per second.

Notice that in equating a full cycle to 2π, I am using *radians* and not degree to measure angles. For those who have not seen a radian, it's just another way to measure angles, wherein a full circle, which we used to think was worth $360°$, now equals 2π radians. Since $2\pi \simeq 6.3$, a radian is roughly $60°$. You will see the advantages of using radians later. For now just remember that a half circle, instead of being $180°$, will now be π radians, and a quarter circle will be $\frac{\pi}{2}$, and so forth.

How fast is this particle moving? It's going around a circle, the angle is increasing at a steady rate ω, and so we know it's going at a steady speed. Let us verify that by computing the velocity

$$\mathbf{v}(t) = \frac{d\mathbf{r}(t)}{dt} \tag{2.51}$$

$$= R\left(\mathbf{i}\frac{d\cos\omega t}{dt} + \mathbf{j}\frac{d\sin\omega t}{dt}\right) \tag{2.52}$$

$$= R\omega(-\mathbf{i}\sin\omega t + \mathbf{j}\cos\omega t). \tag{2.53}$$

At $t = 0$, the velocity is $\mathbf{v} = R\omega\cos 0\,\mathbf{j} = R\omega\mathbf{j}$, so it is moving straight up at speed $v = \omega R$. You may verify that it has the velocity as shown in the figure at later times. The magnitude of the velocity is always ωR although the direction is changing. From the figure we see it remains tangential to the circle. The constancy of the speed v at an arbitrary time may also be established by computing

$$v^2 = (\omega R)^2(\sin^2 \omega t + \cos^2 \omega t) = (\omega R)^2 \tag{2.54}$$

$$v = \omega R. \tag{2.55}$$

Remember the tangential velocity is $v = \omega R$.

Let's take the derivative of the derivative to find the acceleration **a** and its magnitude:

$$\mathbf{a} = -\omega^2 R(\mathbf{i} \cos \omega t + \mathbf{j} \sin \omega t) = -\omega^2 \mathbf{r} \tag{2.56}$$

$$a = \omega^2 R. \tag{2.57}$$

That's a very important result. It tells you that *when a particle moves in a circle of radius R at constant speed v, it has an acceleration, called the* centripetal acceleration, *directed toward the center and of magnitude*

$$a = \omega^2 R = \frac{(\omega R)^2}{R} = \frac{v^2}{R}. \tag{2.58}$$

This acceleration at constant speed reflects the fact that velocity is a vector and you can change the velocity vector by changing its direction. For example, if a car is going on a racetrack and the speedometer says 60 miles per hour, the lay person's view is that the car is not accelerating. But you will say from now on that it indeed has an acceleration equal to $\frac{v^2}{R}$ even though no one's stepping on the accelerator or the brake.

Suppose the particle is not moving fully around a circle but traversing just a quarter of the circle. When it is traveling the quarter of a circle, it has the same acceleration directed toward the center of that quarter circle. In other words, you don't have to be moving actually in a circle to have the acceleration $\frac{v^2}{R}$. At any instant, the curve you are following can be locally approximated as part of some circle, and, in the formula $a = \frac{v^2}{R}$, the acceleration is directed toward the center of that circle, R is its radius and v the instantaneous tangential velocity.

2.7 Projectile motion

I want to consider a particle for which $\mathbf{r}_0$ and $\mathbf{v}_0$ are the position and velocity at $t = 0$ and which has a constant vector acceleration $\mathbf{a}$. What is its location at all future times? By analogy with what I did in one dimension

$$\mathbf{r}(t) = \mathbf{r}_0 + \mathbf{v}_0 t + \frac{1}{2}\mathbf{a}t^2. \tag{2.59}$$

Once you know $\mathbf{r}_0$ and $\mathbf{v}_0$, you can find the position of the object at all future times. Let's take one simple example. Somebody in a car has decided to drive off a cliff as shown in Figure 2.8(a). We want to know when and where the car hits the ground.

We pick our origin $(0,0)$ at the foot of the cliff. Let the height of the cliff be h. The car is traveling with some initial speed v_{0x} in the horizontal direction. Equation 2.59 is really a pair of equations, one along x and one along y with $\mathbf{a} = -\mathbf{j}g$, $\mathbf{v}_0 = v_{0x}\mathbf{i}$, and $\mathbf{r}_0 = h\mathbf{j}$. Separating out the components

$$x(t) = 0 + v_{0x}t + 0 \tag{2.60}$$

$$y(t) = h + 0 - \frac{1}{2}gt^2. \tag{2.61}$$

Notice that the evolution of the two coordinates is completely independent. The time t^* when the car hits the ground $(y = 0)$ satisfies the

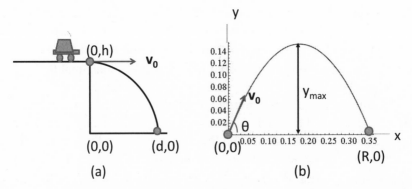

Figure 2.8   (a) A car flies off the cliff at $(0, h)$ and lands at $(d, 0)$. (b) A projectile is launched with initial velocity $\mathbf{v}_0 = (\mathbf{i} + \sqrt{3}\mathbf{j})m/s$. The range is $R = 0.35\ m$ and the maximum height reached is $y_{max} = 0.15\ m$.

equation

$$0 = h - \frac{1}{2}gt^{*2} \tag{2.62}$$

$$t^* = \sqrt{\frac{2h}{g}}. \tag{2.63}$$

This is exactly how long it would take to hit the ground had it simply toppled over the edge from rest. The horizontal velocity does not delay the crash one bit (unless you take into account the curvature of the earth). As to where the car lands, the location is given by $(x(t^*), 0) = (d, 0)$ where

$$d = v_{0x}t^* = v_{0x}\sqrt{\frac{2h}{g}}. \tag{2.64}$$

Finally the problem of projectile motion is depicted in Figure 2.8(b). You fire a projectile from $(0,0)$ with some velocity $\mathbf{v}_0$ at some angle θ. It will go up and then come down, moving horizontally at the same time. Where is it going to land? What is the maximum height y_{max} to which it rises? With what speed will it hit the ground? At what angle should you fire your projectile so it will go the furthest?

Here are the equations that contain all the answers, namely Eqn. 2.59 written out in component form:

$$x(t) = 0 + v_{0x}t = v_0 \cos\theta \cdot t \tag{2.65}$$

$$y(t) = 0 + v_{0y}t - \frac{1}{2}gt^2 = v_0 \sin\theta \cdot t - \frac{1}{2}gt^2. \tag{2.66}$$

You can solve them but it is good to have an idea of what's coming. Imagine you have this monster cannon to fire things. It has a fixed muzzle speed, v_0, but allows you to fire at any angle. How do you aim it so the ball goes as far as possible? There are two schools of thought. One says, aim at your enemy and fire horizontally. Then the ball lands on your foot because it has zero time of flight (assuming the cannon is at zero height). The other school says, maximize the time of flight and point the cannon vertically. It goes up, stays in the air for a very long time, and lands on your head. Then it hits you: the correct answer is somewhere between 0 and $90° = \frac{\pi}{2}$. The naive guess $45° = \frac{\pi}{4}$ turns out to be correct.

Now I want to show you how to use the equations to prove this. What's the strategy for finding the range? You see how long the ball is in the air and multiply that by the constant horizontal velocity $v_0 \cos\theta$. Again, let t^* be the time when it hits the ground. The y-equation

$$0 = t^* \left(v_0 \sin\theta - \frac{1}{2}gt^* \right) \tag{2.67}$$

has two solutions:

$$t^* = 0, \frac{2v_0 \sin\theta}{g}. \tag{2.68}$$

So the cannon ball is on the ground on two occasions. One is initially. We are not interested in that trivial solution. If the time you are interested in is $t^* \neq 0$, you're allowed to divide both sides of Eqn. 2.67 by it and get

$$t^* = \frac{2v_0 \sin\theta}{g} \tag{2.69}$$

and the range

$$R = v_{0x}t^* = v_0 \cos\theta \cdot t^* = \frac{2v_0^2 \sin\theta \cos\theta}{g} = \frac{v_0^2 \sin 2\theta}{g} \tag{2.70}$$

using $\sin 2\theta = 2\sin\theta \cos\theta$. For the greatest range we must make $\sin 2\theta$ as large as possible, which occurs for $2\theta = \frac{\pi}{2}$ or $\theta = \frac{\pi}{4} = 45°$. For any smaller range you will find there are two possible angles that work since $\sin(2\theta) = \sin(\pi - 2\theta)$.

The maximum height is the y-coordinate at the half-way time $\frac{1}{2}t^* = \frac{v_0 \sin\theta}{g}$:

$$y_{max} = 0 + v_{0y}\frac{1}{2}t^* - \frac{1}{2}g\left(\frac{1}{2}t^*\right)^2$$

$$= (v_0 \sin\theta)\frac{v_0 \sin\theta}{g} - \frac{1}{2}g\left[\frac{v_0 \sin\theta}{g}\right]^2 = \frac{v_0^2 \sin^2\theta}{2g}. \tag{2.71}$$

We could equally well find the half-time $\frac{1}{2}t^*$ by setting the vertical velocity to zero:

$$0 = v_0 \sin\theta - g\frac{t^*}{2}. \tag{2.72}$$

How about the velocity at impact? The horizontal part is of course $v_0 \cos\theta$ since there is no acceleration in that direction. The vertical part starts out at $v_0 \sin\theta$ and decreases at a rate g:

$$v_y = v_0 \sin\theta - gt \tag{2.73}$$

so that at t^* it has a value

$$v(t^*) = v_0 \sin\theta - g\frac{2v_0 \sin\theta}{g} = -v_0 \sin\theta, \tag{2.74}$$

which is just the opposite of the initial vertical speed. The magnitude of the final velocity is the same as the initial one since reversing one of the components of $\mathbf{v}$ does not change its length.

There are endless variations. You pick some point off the ground at (X, Y) and want the projectile to arrive there. You are given the launch angle θ and have to find the launch speed v_0. How do you do that problem? You assume the projectile arrives at the destination (X, Y) at some time t^*. You go to the x equation and demand that $x(t^*) = X$ and solve for t^*. Plug this time into y and demand $y(t^*) = Y$ and solve for v_0.

Sometimes the problems are embellished to make everyone feel involved. For example, instead of a ball dropping, nowadays there's a monkey or horse that is falling down, so people in life sciences can say, "Hey, we should learn physics since it seems to have applications to our subject." All those gyrating creatures are very interesting and look great in color, but, in the end, you are told, "Treat the horse as a point particle." If you're going to treat the horse as a point particle, why include its color picture? But I agree there are times when only a horse will do. When the Godfather wants to get the contract for Johnny Fontaine, he doesn't tell his consigliere, "Hey, Tom, put half a point particle on Jack's bed." That would have been a disastrous approximation.

By the way, don't forget to treat the falling car in Figure 2.8 as a point particle.

CHAPTER 3

Newton's Laws I

3.1 Introduction to Newton's laws of motion

This is a big day in your life: you are going to learn Newton's laws, in terms of which you can understand and explain a very large number of phenomena. It's really amazing that so much information can be condensed into three laws.

Your reaction may be that you have already seen Newton's laws, that you have applied them in school. I realized fairly late in life that they are more subtle than I first imagined. It's one thing to plug in all the numbers and say, "I know Newton's laws and I know how they work." But as you get older and you have more spare time, you think about what you are doing. This is something I have had the luxury of doing, and I have realized the laws are more tricky. I want to share some of that understanding.

The first law, or the *law of inertia*, says, "If a body has no forces acting on it, it will maintain its velocity." In other words, in the absence of external forces, a body at rest will remain at rest and a body in motion will retain its velocity. It is not surprising that a body at rest will remain so if not acted upon by a force. We see that all the time. I place an eraser on the table. It will stay put unless I do something to it.

The great discovery that Galileo and Newton made was that you don't need a force for a body to move at constant velocity. You don't see that in daily life—everything seems to come to rest unless you keep pushing or pulling it. But we all know that the reason things come to a halt is that there is always some friction or drag bringing them to rest. If you take a hockey puck on an air cushion and give it a push, it seems it can travel for a very long time. Galileo and Newton abstracted from this an ideal situation in which friction was totally eliminated and the bodies kept moving forever with no help. If you go to outer space, you can check for yourself that, if you throw something, it will go on forever without your intervention. It's in the nature of things to retain a constant velocity. *It is not velocity, but a change in velocity, that calls for a force.*

The law of inertia is not valid for everybody. I'll give an example from your own life. You go on an airplane and then, after the usual delays, the plane begins to accelerate down the runway. At that time, if you leave anything on the floor, you know it's no longer yours. It's going to slide backward and the physicist in the last row is going to collect everything. That is an example of a frame in which the law of inertia does not work: bodies accelerate with no applied force. Once the plane stops accelerating, the law of inertia becomes operative. It fails when it decelerates during landing when everything now slides to the front.

If Newton's law of inertia works for you, you are called an *inertial observer* and your frame of reference is called an *inertial frame.* The plane that's taking off is not an inertial frame, but the one that is cruising is. The earth seems to be a pretty good inertial frame, because if you leave something at some place, it just stays there—unless the thing is your iPod and the place is Grand Central Station. But this is not a violation of Newton's laws, just the laws of New York City.

Although not every frame is inertial, there are plenty of inertial frames to go around. If you find even one inertial observer, namely one person for whom this law of inertia works, I can find for you an infinite number of other people for whom this is true. Who are these people? They are people moving at constant velocity with respect to the first inertial observer. Suppose you are in a train and you're moving past me with velocity **u**, and we both look at some object with no forces on it. We will not agree on its velocity or the velocity of anything: things at rest for me will be moving backward for you at velocity −**u**, and everything at rest in your train will be moving at velocity **u** according to me. In short, you and I will differ on the velocity of any object by our relative velocity. *But we will agree on the acceleration of any object since it is unaffected by adding a constant to velocity.* Adding a constant velocity to objects does not change the fact that those which were maintaining constant velocity still maintain a constant (but different) velocity. Thus neither is every observer inertial, nor is an inertial observer unique.

You must know the earth is not precisely inertial. The earth has an acceleration. Can you see why? Yes, it's spinning around itself and going around the sun, both of which constitute accelerated motion. But the acceleration due to motion around the sun at speed v and radius r is $a = \frac{v^2}{r} = .006 \ ms^{-2}$, which is a very small number, say compared to g. The same goes for the acceleration due to the earth's rotation about its own axis, which is roughly $.03 ms^{-2}$ or roughly $g/300$ at the equator.

The first law might seem tautological because we never see anything that retains its velocity forever, and every time we see velocity change we say that a force is acting. But it's not a big hoax, because you can set up experiments in free space, far from everything, where objects will, in fact, maintain their velocity forever. It's a useful concept even on the earth, because the earth is approximately inertial.

3.2   Newton's second law

Newton's second law says, "If a body has an acceleration **a**, then you need a force

$$\mathbf{F} = m\mathbf{a} \tag{3.1}$$

to produce that acceleration."

In this chapter we will focus on one dimension and write

$$F = ma \tag{3.2}$$

where F and a are along the x-axis.

A few words about units. Acceleration is measured in ms^{-2}. Mass is measured in kilograms or kg. So force has units kilogram meters per second squared. But we get tired of saying that long expression, so we call that a Newton, denoted by N. If you had invented mechanics, we'd be calling it by your name, but it is too late for you now.

Here is a typical problem that you may have solved in your first pass at Newton's laws. A force of $36N$ is acting on a mass $4 \ kg$. What's the acceleration? You divide 36 by 4 and you find it is $a = 9 \ ms^{-2}$. You say, "Okay, I know Newton's laws."

It's actually more complicated than that. Take yourself back to the seventeenth century, when Newton was inventing these laws. You have an intuitive definition of force: when somebody pushes or pulls an object we

say a force is acting on it. Suddenly, you are told there is a law $F = ma$.
Are you better off in any way? Can you do anything with this law? What
does it help you predict? Can you even tell if it's true? Here's a body that's
moving. Is Newton right? How are we going to check? Well, you want to
measure the left-hand side and you want to measure the right-hand side.
If they're equal, you will say the law is working. What can you measure in
this equation?

Let's start with acceleration. What's your plan for measuring acceleration? What instruments will you need? If you say a watch and ruler,
that would be correct provided by *ruler* you don't mean Queen Elizabeth.
Here is your ruler and here is a Rolex. Tell me exactly how you plan to
measure acceleration. Everyone seems to know the answer. First, let it
go a little distance, and take the distance over time. That gives you the
velocity now. Let it go a little more, and repeat the velocity measurement.
Take the difference of the two velocities and divide by the difference of the
two times, and you have got the acceleration. Since the body has moved a
finite distance in a finite time, this gives the average acceleration. You want
to make these three positional measurements more and more rapidly. In
the end, as all the time intervals shrink to 0, you will measure what you
can say is the acceleration *now*, the second derivative $\frac{d^2x}{dt^2} = a(t)$ defined in
calculus.

Back to testing if what Newton told you is right: You see an object in
motion, you measure a, and you get a certain numerical value, say $10 \ ms^{-2}$.
But that's not yet testing the equation, because you still have to find F and
m. What's the mass of this object? One common idea is to take a standard
mass and balance the unknown mass on a seesaw by adjusting its position.
But suppose you were in outer space. There's no gravity. Then the seesaw
will balance even if you put a potato on one side and an elephant on the
other side. What you are doing now is appealing to the notion of mass as
something that's related to the pull of the earth on the object. You have got
to go back and wipe out everything you know. If $F = ma$ is all you have,
there is no mention of the earth in these equations. You only know how
to measure a, but not the other two. So you have a problem. You cannot
say that since $F = ma$, it follows that $m = \frac{F}{a}$; that is circular reasoning since
you have not told me how to measure F either.

Let me give you a hint. How do we decide how long a meter is? You
seem to know that it is arbitrary. A meter is not deduced from anything.
Napoleon or somebody said, "The size of my ego is one meter." That's a

new unit of length. Seriously, at the National Bureau of Standards there used to be a rod made of some special alloy in a glass case, and that *defined* the meter. There are fancier definitions now, but let us stick to this simple one. (See http://physics.nist.gov/cuu/Units/meter.html for more details on definitions of units.) Then I ask you, "What is two meters and what is three meters?" We have ways of handling that. You take the meter and attach it to a duplicate, and that's two meters. You can cut it in half, using dividers and compasses; you can split the meter into any fractions you like. Likewise for mass, we will take a chunk of some material and we will call it a kilogram. That is a matter of convention, just like one second is some convention.

I'm going to give you a glass case that contains a block of some metal *defined* as one kilogram. Then I give you another object, an elephant. What's the mass of the elephant? Here is a hint: I also give you a spring. We cannot do the seesaw experiment because it requires gravity. A spring, on the other hand, will exert a force even in outer space. Here's what we do. We hook one end of the spring to a wall and we pull the other end from its equilibrium position by some amount and we attach the one-kilogram mass to it. We don't know what force it exerts, but it will not matter. We let it go and measure the initial acceleration, a_1. Then we bring the elephant (another point particle) of unknown mass m_E, pull the spring *by the same amount so it can exert the same force*, and find the acceleration a_E of the elephant. Assuming only that the same extension produces the same force in the two cases, we have

$$1 \cdot a_1 = m_E \, a_E \tag{3.3}$$

$$m_E = 1 \cdot \frac{a_1}{a_M}. \tag{3.4}$$

Once you have the mass of the elephant you can use it to measure any other mass m_o by using

$$m_o = m_E \frac{a_E}{a_o} \tag{3.5}$$

where a_E and a_o are produced by the same force.

There are subtleties even here. For example, how do we know that when we pull the spring the second time with the elephant, it will exert the same force as the first time when the 1 *kg* was attached to it? After all, springs wear out. That's why you change the shock absorbers in your car. So first, we have to make sure the spring exerts a fixed force every time (for a given extension). How are we going to check that? We don't have the definition of force yet. But we can do the following. We pull the one-kilogram mass and let it go, and we note the acceleration. Then, we pull it again, by the same amount, and let it go; we do it ten times. If every time we get the same acceleration, we are convinced this is a reliable spring that is producing the same force under the same conditions. On the eleventh time we pull the spring and attach the elephant. With some degree of confidence, we can say we are applying the same force on the elephant as on the one-kilogram mass.

Why is this discussion so important? Because you need to know that everything you or I write down in the notebook or on the blackboard as a symbol is actually a measurable quantity, or, as they say in France, *Les Mesurables*. You should know at all times how you measure anything that enters your theory or calculation. If not, you are just doing math or playing with symbols. You are not doing physics.

This discussion also tells you that the mass of an object has nothing to do with gravitation but with how much it hates to accelerate in response to a force. Newton tells you forces cause acceleration. But the acceleration is not the same on different objects for a given force. Certain objects resist it more than others. They are said to have a bigger mass. We can be precise about how much bigger by saying, "If the acceleration of a body in response to a given force is $\frac{1}{10}$ that of a 1-kilogram mass, then the mass of the body is 10 kilograms."

3.3 Two halves of the second law

We have seen how all objects can be attributed a mass. Now go back to the spring. I want to know how much force $F(x)$ it exerts when I pull it by a certain amount x, which is measured from the point when the spring is neither compressed nor expanded as shown in Figure 3.1. If x is positive, it means the spring is stretched; if it's negative, it means the spring is compressed. *Now I can measure $F(x)$ for any given x because I can measure the acceleration it produces on a known mass m and use $F = ma$.* So I pull it by various amounts, measure F, and draw a graph. It will be a straight line with some slope $-k$ for small values of x:

$$F = -kx \qquad (3.6)$$

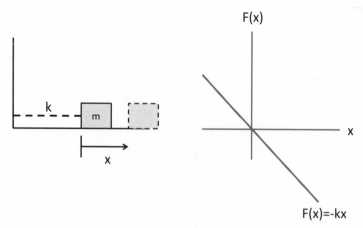

Figure 3.1   (Left) The mass m is attached to a spring (dotted line) of force constant k. The other end of the spring is attached to a wall. The displacement x is measured from the equilibrium position of the spring. (Right) The force $F(x)$ as a function of x.

where k is called the *force constant*. The minus sign says, if you pull it to the right, so that x is positive, the spring will exert a force in the negative direction. If you compress it, then x is negative and the spring will exert a force in the direction of increasing x. All springs will have a graph like this for small enough distortions from equilibrium. Beyond that the line may bend or the spring may even snap. In any case, we now have a way to measure k for any given spring in the regime when $F(x)$ is linear in x.

I want you to think about the two equations $F = ma$ and $F = -kx$. If the first one is Newton's law, then what's the other one? What's the difference between saying $F = -kx$ and $F = ma$? Let me paraphrase the answer I usually hear from students: "$F = ma$ is universally true, independent of the nature of the force acting on a body, while $F = -kx$ is only describing the spring." That is essentially correct, and I will now elaborate.

The cycle of Newtonian dynamics has two parts.

The first one is to find the acceleration a of a body, the force on which is somehow known, using $a = \frac{F}{m}$. The force F is the cause and the acceleration a is the effect and $a = \frac{F}{m}$ is the precise relation between them.

The second is to deduce experimentally what force F will be acting on a given body at any given time. For example, if a mass is attached to a spring that has been extended by x, we must do the experiment to find that the spring force is $F = -kx$. Newton does not tell us that. He never tells us what F is in a given context, with the exception of gravity where he also furnished the left-hand side of $F = ma$ with his law of universal gravitation. If two like charges repel each other, we need Coulomb's law to tell us that the force varies as $1/r^2$. The nuclear force, say between a proton and neutron, falls exponentially with distance. Surely Newton did not tell us this. But once we have experimentally determined a new force F, we can use his second law to deduce the a it would produce, assuming classical mechanics is applicable.

When a new force, like the electric force, is discovered, the F due to it can be measured in one of two ways. One is to compare it to a known force that balances it. For example, to find how the repulsive force between two charged bodies varies with distance, we can glue them to the two ends of a spring of known k and measure by how much it is extended in equilibrium. Another way is to release the charges (of known mass) from rest with some separation, measure the initial acceleration of either, and then compute ma.

So physicists are busy either finding a from F (as when computing the orbit of a satellite given the force law for gravity) or F from a (as when stretching springs to find k or dropping apples from trees to measure the force of gravity).

Now, a small digression on gravity. Consider a body near the surface of the earth, the force of gravity on it being $F = -mg$ where $g = 9.8ms^{-2}$. That's something you find out by dropping things from a tower. Consider a in the field of gravity. We find it is

$$a = \frac{-mg}{m} = -g \qquad (3.7)$$

for all bodies. That's a very remarkable property of the gravitational force— the cancellation of m. If you look at the electrical force, on the proton and electron for example, it's not proportional to the mass of either object. It's proportional to the electric charge of the object. Therefore, when you divide by the mass to get the acceleration, the response of different bodies is inversely proportional to the mass. But gravity has a remarkable property that the pull of the earth is itself proportional to the mass of the object. So, when you divide by the same m to find a, it cancels and everything falls at the same rate on the surface of the earth. In fact, that is a property of gravitational fields anywhere, even in outer space. Everything—gold, silver, diamonds, particles, elephants—all accelerate the same way. This

remarkable fact was known for a long time, but it literally took an Einstein
to figure out why nature behaves in that fashion, why the two masses associated with a body are equal. One is the *inertial mass*, which is how much
a body hates velocity change, how hard it resists acceleration, the mass
in $F = ma$. That quality can be measured far from planets, far from everything. The other is *gravitational mass*, which is the measure of how much it
is attracted to the earth or any other body. There's no reason why these two
attributes had to be equal. Is this just an accident or is it part of a big picture? It turns out that it's part of a big picture called the general theory of
relativity. Here is Albert Einstein's description of gravity. Imagine a stream
and some kids dropping various leaves or paper boats in it. No matter what
they drop (within reason) the object's trajectory will follow the flow lines
of the water. The path of all objects is predetermined. This is what gravity
does to spacetime—it defines trajectories for objects: anything you release
will follow the trajectory etched in spacetime. If you oppose this flow, like
a kid holding on to his paper boat, the resistance you feel is the weight of
that object. What determines the flow lines at each point? The matter and
energy in the universe, as dictated by Einstein's equations.

   Here's a simple example of a complete Newtonian problem. A mass
is attached to a spring. It is pulled by a certain amount x and then released.
What is it going to do? Newton says $F = ma$, which in this case becomes

$$m\frac{d^2x}{dt^2} = -kx. \tag{3.8}$$

To proceed, we must know m and k, and I have already discussed how
these are measured. Now we have a mathematically complete problem:
find the function $x(t)$ whose second derivative is equal to $-\frac{k}{m}$ times the
function $x(t)$. Then, we go to the math department and say, "What's the
solution to this equation?" This is a problem in mathematics and the
answer—that it's going to be oscillating back and forth—will come from
doing the math. Later we will do some of that math ourselves. For now, I
am simply pointing out that once we have stated the laws in mathematical
form, solving for the consequences is a mathematical problem.

   Here is another example. Newton discovers a force of gravity acting
on everything. Here's the sun in Figure 3.2, orbited by a planet. At this
instant, the planet may be moving at some velocity **v**. The acceleration of
the planet is due to the force of gravity between the planet and the sun,
which Newton tells you is directed toward the sun, proportional to the

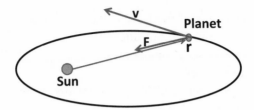

Figure 3.2   The planet is separated from the sun by **r**. The force **F** on it was determined to be always opposite to **r**, i.e., pointing toward the sun and falling as $\frac{1}{r^2}$. (The force is slightly offset from **r** for clarity.)

product of the two masses, and which decreases with distance as $\frac{1}{r^2}$. This completely specifies the left-hand side of $F = ma$. That's the *law of universal gravitation*. Then again, because the second derivative of the position is connected to the position, you go to the mathematicians and say, "What orbit is the solution to that equation?" and they will tell you it is an ellipse.

Of course, Newton did not have mathematicians he could go to. He *was* the math guy. Not only did he formulate laws of gravitation, he also invented calculus and figured out how to solve the differential equation that came out of his $F = ma$. There has been no one like him. Here I speak of Newton the scientist; Newton the man had many flaws.

3.4 Newton's third law

The third law says that if there are two bodies, called 1 and 2, then F_{12}, the force on 1 due to 2, is minus the force on 2 due to 1:

$$F_{12} = -F_{21}. \tag{3.9}$$

Action and reaction are equal and opposite. Coulomb's law and the law of gravity both have this feature. You may assume it for every force in this course.

We are now going to put the laws to work. You have to be good at writing down the forces acting on a body. That's what all these problems are going to boil down to. *Do not forget the existing forces and do not make up your own forces.* I have seen both happen. Every force, with the exception of gravity, is a force due to direct contact with the body: a rope is pulling it, a rod is pushing it, you are pushing it, you are pulling it, one block is pushing another, and so on. Gravity is one force that acts on a body without the source of the force actually touching it. (Later electro-magnetic forces can come in, but not in this book.) That's it. Once more with feeling: With the exception of gravity all forces we will discuss in this book will be contact forces.

We are going to begin with simple problems in mechanics. They will get progressively more difficult. Let's start with our first triumph. There is some object of mass 5 kilograms and I apply $10\ N$ on it. What's the acceleration? Everyone knows it is $a = \frac{10}{5} = 2ms^{-2}$. You may have done this before, but I hope now you understand how we know the force is $10\ N$ and how we know the mass is 5 kg. The algebra is, of course, very trivial here.

Next, I have a 3-kg block placed against a 2-kg block, and I'm push-ing the former with $10\ N$ as shown in Figure 3.3. I want to know what happens. One way is just to use your common sense and realize that these two blocks are going to move together. You know intuitively that if they move together, they will behave like an object of mass 5 kg and the acceler-ation will again be $2ms^{-2}$. What about gravity? What about the force due to the table on which the masses are moving? Imagine that this occurs in outer space where there is no gravity and no need for a table.

There's another way to do this problem, which is to draw *free-body diagrams*. Here you can pick any one body that you like and apply $F = ma$ to it, provided you identify *all* the forces acting *on* that body. We'll first pick the 3-kg mass. My $10\ N$ is certainly acting on it. What other force is acting? The force of the 2 kg, which has a magnitude f acting to the left.

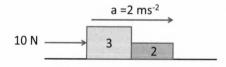

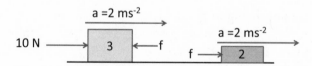

Figure 3.3   Top: A force of 10 N acts on the two blocks, viewed as a single entity. Bottom: The free-body diagram for the two blocks showing all the forces *on* each block. Notice the third law is being invoked.

Do not include the force exerted *by* the 3 *kg* on the 2 *kg*. Next consider the 2-*kg* mass. There is the same f, but acting to the right by the third law. Here is the mistake some people make: they add to that the 10 Newtons. They feel that the 2 *kg* will surely feel it since that is what is behind all the acceleration. That will be a mistake. That's an example of adding a force that you should not be adding. The only force acting on this little guy is this little f.

Now we do $F = ma$ for these two guys:

$$10 - f = 3a \tag{3.10}$$

$$f = 2a \tag{3.11}$$

$$10 = 5a \quad \text{upon adding the previous two equations} \tag{3.12}$$

$$a = 2. \tag{3.13}$$

Notice I'm using the same acceleration for both. I know that if the second mass accelerated faster than the first one, then the picture is completely wrong; it will not feel the force due to the first. If it accelerated less than the first, the first would have plowed into the second. Since that also cannot happen, they're moving with the same acceleration. There's only one unknown a. Once you got $a = 2$, you can go back to Eqn. 3.11 and obtain $f = 4$ N. Now we know the full story: 4 N acting on 2 *kg* gives it an acceleration of 2 ms^{-2}, while $(10 - 4) = 6$ N acting on the 3 *kg* gives the same $a = 2ms^{-2}$.

Here's another variation shown in Figure 3.4. I have a 3-*kg* mass attached by a rope to a 2-*kg* mass, which I pull with a force of 10 N. Again, your common sense tells you that I am pulling something whose effective mass is 5 *kg*; the answer is $2ms^{-2}$. Let's confirm that systematically by using the free-body diagrams in the lower half of the figure. Now, there are really three bodies here: the two blocks and the rope connecting them. In all these examples, I assume the rope is massless. We know there is no such thing as a massless rope, so what we mean is a rope whose mass is negligible compared to the two blocks being pulled. We'll take the idealized limit where the mass of the rope is 0. The 3 *kg* is being pulled by the rope to the right with a force that I'm going to call T, which stands for *tension*. The rope is being pulled backward by the 3 *kg* with a force T by the third law. What is the force on the other end of the rope? What should that be? If you said T, that is right, but not because the rope would

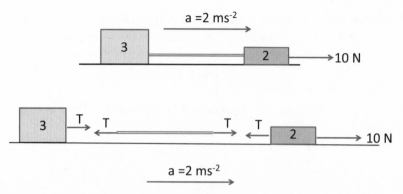

Figure 3.4   (Top) A force of 10 N pulls the two blocks connected by a massless rope, all treated as a single entity. (Bottom) The free-body diagram for the two blocks and the rope, which experiences equal and opposite forces of magnitude T called the tension.

snap otherwise. Something else will be a problem. If the two forces on its ends don't cancel, you have a net force. What are you going to divide by to get the acceleration? Zero, right? So, a massless body cannot have a net force on it, because its acceleration would then be infinite. Massless bodies will always have equal and opposite forces on the two ends. In the case of the rope, this is called the *tension* on the rope. The tension is not 0 just because this T and that $-T$ cancel. Suppose you are being pulled by my favorite animals—the elephants—from both sides by equal and opposite forces. You won't find any consolation in the fact that these forces add up to 0 as you get subdivided.

Now, you can invoke $F = ma$ for the three objects, starting from the left:

$$T = 3a \qquad\qquad\qquad (3.14)$$

$$T - T = 0\,a \qquad\qquad\qquad (3.15)$$

$$10 - T = 2a. \qquad\qquad\qquad (3.16)$$

We add the three equations and we get $10 = 5a$ or $a = 2ms^{-2}$ and $T = 6\,N$. So, the tension on the rope is 6 Newtons. This is very important: when you buy a rope, the specifications will tell you how much tension the rope can take before it will snap; if your plan is to accelerate a 3-kg mass with an acceleration of 2 ms^{-2}, you better have a rope that can take the tension of 6 N.

If you see the rope in isolation you find its acceleration is indeterminate according to Eqn. 3.15. This is correct; it is the non-zero masses that determine a for everybody, and the rope goes along with this a for free.

3.5 Weight and weightlessness

Now let's see what happens when you ride an elevator. In Figure 3.5 you are the stick figure standing on the weighing scale on the floor of an elevator that has a positive (upward) acceleration a. What will the scale register? We will draw free-body diagrams as in the right half of the figure for you and the massless spring. The spring is being pushed down by you with a force W and up by the floor with the same force since it is massless. It will compress by an x such that $W = kx$, and x will somehow be displayed by a needle or digital readout.

(We should note a subtle thing that you may not have realized. Every [massless] spring is pushed or pulled by equal and opposite forces $\pm F$ at the two ends since otherwise it would have $a = F/0 = \infty$. So which of these two appears in $F = -kx$? Recall the mass spring system. One end

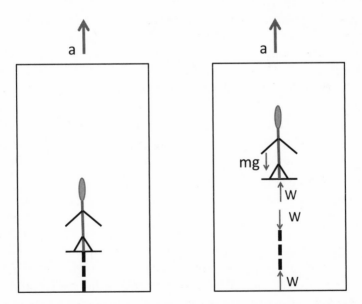

Figure 3.5   (Left) An elevator is accelerating at a rate a and you are standing on a scale whose spring is shown by a dotted line. (Right) The free-body diagram for you and the spring. The spring is compressed at each end by a force W.

of the spring is anchored to the wall. When we pull the other end by $+x$, we apply a force $F = +kx$ to the right. The wall exerts a force $F = -kx$ at the other end. In response to the $+kx$ we apply, the spring exerts a force $F = -kx$ on us or on the mass attached to it, and that is the F we use in writing down $F = ma$ for the mass.)

The equation describing you in the elevator is

$$W - mg = ma, \text{ which means} \tag{3.17}$$

$$W = m(g + a). \tag{3.18}$$

If the elevator is standing still and $a = 0$ we see that $W = mg$ is just your weight. If the elevator is accelerating upward $W = m(g + a)$, and the needle will show a number greater than your weight. You will actually feel heavier. The spring not only has to support you from falling through the floor but also has to accelerate you counter to what gravity wants to do. That's why we have the $g + a$. Say you picked up some speed and are now coasting upward at a steady speed. Then, $a = 0$ again and $W = mg$. As you come to the top of the building, the elevator has to decelerate so that it can lose its positive velocity and come to rest. So a will be negative now. If a is negative, then $g + a < g$. Let us write in this case $a = -|a|$, so that

$$W = m(g - |a|), \tag{3.19}$$

which makes it explicit that $W < mg$. You will feel that your weight is reduced.

On the way down your initial acceleration is negative because you're picking up speed toward the ground. You will feel less heavy. You can see that if $|a| = g$, your downward acceleration is that due to free fall under gravity and you will feel weightless. This can happen when the cable has snapped. You don't feel any weight because your normal weight is the opposition the floor offers to keep you from falling through it; but now the floor is also falling at the same rate. It's wrong to think that when you feel weightless you have escaped the pull of gravity. We all know that in a falling elevator you definitely do not escape the pull of gravity. It's going to catch up with you in a few seconds. The same goes for the people floating around in space stations. They have not escaped the pull of gravity either; they have just stopped fighting it, and they are all accelerating toward the earth at a rate $\frac{v^2}{r} = g^*$, where g^* is the (reduced) acceleration due to gravity at the radius r of the orbit. If they had really escaped the pull of gravity, their spaceship would be off to the far reaches of the universe.

CHAPTER 4

Newton's Laws II

4.1 A solved example

The goal of physics is to be able to predict something about the future, given something about the present. I'm going to provide a very simple example that illustrates how Newton's laws are to be used for this purpose. This treatment will be brief since I will return to this problem in greater detail later.

Figure 4.1 shows a frictionless table, on which is a mass m attached to a spring of force constant k. The other end of the spring is attached to an immobile wall. The dotted outline of the mass shows it when it is displaced by an amount x from equilibrium. I want to pull the mass by some amount A and let it go. So that's the knowledge of the present. What's this guy going to do? That's the typical physics problem. It can get more and more

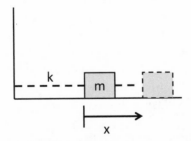

Figure 4.1 A mass m attached to a spring of force constant k in its equilibrium position, when the spring is neither extended not compressed. The dotted outline of the mass shows it when it is displaced by an amount x from equilibrium.

complex. You can replace the mass by a planet; you can replace the spring by the sun, which is attracting the planet; you can bring in many planets; you can make it more and more complicated. But they all boil down to a similar exercise: I have some information now and I want to be able to say what will happen next.

Combining $F = ma$ and $F = -kx$ we obtain the equation that seals the fate of the mass:

$$\frac{d^2x}{dt^2} = -\frac{k}{m}x(t). \tag{4.1}$$

This is a *differential equation*: you have to find a function $x(t)$ whose second derivative is $-\frac{k}{m}$ times itself. A differential equation tells you something about an unknown function $x(t)$ in terms of its derivatives, and you are supposed to find it given this information. Instead of running to the mathematicians, let us solve this equation by guessing, which is a totally legitimate way to solve a differential equation.

Here is how we guess the answer. Let's make our life simple by taking a case where $k = 1, m = 1$. Later on, we can put back any k and m. I'm looking for a function whose second derivative is minus that function. Now, as a word problem, it rings a bell, right? Do you know such a function? Exponential is good. Trigonometric functions are good too, provided you mean sin and cos. I'm going to dismiss the exponential, which is actually a very good guess. If you took $x(t) = e^t$, we have

$$\frac{d^2x}{dt^2} = e^t \neq -x(t). \tag{4.2}$$

It does not help to consider $x(t) = e^{-t}$ since the minus sign will get squared when you take two derivatives, and you will still end up with $+x(t)$. Now $e^{\pm it}$, where $i = \sqrt{-1}$, will work, but we do not want to deal with complex numbers yet.

We just want a function that reproduces itself when we take *two* derivatives. So, I make a guess,

$$x(t) = \cos t, \tag{4.3}$$

and you can check that it works:

$$\frac{d^2x}{dt^2} = \frac{d^2\cos t}{dt^2} = -\cos t = -x(t). \tag{4.4}$$

Did you follow the way this very elementary problem is solved? It's solved by making a guess. But something is not quite okay with this solution. If I put $t = 0$, I get $x = 1$. Why should it be true that I initially pulled it by exactly one meter? I could have pulled it by 2 or 3 or 9 meters. I want to be able to decide how much I pulled it by at $t = 0$. Suppose it was pulled to 5 m and released. I want $x(0) = 5$. This can be arranged by making a choice

$$x(t) = 5\cos t. \tag{4.5}$$

Does the 5 screw up everything? It doesn't, because it just comes along for the ride:

$$\frac{d^2x}{dt^2} = \frac{d^2[5\cos t]}{dt^2} = -5\cos t = -x(t). \tag{4.6}$$

Now I have an answer that does everything I want it to do. At the initial time, it gives me 5 times $\cos 0$, which is 5; and that's what I said was the initial displacement. I find $\frac{dx}{dt} = -5\sin t$, which vanishes at $t = 0$. That too is correct; I pulled it and let it go. So, the instant I released it, it had no velocity. It satisfies Newton's laws, and that's the answer. I want to do this simple example in totality because this is the paradigm. This is the example after which everything else is modeled.

There are two final points about this solution.

First, there is the other option $x(t) = \sin(t)$ that occurs to some students. There is nothing wrong with this option; it just is not needed in this particular example. Equation 4.5 is not the final and most general solution to the problem; it is a solution that certainly works for the one example I had, in which a mass is pulled to 5m and released.

Second, when k and m are not both 1, we go back to the general case

$$\frac{d^2x}{dt^2} = -\frac{k}{m}x(t). \tag{4.7}$$

The solution to this is

$$x(t) = A \cos \sqrt{\frac{k}{m}} t \qquad (4.8)$$

where A, which was 5 in our example, is arbitrary in general and is called the *amplitude*. We will deal with the oscillator in depth later. At this point, we just want to get a feeling for how $F = ma$ is applied in real life, with some help from the mathematicians.

4.2 Never the whole story

Is this the whole story of the mass and spring, or is something missing? Does the mass oscillate forever as predicted by the Eqn. 4.1 we just solved? It does not: it eventually comes to rest. If you never knew about friction, and you solved for $x(t)$ with just the spring force, you would find it didn't work. So, you are at a fork in the road. You can say either that Newton's laws are incorrect or that you are missing some forces. In the latter case, you will look around and eventually deduce the frictional force by insisting that $F = ma$ works with its inclusion. We will see how this is done later in this chapter.

It turns out that even if there were no friction Eqn. 4.1 would not be the correct law, which is given by Einstein's relativistic dynamics. By that I mean, if you really pull a real spring by $5m$ on a frictionless surface and release it, its motion will not be exactly $5 \cos t$; it will be off by a very, very, very tiny amount that you probably will not discover in most laboratories. But if the mass begins to move at a velocity that is comparable to that of light, then this equation will make the wrong predictions. On hearing that even Newton can be wrong in this sense, some people say, "What kind of business are you in? Every once in a while some authority is proven wrong." I'll tell you right now. Everything we know is wrong at some level. Newton didn't try to describe things moving at speeds comparable to light. He dealt with the problems he could deal with at that time. So his laws have a limited domain of validity. You can always push the frontiers of observation until you come to a situation where any given law breaks down. While the special theory of relativity does better than Newtonian mechanics for large velocities, it too fails if the mass becomes very tiny, of atomic dimensions. Then you need the laws of quantum mechanics discovered by Heisenberg, Dirac, and Schrödinger, and of course their laws also have problems in some experimental domain. Sometimes we correctly abandon the formalism; but we should not be too eager to do that. In the present case, the problem is not with formalism but with not including friction. The failure at the quantum level comes not because we did not include some forces, but because the very notions of force or trajectories $x(t)$ seem to be invalid at that scale.

4.3   Motion in $d = 2$

We're going to move on to higher dimensions, where position, velocity, acceleration, and force are all vectors. So let's again start with simple problems in $d = 2$ and make them more and more complex. There is no limit to how difficult mechanics problems can be. If you go back and read some Cambridge University exams from 1700 or 1800, you will find some really difficult problems. Finally, quantum mechanics was invented and life got a lot easier.

Consider a mass m sitting on a table as shown in the left half of Figure 4.2. Because we're in two dimensions, we need to have two axes x and y. Recall that

$$\mathbf{F} = m\mathbf{a} \tag{4.9}$$

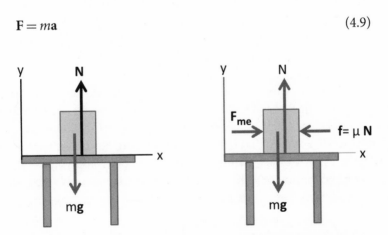

Figure 4.2   A mass m sitting on a table, (left) without and (right) with friction. The coefficient $\mu = \mu_s$ if the block is at rest and $\mu = \mu_k$ if it is moving.

is a vector equation. If two vectors are equal, then their x and y components have to match:

$$F_x = ma_x \qquad\qquad (4.10)$$

$$F_y = ma_y. \qquad\qquad (4.11)$$

I'm going to apply them to this block. There are no known forces acting in the x direction, and the block is not moving in that direction. Therefore it's a case of $0 = 0$. Now I look in the y direction. In the free-body diagram there is the force of gravity mg acting down the y-axis; I will often denote this vector by $m\mathbf{g}$, where $\mathbf{g} = -9.8\mathbf{j}ms^{-2}$. If that's all you had, the block would fall through the table. Because the block is not falling, we know the table is exerting an opposing force denoted by $\mathbf{N}$, where N stands for *normal*. And *normal* is a mathematical term for perpendicular. Evidently $\mathbf{N}$ is a positive force and $m\mathbf{g}$ is a negative force in the y direction. Thus

$$N - mg = ma_y. \qquad\qquad (4.12)$$

In this application of the Newtonian equation, I know the right-hand side. It is 0 because I know this block is neither sinking into the table nor flying off the table. It's sitting on the table; it has no velocity or acceleration in the y-direction. I come to the conclusion

$$N = mg. \qquad\qquad (4.13)$$

4.4 Friction: static and kinetic

Now I'm ready to introduce another force, the force of friction, $\mathbf{f}$. How do we infer there is a force of friction? Consider the mass on the table in the right half of Figure 4.2. What experiment will tell you there is another force called friction? You find that to keep the mass moving at constant velocity, you have to apply a force. That means the force you're applying is canceled by something else, because there is zero acceleration. Here is another good answer: You give the mass a push and soon it stops moving; there must have been a force that produced the deceleration. But even before that, even before it starts moving, you find that, if you push it, it doesn't move. Say I push the podium. If I push it gently, it does not move. But if I push hard enough it begins to move. What's happening before it moves? I'm applying a force and I'm getting nothing in return for it. So I know there is another force exactly balancing mine, and that's the force of *static friction*. Let us call the force I am applying F_{me}. I'm applying it to the right, and so there has to be another force of equal magnitude to the left. Notice it is not a fixed force: it's whatever it takes to keep the mass from moving. It will not be less than what I apply, because then the mass will move; it cannot be more than what I apply because then it will start moving me backward. So, static friction is a force that has a range from 0 to some maximum. The maximum turns out to be

$$f_s = \mu_s N \qquad\qquad\qquad\qquad (4.14)$$

where μ_s is a number called the *coefficient of static friction* and N is the normal force. In our example you may use $N = mg$. So the force of friction seems to depend on how heavy the object is that's sitting on the table. But it doesn't depend on the area of contact. For example, it does not increase if we rest the block on another face with greater area. You might expect more friction because there's more contact. But that is not so, for reasons not readily explained within our elementary treatment. In reality, friction has a subtle origin at the atomic level.

If static friction provides a force equal to the force that I apply, up to a maximum of $\mu_s N$, what happens when I exceed the maximum? The object I am pushing will start moving. Once it starts moving, the frictional force, which is always directed opposite to the velocity, changes to

$$f_k = \mu_k N \qquad\qquad\qquad\qquad (4.15)$$

where μ_k is the *coefficient of kinetic friction*. We find that in all situations $\mu_k < \mu_s$. Let us say $\mu_s = 0.25$ and $\mu_k = 0.2$ for the block on the table. If I apply a force that is 0.1 times its weight, it won't move; at 0.2 times its weight it won't move; at a quarter of its weight it's a tie; at 0.26 of its weight the block will start moving. Once it starts moving, the frictional force will drop to 0.2 times its weight, and the body will accelerate if I maintain the same force as when it first begins moving.

4.5 Inclined plane

Now we are going to do the one problem that has sent more people away from physics than anything else. It is called the inclined plane. A lot of people who do not remember where they were during the Kennedy assassination say, "I remember the day I saw the inclined plane; that's the day I decided I'm not going into physics." This is very bad publicity for our field. You come into a subject hearing about relativity and quantum mechanics, and we hit you with this. So why am I still doing this? Because this is the entry ticket into the business. It is inconceivable to me that you could understand more advanced topics without being able to understand this one. Go ahead and mock the inclined plane, but only after you can prove you have mastered it.

Here is the notorious problem. There is a mass m sitting on a plane, inclined at an angle θ as shown in the left half of Figure 4.3. We know it's going to slide down the hill, but we want to be more precise. The whole purpose of Newton's laws is to quantify things for which you already have an intuition. The novel thing about the inclined plane is that for the first time we are going to pick our x and y axes not along the usual directions, but parallel and perpendicular to the incline. What are the forces on this mass? I have told you, first deal with contact forces. But the only thing in contact with the mass is the plane. In general the inclined plane can exert a force along and perpendicular to its own surface, but I'm first going to take a case where there is no friction. By definition the plane cannot exert a force along its own length, so it can only exert a normal force N. Then there's only one other force, the force of gravity that we agreed we have to remember. Even though the earth is not touching this block, it is able to reach out from down below and pull this block down. These are the two forces, and the mass will do what these forces tell it to do.

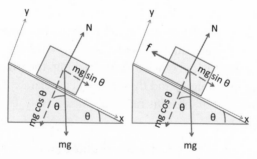

Figure 4.3   A mass m sitting on an inclined plane, without friction (left) and with friction (right).

First I have to take this *mg* pointing down and break it up into components along my *x* and *y* axes. That's called *resolving the force* into various directions. Now, the key to this is to know that this angle θ of the plane is the same as the angle between vertical and the normal to the plane. You have to agree that if I draw a line perpendicular to the horizontal and I draw a line perpendicular to the incline, the angle between those two perpendiculars will be the same as the angle between the two lines with which I began. This is because "make perpendicular" means "rotate by $\frac{\pi}{2}$." If you rotate both lines by $\frac{\pi}{2}$, the angle between the rotated lines will be the same. The vertical is perpendicular to the horizontal, and the normal perpendicular to the inclined plane. Once you understand that, the rest is easy.

So here are my equations:

$$mg \sin \theta = ma_x \qquad\qquad\qquad (4.16)$$

$$N - mg \cos \theta = ma_y. \qquad\qquad\qquad (4.17)$$

Now we know that this block is sliding down the incline. It's not going into the plane, and it is not flying out of the plane. That's the reason we measure our *y* coordinate perpendicular to the plane: it does not change, unlike the traditional *x* and *y*, both of which do. So $a_y = 0$ in Eqn. 4.17, which implies

$$N = mg \cos \theta. \qquad\qquad\qquad (4.18)$$

Then you come to Eqn. 4.16. We cancel *m* and read off the answer

$$a_x = g \sin \theta. \qquad\qquad\qquad (4.19)$$

That's the big result, that the mass will slide down the hill with an acceleration $g \sin \theta$. This provides a good way to measure *g*, because if you drop something vertically, it falls too fast for you to time it. But if you let it go down an incline, by making θ very small, you can reduce the acceleration by a factor $\sin \theta$.

Here's another thing I should tell you right now. The professionals don't put in numbers until the very end. So, if you are told the incline is an angle of 37 degrees, and $g = 9.8 \, msec^{-2}$, don't start putting numbers into the first equation. I know for some of you it's traumatic to work with symbols. There are many reasons to work with symbols rather than with numbers till the end. First of all, if you have already put the numbers in and I suddenly tell you, "Hey, I was wrong about the slope, it's really 39 degrees and not 37," you are forced to do the whole calculation again. But if you use symbols to derive the formula for a_x and then ask me, "What's your θ?" you can just plug that in. Likewise, if I change the value of g because someone made a better measurement, you simply change g in the very last expression for a_x. You can also see if your answer had some mistakes in it. Suppose you got $a_x = g^2 \sin\theta$. You will know it's wrong because the units do not match. Maybe you have the trigonometry wrong. Maybe it's really $a_x = g \cos\theta$? You can do a few tests. For example, as the plane becomes less and less inclined, you have to get less and less acceleration, and when $\theta \to 0$, so must a_x, which must then be proportional to $\sin\theta$. Or, if you make the incline almost vertical, the block is just falling under gravity with $a_x = g$. The result $a_x = g \cos\theta$ does not do that when we set $\theta = \frac{\pi}{2}$. Finally, $a_x = g \sin\theta$ is independent of m. That's an interesting property of the result you would not notice if you kept numbers everywhere. So we will agree to work with symbols until the end.

Now we are going to make life a little more complicated by adding friction, as shown in the right half of Figure 4.3. Let the block be at rest and imagine there are some hinges on the inclined plane that allow me to increase θ from 0 to $\frac{\pi}{2}$. I want to know when the block will begin to slide down. Let me cut to the chase and write the equations:

$$N - mg\cos\theta = ma_y \qquad\qquad (4.20)$$

$$mg\sin\theta - f = ma_x. \qquad\qquad (4.21)$$

The first equation is the same as before and tells us that since $a_y = 0$,

$$N = mg\cos\theta. \qquad\qquad (4.22)$$

In the second equation, do not write $f = \mu_s N$ because the frictional force is not always $\mu_s N$; it is whatever it takes to keep the block still, up to a maximum of $\mu_s N$. In other words, if θ is very small, the frictional force will in fact be a much smaller amount $f = mg\sin\theta$, which you get by setting $a_x = 0$ in the right-hand side.

Let us crank up the angle to θ^*, beyond which it cannot stay still. At this angle the friction is maximum and we can assert

$$mg \sin \theta^* = \mu_s N = \mu_s mg \cos \theta^*, \qquad (4.23)$$

which gives us a way to measure μ_s:

$$\mu_s = \tan \theta^*. \qquad (4.24)$$

Again, the mass cancels. So, it doesn't matter how heavy a car you parked on the slope, as long as μ_s was the same. But g cancels too. So, whether you park your car on the earth or park it on another planet, the same restriction $\tan \theta \leq \mu_s$ applies.

When we pass this limit, when $\tan \theta > \mu_s$, the block will begin to slide down. It will now have a non-zero downhill acceleration because it is kinetic friction $\mu_k N$ that now opposes the velocity and $\mu_k < \mu_s$. The operative equations are

$$N - mg \cos \theta = ma_y \qquad (4.25)$$
$$mg \sin \theta - \mu_k N = ma_x. \qquad (4.26)$$

What's the acceleration now? I take the y equation that says as before that $N = mg \cos \theta$ since $a_y = 0$, and plug that into the x equation to obtain the downhill acceleration of the block:

$$mg \sin \theta - \mu_k mg \cos \theta = ma_x \qquad (4.27)$$
$$g(\sin \theta - \mu_k \cos \theta) = a_x. \qquad (4.28)$$

4.6 Coupled masses

Next is a problem of two masses m and M connected by a string that goes over a massless pulley mounted at the top of a frictionless inclined plane, as shown in Figure 4.4. What can we say before doing the math? If $M > m$ am I assured it'll go downhill? No, it depends on the angle θ because only part of its weight, $Mg \sin \theta$, is helpful in going downhill while all of mg is pulling m down. All we know for sure is that for a fixed θ, the bigger mass M will go down the slope as $M \to \infty$ while the smaller mass m will go

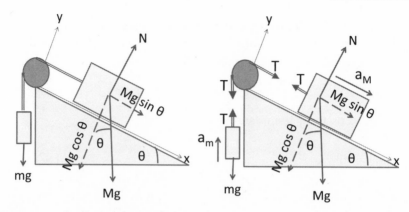

Figure 4.4   (Left) Two masses linked by a massless rope and massless pulley. (Right) The free-body diagram.

straight down as $\theta \to 0$. Let us figure out now when exactly the balance shifts.

Since we have two masses here, no choice of axes is going to make all their motions simple: m goes straight up and down while M glides along the slope. Let us ignore the boring $N = Mg\cos\theta$ equation for the y coordinate and focus on the motion along the plane. In the free-body diagram I have shown the same tension T wherever the rope is involved. For a straight massless rope it is clear the opposing forces at the ends must have the same magnitude T since there can be no net force on a massless object. For a rope that curves around the pulley, the explanation is more subtle, as I will explain momentarily.

For M, the equation along the x-direction is

$$Mg\sin\theta - T = Ma. \tag{4.29}$$

For m the equation in the up direction is

$$T - mg = ma \tag{4.30}$$

where I have deliberately used the *same a* as in Eqn. 4.29. I am not saying the acceleration of M and m are equal as vectors. They do not even have the same direction: m moves up and down while M moves along the plane. By a I mean the components of the two accelerations in the allowed directions of motion, defined to be positive if M is moving downhill and m is moving up. This equality reflects the inelasticity of the rope: if m moves up one inch, M will slide downhill one inch along the incline. If the displacement is the same, so are the velocity and acceleration. We can solve for a by adding Eqns. 4.29 and 4.30 to obtain

$$Mg\sin\theta - mg = Ma + ma \quad \text{which means} \tag{4.31}$$

$$a = g\left[\frac{M\sin\theta - m}{m + M}\right]. \tag{4.32}$$

Does this solution make sense? Now you notice that for a to be positive and for M to go downhill, it's not enough that $M > m$; we need $M\sin\theta > m$ because only the force $Mg\sin\theta$ is pulling M downhill: the force $Mg\cos\theta$ is trying to ram it into the inclined plane, and that's being countered by the normal force N.

Can we use this formula when $a < 0$, that is, when M moves uphill? After all, I did all my analysis assuming it's going down. Yes, we can. All the forces I drew here—N, Mg, and so on—are not going to change when you change the body's direction of motion. Gravity is always going to pull down whether the block moves with it or against it. So once you have a formula for positive a, you can apply it to negative a as well. However, if friction is present, you cannot do that because you have to assume a particular direction of motion before you can assign a direction to the frictional force in the equations of motion.

Now let's consider why the forces at the two ends of the rope emerging from the pulley have equal magnitudes T. Assume that the pulley is massless and that the rope does not slip; that as the masses move and rope moves over the pulley, the pulley is forced to rotate without relative slippage. In this case the part of the rope instantaneously in contact with the pulley and the pulley may be viewed as one rigid body. (It is like a bicycle chain whose links mesh with the teeth on the wheel the pedal is turning.) The tensions on the rope at the two ends are clearly trying to rotate this body in opposite directions, as is clear in the free-body diagram in Figure 4.4. If these tendencies are not exactly balanced—that is, if these forces are not equal in magnitude—they would produce an infinite *rotational* acceleration of this massless rigid body, just like a non-zero force would produce infinite *linear* acceleration on a massless object. By the no-slip condition, this means an infinite linear acceleration for the two masses, which is impossible because the masses are non-zero and have finite forces acting on them. So the forces at the two ends of the rope emerging from the pulley must have equal magnitudes. (We will return to a proper study of rotations and the case of the massive pulley in Chapter 10.) Note that even though the rotational effects of the two non-parallel forces cancel, their non-zero vector sum can produce linear acceleration of the pulley. The axle about which the pulley rotates provides an equal and opposite force to prevent this. Luckily this last force does not contribute to rotations around the axle.

4.7 Circular motion, loop-the-loop

Now we'll turn to some interesting problems in circular motion of various kinds. The first, shown in part A of Figure 4.5, describes a ride in an amusement park. You sit in these baby rockets hanging from a rope along with other petrified victims. The whole thing begins to spin, and the rope, instead of remaining vertical, starts tilting at an angle θ, which we want to determine as a function of the tangential speed v and radius R of the circular orbit. We apply $\mathbf{F} = m\mathbf{a}$ and start listing the forces on the baby rocket. Gravity provides the usual mg pointing down. The rope can only exert a force T along its length. Let us trade the tension along the rope for the sum of two equivalent forces, a vertical part, $T_y = T\cos\theta$, and a horizontal part, $T_x = T\sin\theta$ directed toward the center of the circular orbit. The equations are

$$T\cos\theta - mg = 0 \quad \text{vertical} \tag{4.33}$$

$$T\sin\theta = ma = m\frac{v^2}{R} \quad \text{radial or horizontal.} \tag{4.34}$$

In the first equation, we recall that by assumption the rocket's orbit is a horizontal circle and it has no net acceleration in the vertical direction. In the second, we recall from Section 2.7 that a body moving in a circle of radius R at speed v has a centripetal acceleration $\frac{v^2}{R}$. So this is a case where we know $\mathbf{a}$ in $\mathbf{F} = m\mathbf{a}$. Eliminating T we find

$$\tan\theta = \frac{v^2}{Rg}. \tag{4.35}$$

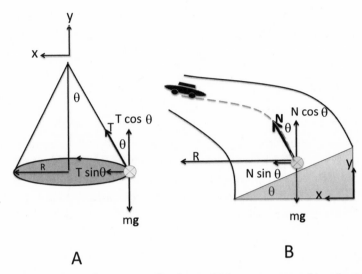

Figure 4.5   (Left) An "amusement" ride in which you go around in a horizontal circle of radius R in a baby rocket. The rope supporting you necessarily makes an angle $\theta = \tan^{-1} \frac{v^2}{Rg}$ with the vertical. (Right) A car going around a circular racetrack of radius R at speed v. The road has a banking angle that obeys $\tan \theta = \frac{v^2}{Rg}$. The normal force N has a horizontal part that gives the necessary force $\frac{mv^2}{R}$ to bend the path into a circle. The symbol $\otimes$ is used in both parts to indicate that the rocket or car is going into the page, away from you. The convention is based on how an arrow with feathers would appear going away from you. Likewise $\odot$ indicates an arrow coming out of the page toward you.

It is worth finding the tension on the rope since our lives may depend on it. It is

$$T = mg\sqrt{1 + \left(\frac{v^2}{Rg}\right)^2} \qquad\qquad (4.36)$$

using $T = \sqrt{T_x^2 + T_y^2}$.

Here is another interesting problem. You are driving on a circular racetrack of radius R at a speed v. Suppose the plane of the road is strictly horizontal, perpendicular to $\mathbf{g}$. Some agency has to exert a force $m\frac{v^2}{R}$ on the car to bend it into this circle. It is of course the road that does this, thanks to the frictional force $f \leq \mu_s mg$ directed toward the center. (I use

μ_s, the static coefficient, and not μ_k, the kinetic one, even though the car is moving, because we are discussing the force in the radial direction and the car has no velocity in that direction, unless it is skidding. Note also that the car does not really have to travel in a circle; all we need is that at this instant the trajectory is part of some circle of radius R.)

If you don't have the requisite static friction, if $\mu_s mg < m\frac{v^2}{R}$, your car will not be able to make the curve; it will fly off. But there is a clever way in which you can still make the turn without any friction, and that is to bank your road by an angle θ as shown in the right half of Figure 4.5. Imagine now you're going into the paper. The frictionless road only exerts a normal force N. Let us resolve that force into a vertical part $N \cos \theta$ and a horizontal one $N \sin \theta$ directed toward the center of the circle. The equations are

$$N \cos \theta = mg \qquad\qquad (4.37)$$

$$N \sin \theta = m\frac{v^2}{R}. \qquad\qquad (4.38)$$

Eliminating N we find the banking angle to be

$$\tan \theta = \frac{v^2}{Rg}. \qquad\qquad (4.39)$$

Let me elaborate. You want the car to go around the bend at a certain speed. If you bank your road at that angle, you don't need any friction to make the turn. Even though the frictionless road can only exert a normal force, thanks to banking, a part of the normal force points toward the center, providing the requisite centripetal force. Of course, when you drive on a real road, you do not have to travel at exactly this speed for a given R, for any small differences will be made up by the frictional force of the tire. It is just that you do not want to rely on friction for the entire radial force.

Finally, the famous loop-the-loop problem, which defies common sense, is shown in Figure 4.6. You come down on a roller-coaster track from some height H, you go on a vertical circle, and for a while you are upside down. The eternal question is, "Why don't you fall down?" We'll find we can understand this phenomenon fully with Newton's laws. The forces on the coaster are mg acting down and the force of the track, which has to be normal to it since it is assumed to be frictionless. But

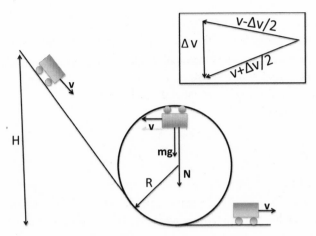

Figure 4.6   The roller coaster comes down from a height H and goes into a loop in the vertical plane. Why does it not not fall down? The forces on it are mg acting down and the track force N also acting down! It does fall, as explained in the text. The three arrows forming a triangle in the inset show the initial velocity $\mathbf{v} - \Delta\mathbf{v}/2$ just before it reaches the top, the change $\Delta\mathbf{v}$ in a small interval near the top, and the final velocity $\mathbf{v} + \Delta\mathbf{v}/2$ just after it passes the top.

it too points down! We are doomed! Why don't we fall? The answer is that we do fall, that is, we accelerate downward, but this does not mean we get any closer to the center. The two forces mentioned combine to bend the coaster into a circular path and produce the requisite downward centripetal acceleration:

$$N + mg = m\frac{v^2}{R}.$$
(4.40)

Solving for N we find

$$N = m\left(\frac{v^2}{R} - g\right).$$
(4.41)

If N comes out positive, that is, points down in our convention, which happens if $\frac{v^2}{R} > g$, we are safe. If it comes out negative, that is, if

$$\frac{v^2}{R} < g,$$
(4.42)

it means the track exerts an *upward* force, which is impossible, unless there is some other mechanism, like a T-bracket, that goes under the track and supports the coaster even if it is just hanging upside down. I believe such things exist in real roller coasters, in case they get stuck at the top or do not go fast enough. In our idealized coaster, without any of this backup, the speed v must obey

$$v^2 > Rg \qquad\qquad (4.43)$$

to safely make the loop.

We will figure out the minimum height from which it must be released to satisfy this condition when we derive the law of conservation of energy.

Let us be sure to understand again why accelerating down does not always mean gaining speed toward the earth. If you drop an apple, accelerating down means really picking up speed toward the ground, toward the center of the earth. It starts with zero vertical speed and picks up speed. In our example, the coaster is also accelerating, but the tiny change in velocity in a tiny time, which points *radially down* in Figure 4.6 near the top of the loop, is now added to a huge *horizontal* velocity pointing to the left.

We will now see that this implies the velocity has a constant magnitude and changing direction. Consider the impact of adding a tiny change $\Delta \mathbf{v}$ to a velocity $\mathbf{v}$ on the magnitude of velocity. Assume $\Delta \mathbf{v}$ can be in any direction relative to $\mathbf{v}$. The resultant velocity has a magnitude squared given by

$$|\mathbf{v} + \Delta \mathbf{v}|^2 = (\mathbf{v} + \Delta \mathbf{v}) \cdot (\mathbf{v} + \Delta \mathbf{v}) \qquad\qquad (4.44)$$

$$= \mathbf{v} \cdot \mathbf{v} + 2\mathbf{v} \cdot \Delta \mathbf{v} + \Delta \mathbf{v} \cdot \Delta \mathbf{v} \qquad\qquad (4.45)$$

$$= v^2 + 2\mathbf{v} \cdot \Delta \mathbf{v} + |\Delta \mathbf{v}|^2 \qquad\qquad (4.46)$$

$$\Delta v^2 = 2\mathbf{v} \cdot \Delta \mathbf{v} + |\Delta \mathbf{v}|^2 \qquad\qquad (4.47)$$

$$\frac{dv^2}{dt} = 2\mathbf{v} \cdot \frac{d\mathbf{v}}{dt} = 2\mathbf{v} \cdot \mathbf{a}. \qquad\qquad (4.48)$$

(To find $\frac{dv^2}{dt}$, we need to keep just the term linear in $\Delta \mathbf{v}$.) This equation generally implies a non-zero rate of change of the magnitude of $\mathbf{v}$, *unless $\Delta \mathbf{v}$ and the acceleration $\mathbf{a}$ are perpendicular to $\mathbf{v}$*, as is the case at all times in circular motion.

Figure 4.7   You fire two bullets from a tower at increasing speeds, which land farther and farther away (points 1 and 2). Beyond a critical speed, the bullet would go into orbit. It has of course never ceased to accelerate toward the earth.

So a little later, the total velocity vector merely gets rotated with no change in length and becomes tangent to the circle at a slightly different point. Going around in a circle is an example of constantly accelerating toward the center but not getting any closer.

Here is another example of this phenomenon. Suppose you are on a tower and you fire a gun horizontally as shown in Figure 4.7. The bullet hits the ground at point 1, under the pull of gravity. If you fire another bullet at a greater velocity it will land a little further away, at point 2. While greater initial speed will extend the time of flight even on a flat earth, the flight is further enhanced by the earth curving under the bullet. There will be a certain speed at which the bullet will keep falling but will not get any closer to the center, because the earth is falling under it just as fast. It is in orbit, as shown by the circle. That is in fact how you launch a bullet into orbit. So, what's the first thing you should do when you fire this gun? Move away, because it's going to come back in about 84 minutes and get you from behind at 17,650 miles per hour. This calculation assumes that there is no atmosphere, which of course leaves you with an even more pressing problem.

CHAPTER 5

Law of Conservation of Energy

5.1 Introduction to energy

The law of conservation of energy is a robust and powerful one. When the laws of quantum mechanics were discovered in the subatomic world, many cherished notions were abandoned. You must have heard the ugly rumors: particles do not have a definite position and definite velocity at a given time. They don't move along continuous trajectories. You might think the particle must have had an interpolating trajectory connecting two sightings, but it does not, and to assume it does causes conflict with experiment. While many of the ideas of Newtonian mechanics were abandoned, the notion of a conserved energy survived the quantum revolution. There was a period when people were studying nuclear reactions, and the energy they began with didn't seem to be the energy they ended up with. Niels Bohr, the father of the atom, suggested that maybe the law of conservation of energy was not valid in quantum theory. Then in 1931 Wolfgang Pauli decided to put his money on the law of conservation of energy; he postulated that some other tiny electrically neutral particle, which escaped detection, was carrying away the missing energy. That was a radical position to take in those days, when people did not lightly postulate new particles, as compared to today when if you don't postulate a few new particles you don't get your Ph.D in particle physics. Pauli's particle, called a *neutrino*, was detected after many, many years in 1959 by Clyde Cowan and Frederick Reines. Nowadays, neutrinos are one of the most exciting, elusive, and mysterious things one could study, and they hold the key to many puzzles concerning the universe.

Let's see how the notion of energy conservation arises, starting with one dimension. When a force acts on a body, it changes its velocity. In one dimension, this simply means speeding up or slowing down. We're going to find the relation between the speed accumulated when a force acts on a body for some time and the distance the body has traveled in that time.

Consider the case when the force F is constant, not varying with time. That produces an acceleration $a = \frac{F}{m}$. In the first chapter we learned that if a body has constant acceleration a,

$$v^2 = v_0^2 + 2a(x - x_0) \tag{5.1}$$

where v and x are the current velocity and position, and v_0 and x_0 the corresponding initial values. In those days of kinematics, we didn't ask, "Why does it have a constant acceleration?" We were just told, "It has a constant acceleration; just find out what happens." Now that we have learned dynamics, we know acceleration has a cause, namely some force. So I'm going to replace a by $\frac{F}{m}$ and make one more cosmetic change in notation. All the initial (final) quantities will carry a subscript 1 (2). Then Eqn. 5.1 takes the form

$$v_2^2 = v_1^2 + 2\frac{F}{m}d \tag{5.2}$$

where $d = x - x_0 \equiv x_2 - x_1$ is the distance traveled during this interval.

5.2   The work-energy theorem and power

Equation 5.2 says that when the force acts on a body, it changes the velocity, and the change depends on how far the force has been acting, on how many meters it has been pushing the object. The change is not simply in velocity but in velocity squared. Let us move everything involving the force to one side and the particle to the other side:

$$\frac{1}{2}mv_2^2 - \frac{1}{2}mv_1^2 = Fd. \tag{5.3}$$

The combination $\frac{1}{2}mv^2$ is called *kinetic energy* and is denoted by K. The product Fd is called the *work done by the force* and denoted by W. The units of W are Newton-meters and we replace that by joules or J.

We have found the simplest version of the work-energy theorem:

$$K_2 - K_1 = Fd = W. \tag{5.4}$$

What if there are 36 forces acting on the body? Which one should I use? Say I'm pulling and you're pushing. Then, F has got to be the net force, because Newton's law connects the net force to the acceleration. If you and I have a tug of war, and we cancel each other out to zero, then there's no acceleration, and the body with some initial velocity will maintain its initial velocity. The work-energy theorem says, *The change in kinetic energy is equal to the work done by all the forces.* In this case, there is no change in energy, though there are two forces at work that cancel. It turns out that it is sensible to define the work done by me, which is equal to F times the distance traveled, and the work done by you, which will be given an opposite sign. When do we attach a plus sign and when do we attach a minus sign? If you go back and review the whole derivation, you will see it was understood that a was a positive quantity. Then everything works. If the body is moving to the right, and I'm pushing to the right, then the work done by me is positive. And if you were pushing to the left and the body still moved to the right, the work done by you is negative. In other words, if you get your way, if things move the way you're pushing, the work done by you is positive. If the object is moving counter to your will, in the opposite direction to your force, the work done by you is negative. I'm lifting this piece of chalk at constant velocity from the ground. Its kinetic energy is not changing. That means the total work done on the chalk is zero, but not because there are no forces on it. There is gravity acting down, and I am countering gravity with exactly mg. The work done by me is positive because I want the chalk to go up, and it does. If it goes up by an amount h, the work done by me is $W = mgh$. The work done by gravity in the meantime is $-mgh$ and the total work done is zero.

Let us imagine all of this occurs in a time Δt and the distance moved is $d = \Delta x$. Then if you divide both sides of Eqn. 5.4 by Δt and take all the usual limits

$$\frac{dK}{dt} = F\frac{dx}{dt} = Fv \equiv P \tag{5.5}$$

where P is defined as the *power*. So power is the rate at which work is done. For example, if I climb a twelve-story building, I have done some work, my mg times the height of the building. I can climb the building in one minute; I can climb the building in one hour. The work done is the same, but the power is a measure of how rapidly work is done. That's why it's the product of force and velocity or work divided by time. The units for power are joules per second. That also has a new name, which is *watts* or just W. You may use a kilowatt, or kW, which is a thousand watts. So, if you have a 60 W bulb, it's consuming energy at the rate of 60 joules per second.

Now let's turn to the next generalization, when the force is not a constant but varies with x. Do we know any example of a force that varies with x? A spring is a good example, with $F(x) = -kx$. Even gravity is a good example, if we consider large distances. I think it is no secret that the force of gravity is mg only near the earth and that if you go sufficiently far, you will notice gravity itself is getting weaker. It will still look like mg' locally but g' won't be a fixed $9.8ms^{-2}$; it will be decreasing as we move away from the earth.

What is the work-energy theorem when the force varies? Let's draw ourselves a force $F(x)$ that varies with x, as in Figure 5.1. I'm just taking any function of x that I want. Now the acceleration is not constant because the force is not a constant. We cannot apply the formula $W = Fd$ because F varies over the distance d. So we resort to the usual trick in calculus: find an interval of width dx that is as narrow as you want, so that during that

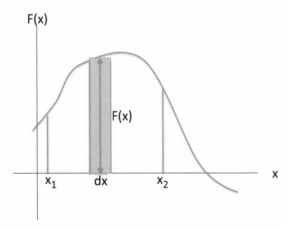

Figure 5.1   When a variable force $F(x)$ acts on a body that moves by a distance dx, it does work $dW = F(x)dx$ shown by the shaded rectangle.

period F is essentially a constant equal to $F(x)$, the value of F at that x. For that tiny interval I can still say that the change in K is

$$dK = F(x)dx. \tag{5.6}$$

Geometrically, $F(x)dx$ is the area of the thin rectangle whose base is dx and whose height is the function F at that x. (If $F(x) < 0$, the area is counted as negative.) If you eventually went from x_1 to x_2, then the work done by the force is given by the area under that graph in the sense of calculus. In every segment you pick up the change in kinetic energy dK, you add it all up to get $K_2 - K_1$ from the left-hand side. The right-hand side is the integral of the function $F(x)$ between x_1 and x_2. The general work-energy theorem now says

$$K_2 - K_1 = \int_{x_1}^{x_2} F(x)dx \equiv W. \tag{5.7}$$

Even if you have never heard of an integral, if I give you a function you can still deal with this problem. You'll come to me and say, "Give me your function. I'm going to plot it on some kind of graph paper with a grid on it, and I'm just going to count the number of tiny squares enclosed. That's the area and that's the change in kinetic energy." So integration is just finding the area bounded by the function at the top, the x axis below, and two vertical lines at the starting and ending points x_1 and x_2.

Now here is a little digression, a three-minute introduction to a great secret for finding the area. If you give me a function and you tell me to find the area under it from x_1 to x_2, I can show you a trick. You don't have to draw anything on graph paper. First, I find a function $G(x)$ specified by its derivative

$$F(x) = \frac{dG}{dx}. \tag{5.8}$$

Then I claim (and will prove shortly) that

$$\int_{x_1}^{x_2} F(x)dx = G(x_2) - G(x_1). \tag{5.9}$$

This is the opposite of taking derivatives: now we want a function whose derivative is given to be F. If I say $F(x) = x^3$, then $G(x)$ is that function whose derivative is x^3. Now, I know I have to start with x^4 because when I take derivatives, I will lose a power. But I will also get an unwanted 4 in front. I fight that by putting a 4 downstairs and find $G(x) = \frac{x^4}{4}$. You might point out that if you have a function whose derivative is something, adding a constant doesn't change the derivative. Then you can say, "Well, we are in trouble now because the world cannot agree on what G is, because if I have one G, you can get another one with a different constant." But the beauty is that when you take $G(x_2) - G(x_1)$, this difference in the choice of c goes away. So most of the time we don't bother with the constant. Sometimes we make a special choice that recommends itself, as you will see when we study gravity on a celestial scale. The choice of constant is like the choice of origin in the projectile problem: any origin will do, but it is convenient to choose it as the point of take-off.

Why is $G(x)$ the function whose derivative is $F(x)$? Let us call the area from some arbitrary point x_0 to the point x as $G(x)$. If I add a little more area, out to $x + dx$, the extra area is $F(x)dx$. This, by definition, is dG, the change in G. Dividing by dx and taking the limit, we see $F = \frac{dG}{dx}$. How do different G's differing by a constant arise? By choosing different starting points x_0 from which to reckon the area. But whatever we choose for x_0, the *change* in the area due to changing the upper limit x is always given by $dG = F(x)dx$.

5.3 Conservation of energy: $K_2 + U_2 = K_1 + U_1$

Let us now incorporate what we have discussed so far and write

$$K_2 - K_1 = \int_{x_1}^{x_2} F(x)dx = G(x_2) - G(x_1) \equiv G_2 - G_1, \qquad (5.10)$$

which we can rearrange to give

$$K_2 - G_2 = K_1 - G_1. \qquad (5.11)$$

We now make a little cosmetic change, and introduce the function

$$U(x) = -G(x) \qquad F(x) = -\frac{dU}{dx} \qquad (5.12)$$

in terms of which we obtain the standard form

$$E_2 \equiv K_2 + U_2 = K_1 + U_1 \equiv E_1. \tag{5.13}$$

This is the law of conservation of energy; $E = K + U$ is called the *total mechanical energy* and U is called the *potential energy*. Conservation of energy in physics has a totally different meaning from "Turn the lights off when you leave the room!" Here it means that when a body is moving under the effect of this force $F(x)$, even though it is speeding up and slowing down, a certain special combination

$$E = \frac{1}{2}mv^2 + U(x) \tag{5.14}$$

does not change with time. If you know the value of E at one time, you know E at all times.

Let's consider a simple example. We take a rock, and we drop it. We know it's picking up speed; we know it's losing height. So, you may expect there is some quantity that is a combination of height and speed, a combination which does not change in this exchange. We can find that combination by this law. In the case of gravity where $F = -mg$ the expression for U is

$$U = mgy \quad \text{since it obeys} \tag{5.15}$$

$$-\frac{dU}{dy} = -mg = F. \tag{5.16}$$

The energy conservation law takes the form

$$E_2 = \frac{1}{2}mv_2^2 + mgy_2 = \frac{1}{2}mv_1^2 + mgy_1 = E_1. \tag{5.17}$$

In the mass and spring system, the corresponding relations are

$$U(x) = \frac{1}{2}kx^2 \quad \text{since it obeys} \tag{5.18}$$

$$-\frac{dU}{dx} = -kx = F(x) \tag{5.19}$$

and the conservation law assumes the form

$$E_2 = \frac{1}{2}mv_2^2 + \frac{1}{2}kx_2^2 = \frac{1}{2}mv_1^2 + \frac{1}{2}kx_1^2 = E_1. \qquad (5.20)$$

Let us put that to work. I'm going to pull the mass by an amount A and let it go. I want to know how fast it will be moving when it comes to some point, say, $x = 0$. If you go back to Newton's laws, this is a pretty complicated problem. Think about why. You start with a mass at rest. If you pull it by an amount A, a force $-kA$ initially acts on it. That will produce an acceleration, $-kA/m$, which will give it a small negative velocity by the time it moves a distance Δx to the left. But once it comes to the new location, a different force will be acting on it, because the x is now different from A. So the acceleration during the next tiny interval is different, and the gain and the velocity during that second interval will be different from that in the first. You have to add all these changes to find the velocity at $x = 0$. That's a difficult proposition, but with the law of conservation of energy you'll do that in no time. Let's do it, not just for $x = 0$, but an arbitrary x. We set $x_1 = A$, $v_1 = 0$ at the start and drop the subscript 2 on x and v at the generic point to obtain

$$\frac{1}{2}mv^2 + \frac{1}{2}kx^2 = 0 + \frac{1}{2}kA^2. \qquad (5.21)$$

At the initial time, all the energy is potential energy; there is no kinetic energy because there is no motion. At any subsequent time, we can solve for the velocity at any location x. If you want $x = 0$, that's very easy:

$$\frac{1}{2}mv^2 + \frac{1}{2}k \cdot 0^2 = 0 + \frac{1}{2}kA^2 \qquad (5.22)$$

$$v^2 = \frac{kA^2}{m} \qquad (5.23)$$

$$v = \pm\sqrt{\frac{kA^2}{m}} = \pm A\sqrt{\frac{k}{m}} = \pm\omega A. \qquad (5.24)$$

We get two answers because the mass may be going past the origin in either direction. If it is passing the origin for the first time, it will be moving to the left with a negative velocity.

For the case of general x, Eqn. 5.21 tells us

$$v(x) = \pm\sqrt{\frac{k}{m}}\sqrt{A^2 - x^2}.$$ (5.25)

You can see why finding the velocity is so much easier now; it is because we have calculated the kinetic energy change produced by the spring between two points x_1 and x_2 once *and for all* by doing the integral of the spring force and encoding that in $U(x_1) - U(x_2)$.

Now, let us take another problem: a mass is hanging from the ceiling. Let us choose the origin of the vertical coordinate y at the lower end of the undistorted spring, so that the force due to the spring is $-ky$. There will now be two kinds of potential energy, the gravitational and spring-based, because there are two forces:

$$K_2 - K_1 = \int_{y_1}^{y_2} F(y)dy$$ (5.26)

$$= \int_{y_1}^{y_2} (-mg - ky)\, dy$$ (5.27)

$$= mg(y_1 - y_2) + \frac{1}{2}k\left(y_1^2 - y_2^2\right),$$ (5.28)

which can be written as

$$K_2 + mgy_2 + \frac{1}{2}ky_2^2 = K_1 + mgy_1 + \frac{1}{2}ky_1^2.$$ (5.29)

5.4 Friction and the work-energy theorem

One bad apple ruins the whole thing: friction. Let's take a body with the spring force $-kx$ and a frictional force f acting on it. I will try as usual to get a law of conservation of energy when friction is acting, and you will see that I will not succeed. It looks like everything here is water-tight, right? You tell me all the forces acting on a body. I integrate each force from start to finish and call that the difference of the corresponding potential energies, as I did in the case of a vertical spring-mass system subject to the spring and gravity. Then I have got my result. So here we go:

$$K_2 - K_1 = \int_{x_1}^{x_2} F(x)dx = \int_{x_1}^{x_2} (-kx)dx + \int_{x_1}^{x_2} f(x)dx$$

$$(5.30)$$

$$= \frac{1}{2}kx_1^2 - \frac{1}{2}kx_2^2 + \int_{x_1}^{x_2} f(x)dx \qquad (5.31)$$

$$\left(K_2 + \frac{1}{2}kx_2^2\right) - \left(K_1 + \frac{1}{2}kx_1^2\right) = \int_{x_1}^{x_2} f(x)dx. \qquad (5.32)$$

If we do the integral over $f(x)$ and call that the difference of potential energies due to friction, we can take that to the left-hand side and get a formula for energy as a sum of K and two potential energies. What is stopping me? The answer is that the force of friction is not a function of just x. You might say, "What do you mean? I'm pushing this mass, and I know how hard it's pushing me back." But then I tell you, "Push it the opposite way!" Then, you'll find the force of friction, at the very same location, is pointing in the opposite direction. This is unlike the force of a spring, which is $-kx$ whether you're going toward the origin or away from it. The same is true for gravity. Gravity is pulling down with a force $-mg$ and it doesn't care whether the object is going up or coming down; at a given location there is a fixed force. So the problem is not that the frictional force varies with x but that the force is also a function of the direction of velocity, which can have either sign. Thus $f = f(x, v(x))$. Let us postpone the evaluation of the integral over f and write

$$\left(K_2 + \frac{1}{2}kx_2^2\right) - \left(K_1 + \frac{1}{2}kx_1^2\right) = \int_{x_1}^{x_2} f(x, v(x))dx \qquad (5.33)$$

$$E_2 - E_1 = \int_{x_1}^{x_2} f(x, v(x))dx. \qquad (5.34)$$

I cannot evaluate the integral of f given just x_1 and x_2, because it can go from x_1 to x_2 directly or after one or more oscillations. During these oscillations the velocity will change sign repeatedly and so will the force of friction at any given x. *We must break down the total journey into segments wherein the velocity and f have a definite sign at each x.*

Here is an example where we consider just one segment within which v has a definite sign and hence so does f. I pull a mass to $x = A$ and let go. First assume a small amount of friction. I know that when the mass returns to the center it won't be going quite as fast as without friction. And I also know that when it overshoots to the other side, it will not go all the way back to $x = -A$. I want to find A', the coordinate of the left-most point. For this leftward journey from $x = A$ to $x = A'$ the force is a well-defined function: $f = +\mu_k mg$ pointing to the right. The displacement is negative: $d = -(A - A')$ and the work done by friction is

$$W_f = -\mu_k mg(A - A') \equiv -f \cdot (A - A') \tag{5.35}$$

and the energy equation is

$$E_2 - E_1 = \int_{x_1}^{x_2} f(x, v(x))dx = -f \cdot (A - A') \tag{5.36}$$

$$\frac{1}{2}kA'^2 - \frac{1}{2}kA^2 = -f \cdot (A - A') \tag{5.37}$$

where we have set $K_1 = K_2 = 0$ at the start and finish. This quadratic equation for A' may be rearranged as follows:

$$\frac{k}{2}(A' - A)\left[A' + A - \frac{2f}{k}\right] = 0. \tag{5.38}$$

The root $A' = A$ is trivial: it corresponds to the initial point where the kinetic energy is indeed zero and the total energy equals the initial energy. Let us analyze the nontrivial solution

$$A' = -A + \frac{2f}{k}. \tag{5.39}$$

As we crank up $\frac{f}{k}$ from 0 to A, A' starts moving rightward from $x = -A$ and turns positive when we pass $\frac{f}{k} = \frac{A}{2}$. That is, the final point now lies at positive x: the oscillator never even makes it back to the equilibrium point. Raising f further, we reach the limiting case $\frac{f}{k} = A$ when the nontrivial solution coalesces with the trivial one: the mass never leaves $x = A$ since the spring force, at its very maximum of kA, exactly equals the frictional force. (In this discussion we assume $\mu_s = \mu_k$.)

Here is the bottom line for the law of conservation of energy. You take all the forces acting on a body and equate $K_2 - K_1$ to the work done, the integral of all the forces. Divide the forces into *conservative* forces (gravity, spring) that depend on location only and the non-conservative like friction (which is the only non-conservative force we will consider). The integral over each conservative force will turn into an associated potential energy difference. Let W_f be the work done by friction. The final result will be

$$E_2 - E_1 = W_f \quad \text{with} \tag{5.40}$$

$$E = \frac{1}{2}mv^2 + U_s + U_g + \dots \tag{5.41}$$

where the potential energy U_s is due to the spring, U_g is due to gravity, and so on.

CHAPTER 6

Conservation of Energy in $d=2$

6.1 Calculus review

We begin with some mathematical preparation for what I'm going to do next. Let's take some function $f(x)$ shown in Figure 6.1. I start at some point x with a value $f(x)$. When I go to a nearby point, $x + \Delta x$, the function changes by $\Delta f = f(x + \Delta x) - f(x)$. All these tiny quantities are exaggerated in the figure so you can see them. We are going to need approximations to the change in the function as $\Delta x \to 0$. A common one is to pretend the function is linear with the local value of the slope $f'(x) = \frac{df}{dx}$, as depicted by

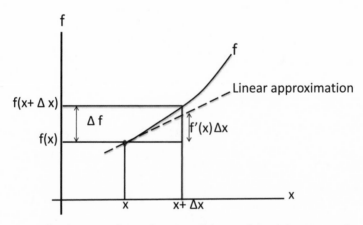

Figure 6.1   The change, Δf, in a function $f(x)$ as x changes by Δx may be approximated by $\Delta f \simeq f'(x)\Delta x$, where $f'(x) = \frac{df}{dx}$. The solid line is the actual function and the dotted line is the approximation by a straight line of slope $f'(x)$.

the dotted line. The change in f along the straight line is $f'(x)\Delta x$. It differs from the actual Δf by a tiny amount because the function is not following the same slope that you have to begin with; it's curving up. Take a concrete example:

$$f(x) = x^2 \tag{6.1}$$

$$f(x + \Delta x) = x^2 + 2x\Delta x + (\Delta x)^2 \tag{6.2}$$

$$\Delta f = 2x\Delta x + (\Delta x)^2 \tag{6.3}$$

$$\Delta f = f'(x)\Delta x + (\Delta x)^2. \tag{6.4}$$

This result is valid for Δx of any size. We see that the exact change is $f'(x)\Delta x$ plus something quadratic in Δx. If we are interested in very small Δx, we may start ignoring all but the term linear in Δx:

$$\Delta f = f'(x)\Delta x + \mathcal{O}(\Delta x)^2 \tag{6.5}$$

where $\mathcal{O}(\Delta x)^2$ signifies that the neglected terms are of order $(\Delta x)^2$ and higher.

Often we will use

$$\Delta f \simeq f'(x)\Delta x \tag{6.6}$$

as an approximation for small Δx.

Consider, for example, $f(x) = (1 + x)^n$ and its values near $x = 0$. Clearly $f(0) = 1$. Suppose you want the function at a point x very close to the origin. In this case $\Delta x = x - 0$ is just x itself and the approximate value will be

$$f(x) = f(0) + f'(0)x + \ldots = 1 + n\,(1 + x)^{n-1}\big|_{x=0} x + \ldots = 1 + nx + \ldots, \tag{6.7}$$

a result we will exploit mercilessly.

On other occasions, we will take the limit $\Delta x \to 0$ in the end and write the equality

$$df = f'(x)dx \tag{6.8}$$

with the understanding that both sides are to be integrated to obtain

$$\int_1^2 df = f(x_2) - f(x_1) = \int_{x_1}^{x_2} f'(x)dx. \tag{6.9}$$

6.2 Work done in $d = 2$

Now we are going to derive the work-energy theorem and the law of conservation of energy in two dimensions. I am hoping I will get some relation like $K_1 + U_1 = K_2 + U_2$, where $U = U(x, y)$. How do you visualize the function of two variables $f(x, y)$? On top of each point (x, y) you measure the value of $f(x, y)$ in the third perpendicular direction. The function defines a surface over the $x - y$ plane and the distance from the plane to the surface is the value of f at the point (x, y). For example, (x, y) could be coordinates of a point in the United States and the function could be the temperature $T(x, y)$ at that point. So you plot on top of each point in the United States the local temperature.

Once I have got the notion of a function of two variables, I want to move around the plane and ask how the function changes. But now I have an infinite number of options. I can move along x, I can move along y, I can move at some intermediate angle. Consider derivatives along the two principle directions x and y. We're going to define a *partial derivative* as follows. You start at the point (x, y), go to the point $(x + \Delta x, y)$, subtract the function at the starting point, divide by Δx, and take $\Delta x \to 0$. This defines the *partial derivative with respect to x*:

$$\frac{\partial f}{\partial x} = \lim_{\Delta x \to 0} \frac{f(x + \Delta x, y) - f(x, y)}{\Delta x}. \tag{6.10}$$

The curly ∂ instead of d tells you it's the partial derivative. As you move horizontally, you notice you don't do anything to y. We could make it very explicit by using a subscript y as follows:

$$\left.\frac{\partial f}{\partial x}\right|_y = \lim_{\Delta x \to 0} \frac{f(x + \Delta x, y) - f(x, y)}{\Delta x}. \tag{6.11}$$

We will not do that: if one coordinate is being varied, all the others (of which there is just one in $d=2$) will be assumed fixed. In the same notation

$$\left.\frac{\partial f}{\partial y}\right|_x \equiv \frac{\partial f}{\partial y} = \lim_{\Delta y \to 0} \frac{f(x, y + \Delta y) - f(x, y)}{\Delta y}. \tag{6.12}$$

Let's get some practice with $f = x^3 y^2$. To find $\frac{\partial f}{\partial x}$ we see how f varies with x keeping y constant. That means we treat y like a number such as 5 when we encounter it. So we have

$$\left.\frac{\partial f}{\partial x}\right|_y = 3x^2 y^2 \tag{6.13}$$

$$\left.\frac{\partial f}{\partial y}\right|_x = 2x^3 y. \tag{6.14}$$

We know from the calculus of one variable that you can take the derivative of the derivative. Here are the four possible second derivatives and their explicit values for $f = x^3 y^2$:

$$\frac{\partial}{\partial x}\left(\frac{\partial f}{\partial x}\right) \equiv \frac{\partial^2 f}{\partial x^2} = 6xy^2 \tag{6.15}$$

$$\frac{\partial}{\partial y}\left(\frac{\partial f}{\partial y}\right) \equiv \frac{\partial^2 f}{\partial y^2} = 2x^3 \tag{6.16}$$

$$\frac{\partial}{\partial x}\left(\frac{\partial f}{\partial y}\right) \equiv \frac{\partial^2 f}{\partial x \partial y} = 6x^2 y \tag{6.17}$$

$$\frac{\partial}{\partial y}\left(\frac{\partial f}{\partial x}\right) \equiv \frac{\partial^2 f}{\partial y \partial x} = 6x^2 y. \tag{6.18}$$

Notice that the *mixed or cross derivatives* are equal:

$$\frac{\partial^2 f}{\partial y \partial x} = \frac{\partial^2 f}{\partial x \partial y}. \tag{6.19}$$

That's a property of the generic functions that we will encounter. I'd like to give you a feeling for why that is true. For what follows, bear in mind that when you make a small displacement in the plane, the change in any function is approximately

$$\Delta f \simeq \frac{\partial f}{\partial x} \Delta x + \frac{\partial f}{\partial y} \Delta y, \tag{6.20}$$

which becomes an equality in the limit $\Delta x \to 0$, $\Delta y \to 0$ and $\Delta f \to 0$:

$$df = \frac{\partial f}{\partial x} dx + \frac{\partial f}{\partial y} dy. \tag{6.21}$$

These limits appear naturally when we plan to sum over the infinitesimal changes to get the corresponding integrals.

Armed with this, let us ask how much the function changes when we go from some point (x, y) to $(x + dx, y + dy)$ in Figure 6.2. We're going to

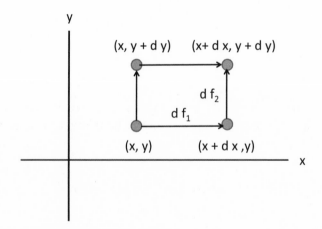

Figure 6.2   Two ways to go from (x, y) to $(x + \Delta x, y + \Delta y)$: move horizontally and then vertically or vice versa. That the change in f must be the same both ways becomes the requirement that $\frac{\partial^2 f}{\partial y \partial x} = \frac{\partial^2 f}{\partial x \partial y}$.

make the move in two stages. We go via an intermediate point $(x+dx,y)$ and add the changes df_1 and df_2 in each step:

$$df_1 = \left.\frac{\partial f}{\partial x}\right|_{(x,y)} dx \tag{6.22}$$

$$df_2 = \left.\frac{\partial f}{\partial y}\right|_{(x+dx,y)} dy \tag{6.23}$$

$$df = \left.\frac{\partial f}{\partial x}\right|_{(x,y)} dx + \left.\frac{\partial f}{\partial y}\right|_{(x+dx,y)} dy. \tag{6.24}$$

Notice that the second step requires the y partial derivative at $(x+dx,y)$. Because the partial derivative is itself just another function of x and y, we may write to leading order in dx

$$\left.\frac{\partial f}{\partial y}\right|_{(x+dx,y)} = \left.\frac{\partial f}{\partial y}\right|_{(x,y)} + \frac{\partial^2 f}{\partial x \partial y} dx. \tag{6.25}$$

Upon feeding this into Eqn. 6.24 we find

$$df = \left.\frac{\partial f}{\partial x}\right|_{(x,y)} dx + \left.\frac{\partial f}{\partial y}\right|_{(x,y)} dy + \left.\frac{\partial^2 f}{\partial x \partial y}\right|_{(x,y)} dx dy. \tag{6.26}$$

If we first moved up to $(x,y+dy)$ and then to $(x+dx,y+dy)$, we would get a change in f with x and y interchanged. Equating the results from the two ways to find the change in f between (x,y) and $(x+dx,y+dy)$ we find

$$\left.\frac{\partial^2 f}{\partial x \partial y}\right|_{(x,y)} dx dy = \left.\frac{\partial^2 f}{\partial y \partial x}\right|_{(x,y)} dy dx. \tag{6.27}$$

Canceling the products of the infinitesimals, we get the equality of the mixed derivatives.

6.3   Work done in $d=2$ and the dot product

Let us come back to deriving the law of conservation of energy in two dimensions. In one dimension we found that if $K = \frac{1}{2}mv^2$,

$$\frac{dK}{dt} = mv\frac{dv}{dt} = mva = Fv = F\frac{dx}{dt} \tag{6.28}$$

$$dK = Fdx \quad \text{upon canceling } dt \text{ above} \tag{6.29}$$

$$K_2 - K_1 = \int_{x_1}^{x_2} F(x)dx \quad \text{upon integrating both} \tag{6.30}$$
$$\text{sides above}$$

$$\therefore F = -\frac{dU}{dx}$$

$$= U(x_1) - U(x_2), \quad \text{which can be} \tag{6.31}$$
$$\text{rearranged to give}$$

$$K_2 + U_2 = K_1 + U_1 \tag{6.32}$$

provided F did not depend on anything else besides x, such as $v(x)$.

We want to try the same thing in two dimensions. What expression should I use for the work done in two dimensions, given that the force and displacement are both vectors with two components each? How should I multiply all these parts in generalizing $dW = Fdx$? Here is the solution. I'm going to find $\frac{dK}{dt}$ for a body moving in two dimensions and call that the power $P = \frac{dW}{dt}$ just as in $d=1$. For that I need a formula for kinetic energy. The obvious choice that reduces to what we know is correct for motion along just x or y is

$$K = \frac{1}{2}mv^2 = \frac{1}{2}m(v_x^2 + v_y^2). \tag{6.33}$$

Now we find

$$\frac{dK}{dt} = m\left(v_x\frac{dv_x}{dt} + v_y\frac{dv_y}{dt}\right) \tag{6.34}$$

$$= F_xv_x + F_yv_y = F_x\frac{dx}{dt} + F_y\frac{dy}{dt} \tag{6.35}$$

$$dK = F_xdx + F_ydy \tag{6.36}$$

where I have used Newton's second law $\mathbf{F} = m\frac{d\mathbf{v}}{dt}$ and multiplied both sides of Eqn. 6.35 by dt, which is allowed in the sense explained earlier. If I define the work done as

$$dW = F_x dx + F_y dy, \tag{6.37}$$

I find, just as in $d=1$, that

$$dK = dW = F_x dx + F_y dy. \tag{6.38}$$

The force and displacement are both vectors

$$\mathbf{F} = \mathbf{i}F_x + \mathbf{j}F_y \tag{6.39}$$

$$d\mathbf{r} = \mathbf{i}dx + \mathbf{j}dy \tag{6.40}$$

and their components enter dW in the combination $dW = F_x dx + F_y dy$. Likewise the power P may be written as

$$P = \frac{dK}{dt} = F_x v_x + F_y v_y. \tag{6.41}$$

Given two vectors

$$\mathbf{A} = \mathbf{i}A_x + \mathbf{j}A_y \tag{6.42}$$

$$\mathbf{B} = \mathbf{i}B_x + \mathbf{j}B_y, \tag{6.43}$$

we see that the combination $A_x B_x + A_y B_y$ appears very naturally. It has a name: *the dot product of* $\mathbf{A}$ *and* $\mathbf{B}$, denoted by $\mathbf{A} \cdot \mathbf{B}$. That is, by definition,

$$\mathbf{A} \cdot \mathbf{B} = A_x B_x + A_y B_y. \tag{6.44}$$

In this notation

$$dW = \mathbf{F} \cdot d\mathbf{r} \tag{6.45}$$

$$P = \mathbf{F} \cdot \mathbf{v}. \tag{6.46}$$

A few factoids about $\mathbf{A} \cdot \mathbf{B}$. First

$$\mathbf{A} \cdot \mathbf{A} = A_x^2 + A_y^2 = A^2 \tag{6.47}$$

where A is the length of $\mathbf{A}$.

Next if θ_A and θ_B are the angles $\mathbf{A}$ and $\mathbf{B}$ make with the x-axis, then

$$\mathbf{A} \cdot \mathbf{B} = A_x B_x + A_y B_y \tag{6.48}$$

$$= A \cos \theta_A B \cos \theta_B + A \sin \theta_A B \sin \theta_B \tag{6.49}$$

$$= AB \left[\cos \theta_A \, \cos \theta_B + \sin \theta_A \, \sin \theta_B \right] \tag{6.50}$$

$$= AB \cos \left[\theta_B - \theta_A \right] = AB \cos \left[\theta_A - \theta_B \right], \tag{6.51}$$

which is usually written more compactly as

$$\mathbf{A} \cdot \mathbf{B} = AB \cos \theta, \tag{6.52}$$

where it is understood θ is the angle between the vectors. It can be measured from $\mathbf{A}$ to $\mathbf{B}$ or the other way since $\cos \theta$ is unaffected by a sign change in θ.

Equation 6.52 works even in $d=3$ because $\mathbf{A}$ and $\mathbf{B}$ can still be made to lie in a plane and θ defined as the angle between them in this plane. However, in terms of components we must bring in all three components:

$$\mathbf{A} \cdot \mathbf{B} = A_x B_x + A_y B_y + A_z B_z, \tag{6.53}$$

a result that seems reasonable and one which can be verified after some messy trigonometry.

The dot product is symmetric since $\cos \theta = \cos(-\theta)$:

$$\mathbf{A} \cdot \mathbf{B} = \mathbf{B} \cdot \mathbf{A}. \tag{6.54}$$

Note that if we set $\mathbf{A} = \mathbf{B}$, then $\mathbf{A} \cdot \mathbf{A} = AA \cos 0 = A^2$.

The two definitions of the dot product, Eqns. 6.44 and 6.52, are fully equivalent. If you are thinking in terms of the components, $A_x B_x + A_y B_y$ is more natural, while if you are thinking in terms of arrows of some lengths and angles, $AB \cos \theta$ is preferred. Which one you use depends on your goals.

For example, to establish an important property that the dot product is distributive:

$$\mathbf{A} \cdot (\mathbf{B} + \mathbf{C}) = \mathbf{A} \cdot \mathbf{B} + \mathbf{A} \cdot \mathbf{C} \tag{6.55}$$

it is easier to proceed as follows

$$\mathbf{A}\cdot(\mathbf{B}+\mathbf{C})=A_x(B_x+C_x)+A_y(B_y+C_y) \tag{6.56}$$
$$=A_xB_x+A_yB_y+A_xC_x+A_yC_y=\mathbf{A}\cdot\mathbf{B}+\mathbf{A}\cdot\mathbf{C}. \tag{6.57}$$

On the other hand, using $\mathbf{A}\cdot\mathbf{B}=AB\cos\theta$, it is easier to establish the following very useful results:

- If $\mathbf{A}$ and $\mathbf{B}$ are parallel, i.e., $\theta=0$, their dot product is a maximum.
- If $\mathbf{A}$ and $\mathbf{B}$ are perpendicular, their dot product is zero.
- Under a rotation of axes, $\mathbf{A}\cdot\mathbf{B}$ is invariant or unchanged, because the lengths and the *relative angle* are unchanged by a rotation of axes.

Of course, for every proof with one definition, a possibly more cumbersome one, which uses the other definition, also exists.

Let us return to the work-energy theorem using the dot product notation:

$$dK=\mathbf{F}\cdot d\mathbf{r}=dW. \tag{6.58}$$

The work done by a force when it moves a body by a vector $d\mathbf{r}$ is the length of the force vector times the distance traveled, times the cosine of the angle between the force vector and the displacement vector. That is also the change in kinetic energy dK. Let us make a big trip in the $x-y$ plane, shown in Figure 6.3, starting from a point $\mathbf{r}_1\equiv1$ and ending at $\mathbf{r}_2\equiv2$, and made up of a sequence of little segments $d\mathbf{r}$ in each one of which I calculate $\mathbf{F}\cdot d\mathbf{r}$. When I add their contributions to the change in K and the work done, I get, *as the segments' sizes tend to zero,*

$$\int_1^2 dK=K_2-K_1=\int_1^2\mathbf{F}\cdot d\mathbf{r}. \tag{6.59}$$

The right-hand side is called the *line integral of the force* $\mathbf{F}$ *between* 1 *and* 2 *along a path P.*

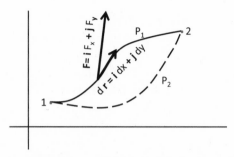

Figure 6.3   The line integral of a force between points 1 and 2 along a path P_1 is the sum of dot products $\mathbf{F} \cdot d\mathbf{r}$ over tiny segments that make up the path, in the limit $d\mathbf{r} \to \mathbf{0}$. Also shown by a dotted line is another path P_2 between the same end points.

6.4   Conservative and non-conservative forces

Suppose it is true, just like in one dimension, that the line integral of the force is something that depends only on the end points. Let us call the answer $U(1) - U(2)$, just like we did in one dimension. I am done, because then I have

$$K_2 + U_2 = K_1 + U_1. \tag{6.60}$$

To make sure this is correct, I ask the mathematicians a question: "You told me the integral of $F(x)$ from start to finish is really the difference of another function G at the upper limit minus G at the lower limit, with G related to F by $F = \frac{dG}{dx}$. Is there a similar result in $d = 2$?" Sadly, this is not the case. What could go wrong? Yes, friction will do it, but let us assume there is no friction, and that $\mathbf{F}$ depends only on $\mathbf{r}$. Can something still be wrong? Well, let me ask you the following question. Suppose I go from 1 to 2 along path P_1 and another person goes along path P_2. Do you think that person will do the same amount of work, even though the force is now integrated on a longer path? In two dimensions, there are thousands of ways to go from one point to another point. Therefore, this integral is not specified by just the end points; it depends on the entire path, which needs to be specified. If the work done depends on the path, then the answer cannot be of the form $U(1) - U(2)$, which depends only on the end points.

I digress to point out that even in $d = 1$, there are many ways to go from x_1 to $x_2 > x_1$. For example, we can go directly to x_2 or we can overshoot to x_3 and swing back to x_2. The answer will be the same, because for every segment from x_2 to x_3 that makes a contribution $F(x)dx$, an equal and opposite contribution exists on the way back to x_2, because $F(x)$ remains the same, and dx changes sign. In this sense, every force $F(x)$ in $d=1$ is conservative. Of course, if it is friction we are talking about, the two canceling pieces now add, because $F = F(x, v(x))$ reverses sign along with dx.

Returning to $d=2$, I am going to show that the work done by a generic force will be path-dependent. To generate a random force, I asked my class to give me numbers from 1 to 3, and I got the following list: 2, 2, 2, 1, 1, and 2. Using these randomly generated numbers as coefficients and exponents, I wrote down a force:

$$\mathbf{F}(x, y) = \mathbf{i}2x^2y^2 + \mathbf{j}xy^2 \tag{6.61}$$

For example, the $2x^2y^2$ is from the first three 2's chosen by the class.

Is it true for this generic force, essentially picked out of a hat, that the work done in going from one point to another depends only on the end points, or does it depend in detail on how you go between the end points? We will find that the work done along two paths, joining the same two end points, will give two different answers.

Let's find the work done in moving from the origin, $(0,0)$, to the point $(1,1)$. I will take the two paths shown in Figure 6.4. In one path I go horizontally until I'm at $(1,0)$, below the point $(1,1)$, and then straight up to $(1,1)$. In the other path, I'm going straight up to $(0,1)$ and then on horizontally to $(1,1)$. So, let's find the work done when I go the first way. I'm

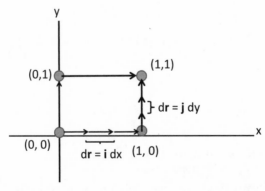

Figure 6.4   The line integral of a vector from $(0,0)$ to $(1,1)$ along two paths.

going to integrate $\mathbf{F} \cdot d\mathbf{r}$ first on the horizontal segment, then on the vertical segment. On the x-axis if I move a little bit I have

$$dr = idx \tag{6.62}$$

$$\mathbf{F}(x,y) = i2x^2y^2 + jxy^2 = 0 \text{ because } y = 0 \text{ on the } x\text{-axis} \tag{6.63}$$

$$\mathbf{F} \cdot d\mathbf{r} = 0. \tag{6.64}$$

In other words, the work done in this segment is zero because $\mathbf{F}$ itself vanishes when $y = 0$. In the vertical segment from $(1,0)$ to $(1,1)$,

$$dr = jdy \tag{6.65}$$

$$\mathbf{F}(x,y) = i2x^2y^2 + jxy^2 = i2y^2 + jy^2$$

$$\text{because } x = 1 \text{ on this segment} \tag{6.66}$$

$$\mathbf{F} \cdot d\mathbf{r} = y^2 dy \tag{6.67}$$

$$\int \mathbf{F} \cdot d\mathbf{r} = \int_0^1 y^2 dy = \frac{1}{3}. \tag{6.68}$$

So the work done on this path is $W_1 = 0 + \frac{1}{3} = \frac{1}{3}$.

On the second path, we have no contribution from the vertical segment because $\mathbf{F} = 0$ for $x = 0$. In the horizontal segment at $y = 1$, we have

$$dr = idx \tag{6.69}$$

$$\mathbf{F}(x,y) = i2x^2y^2 + jxy^2 = i2x^2 + jx$$

$$\text{because } y = 1 \text{ on this segment} \tag{6.70}$$

$$\mathbf{F} \cdot d\mathbf{r} = 2x^2 dx \tag{6.71}$$

$$\int \mathbf{F} \cdot d\mathbf{r} = \int_0^1 2x^2 dx = \frac{2}{3}. \tag{6.72}$$

So the work done on this path is $W_2 = 0 + \frac{2}{3} = \frac{2}{3}$.

The answer is path-dependent.

I have shown you that if we took a random force, the work done is dependent on the path. For this *non-conservative force*, you cannot define a potential energy, whereas in one dimension any force other than friction allowed you to define a potential energy.

Our quest for a conserved energy leads us to search for a *conservative force*, a force for which the work done in going from 1 to 2 is path-independent.

6.5 Conservative forces

At first sight a conservative force looks miraculous. A randomly generated force was seen to have a line integral that depended on the path. How can the path dependence ever go away? Do conservative forces exist, and, if yes, how are we to find them?

Do not despair. Here is an algorithm that will produce any number of conservative forces.

- Take any function $U(x,y)$.
- The corresponding conservative force is

$$\mathbf{F} = -\mathbf{i}\frac{\partial U}{\partial x} - \mathbf{j}\frac{\partial U}{\partial y}. \tag{6.73}$$

- The potential energy associated with this conservative force will be U itself.

Here is an example.

$$U(x,y) = xy^3 \tag{6.74}$$

$$\frac{\partial U}{\partial x} = y^3 \tag{6.75}$$

$$\frac{\partial U}{\partial y} = 3xy^2 \tag{6.76}$$

$$\mathbf{F} = -\mathbf{i}y^3 - \mathbf{j}3xy^2. \tag{6.77}$$

Let me prove to you that the recipe works. The change in the function U, due to a small deviation from (x,y) to $(x+dx, y+dy)$, is

$$dU = \frac{\partial U}{\partial x}dx + \frac{\partial U}{\partial y}dy \tag{6.78}$$

in the limit as all changes go to zero. Writing this in terms of $\mathbf{F}$

$$dU = -F_x dx - F_y dy = -\mathbf{F} \cdot d\mathbf{r}. \tag{6.79}$$

Adding all the little pieces and changing the sign of both sides, we get

$$U(1) - U(2) = \int_1^2 \mathbf{F} \cdot d\mathbf{r} = K_2 - K_1, \tag{6.80}$$

which is the law of conservation of energy with U as the potential energy.

So I cooked up a force such that $\mathbf{F} \cdot d\mathbf{r}$ was a change in a certain function U. If I add all the $\mathbf{F} \cdot d\mathbf{r}$'s, I'm going to get the change in the function U from start to finish. We are beginning to see why certain integrals do not depend on the path. Here is an analogy. Forget about integrals. Imagine I am on some hilly terrain. I start at one point, and I walk to another point. At every portion of my walk, I keep track of my change in altitude, with uphill as positive and downhill as negative. That is like my dU. I add them all up. The total height change will be the difference in the heights of the end points. Now, you start with me but go on a different path. You wander all over the place but finally stop where I stopped. If you kept track of how long you walked, it won't be the same as my walk. But if you also kept track of how many feet you climbed at each step and added them all up, you would get the same answer I got. I repeat: if what you were keeping track of was the height change in a function, then the sum of all the height changes will simply be the height at the end minus the height at the beginning, independent of the path. Conversely, starting with the height function, if you manufacture a force $\mathbf{F}$ whose components are its partial derivatives, $\mathbf{F} \cdot d\mathbf{r}$ will measure the height change in each segment, and the line integral will yield the total height change between start and finish, independent of the path.

Consider the line integral of a conservative force on a *closed loop*, that is, when the starting and ending points 1 and 2 in Figure 6.3 coincide. Because this represents the change in U between some point and the *same* point, it vanishes for any loop. This is expressed as follows:

$$\oint \mathbf{F} \cdot d\mathbf{r} = 0 \quad \text{if } \mathbf{F} \text{ is conservative.} \tag{6.81}$$

Are there other ways to manufacture the conservative force? No! One can show that every conservative force can be obtained by differentiating some corresponding U.

Remember how I went out on a limb with the randomly chosen force the class generated and promised I was going to do the integral along two paths and get two different answers? What if the force had been a conservative force? Then I would have been embarrassed, because I would find, after all the work, that both paths gave the same answer. So, I had to make sure right away that the force was not conservative. How could I tell? I asked myself, "Could there be some function U (the negative of) whose x and y derivatives could equal $\mathbf{i}2x^2y^2 + \mathbf{j}xy^2$?" I knew the answer was no because if I took a y derivative of such a U to get F_y, then F_y should have one less power of y than F_x, but in our example the powers of y were the same in both. I will describe shortly a better way to analyze this question.

While it is true that even one example of path dependence (as illustrated above) is enough to show a force is non-conservative, getting the same answer on two or even two thousand paths between any number of fixed end points does not mean the force is conservative. It could be accidental. Some other path or some other end points may show the force is non-conservative. Conversely it could happen that a non-conservative force, like the one I just worked with, has the same integral for two particular paths joining two particular end points by pure accident. I took that gamble and lucked out.

But if the force is really conservative, how are we to show that? Here is the wonderful *test* I promised. If $\mathbf{F}$ is conservative, it must come from a U by taking partial derivatives, as per Eqn. 6.73. It follows that

$$\frac{\partial F_x}{\partial y} = -\frac{\partial^2 U}{\partial y \partial x} \tag{6.82}$$

$$\frac{\partial F_y}{\partial x} = -\frac{\partial^2 U}{\partial x \partial y} \quad \text{which means} \tag{6.83}$$

$$\frac{\partial F_x}{\partial y} = \frac{\partial F_y}{\partial x} \quad \text{because the cross derivatives are equal.} \tag{6.84}$$

If I give you a force and ask you, "Is it conservative?" you simply see if

$$\frac{\partial F_x}{\partial y} = \frac{\partial F_y}{\partial x}. \tag{6.85}$$

If it is, you know the force is conservative; if not, it is not.

The example we considered,

$$\mathbf{F}(x,y) = \mathbf{i}2x^2y^2 + \mathbf{j}xy^2, \tag{6.86}$$

fails the test:

$$\frac{\partial F_x}{\partial y} = 4x^2y \neq \frac{\partial F_y}{\partial x} = y^2. \tag{6.87}$$

The two most ubiquitous forces, gravitational and electrostatic, are conservative.

For longer discussion of this topic that fills in many blanks, see my *Basic Training in Mathematics*.

6.6 Application to gravitational potential energy

Let's take the most popular example: the force of gravity near the surface of the earth given by $\mathbf{F}_g = -\mathbf{j}mg \equiv m\mathbf{g}$ where $\mathbf{g} = -\mathbf{j}g$. It is conservative because the x derivative of F_y vanishes, and there is no F_x to differentiate, so that $\frac{\partial F_x}{\partial y} = \frac{\partial F_y}{\partial x} = 0$. What is the potential U that led to this? You can easily guess that $U = mgy$ will obey $F_y = -\frac{\partial U}{\partial y}$. You can also have $U = mgy + 96$, but we will not add those constants. In the law of conservation of energy, $K_1 + U_1 = K_2 + U_2$, adding a 96 to the U on both sides doesn't do anything. You already knew this from our study of motion in one dimension, and I am pointing out that this is also true in two dimensions.

Consider an application. Figure 6.5 shows a roller-coaster track that has a wiggly shape. At every x, there's a certain height $y(x)$ and a potential energy $U(x) = mgy(x)$, which is essentially just the profile of the roller-coaster track. If a coaster begins at rest at point A at the top, what is its total energy? It has a potential energy given by the height h, it has no kinetic energy, and so the total energy is just $E_1 = mgh$. But the total energy cannot change as the coaster goes up and down. So, you draw a line at height E_1 to represent this total energy. If the coaster is at some point x, then $U_1(x)$ is its potential energy and the rest of E_1 is its kinetic energy $K_1(x)$ as shown. As it oscillates up and down during its ride, the coaster gains and loses kinetic and potential energies, which always add up to the same E_1.

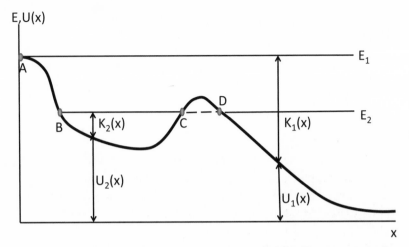

Figure 6.5   The roller-coaster ride. The total energy is fixed at E_1 or E_2 in the two examples discussed. At every point x, the sum of the potential energy $U(x)$ and the kinetic energy $K(x)$ equals a constant E. If the energy is E_2, the coaster can only be found between B and C or to the right of D. It is disallowed in the region CD where the potential energy exceeds the total energy, and K would have to be negative.

Consider a roller coaster whose total energy is E_2. We release it from rest at point B. It'll come down, pick up speed, slow down, stop, and turn around at C, because, at that point, the potential energy is equal to the total energy and there is no room for any kinetic energy. It'll rattle back and forth between B and C. If we release it from rest at D, it will have the same energy E_2, and it will coast down to the end of the ride. But it can never go from C to D because in the region CD it would have more potential energy than total energy, and hence negative kinetic energy, which is impossible.

However, according to laws of quantum mechanics, a particle with energy E_2 can disappear from the region BC and tunnel to D with the same energy. I use the word *tunnel* because in classical mechanics, the particle cannot cross the potential energy barrier in the interval CD. In quantum theory you cannot raise this objection because particles do not move along continuous, interpolating trajectories between two observed locations.

Back to the coaster: We can use energy conservation to find the speed at any point along the track. We can use it to determine the minimum height H from which the coaster in Figure 4.6 must be released so as to reach the top of the loop (at a height $2R$ measured from the ground) at the minimum requisite speed of $v = \sqrt{Rg}$. We write

$$\frac{1}{2}m \cdot 0^2 + mgH = \frac{1}{2}mRg + mg(2R) \qquad (6.88)$$

$$H = \frac{5}{2}R. \qquad (6.89)$$

A final note. The law of conservation of energy for the coaster as I stated it is incomplete, because gravity is not the only force acting. There is $\mathbf{F}_T$, the normal force of the track. Look, if I didn't want to have any force but gravity, I could take this roller coaster and just push it over the edge of a cliff. That converts potential to kinetic energy, but the outcome is not going to be good for the riders. Park designers build a track because they want the customers to survive the ride and come back for more. So the track should exist, and the consequent $\mathbf{F}_T$ should be included in computing the work. Luckily, this normal force does no work, because $\mathbf{F}_T \cdot d\mathbf{r} = 0$ in every portion. So the correct thing to do would be to say $K_2 - K_1$ is the integral of all the forces, divide them into $\mathbf{F}_T$ due to the track and $\mathbf{F}_g$ due to gravity, and drop $\mathbf{F}_T$ for the reason mentioned.

CHAPTER 7

The Kepler Problem

7.1 Kepler's laws

Next we discuss one of the most famous problems involving a conservative force: celestial motion under the influence of Newtonian gravity. We're going to make a big leap beyond inclined planes, pulleys, and whatnot; we are going to understand how planets move around the sun. That's a mega problem, right? The little m's you put in the equation are not the masses of a pulley or a block, but the mass of Jupiter or the sun. You're doing something of cosmological proportions. And you don't need to know too much more to do that. You're almost there.

The situation was as follows at Newton's time. Nicolaus Copernicus had proposed that the way to think about our solar system is to put the sun S at a fixed point as shown in Figure 7.1 and let the planets move around it. I have shown just one planet called P. You have to agree that Copernicus's contribution was truly remarkable. First of all, that was an era when it was not "publish or perish" but "publish and perish." It was not a good idea to come out and say what you thought about celestial objects. More than that, how did Copernicus ever figure this out? Even today, even when I know the heliocentric model is correct, when I look around, it doesn't at all look like this to my eyes. I concede that if you looked at our solar system from far away, it would look very simple. But what is our vantage point? We are sitting on the third planet from the sun, spinning around our own axis, and going around the sun. We think we are at rest, and everything is spinning around the opposite way. To deduce from that chaos this simple picture was quite remarkable. This shift in thinking is properly called the Copernican revolution.

After the Copernican revolution, people decided to take the data we get with the earth as the center, transcribe the data with the sun as the center, and analyze the new results. Tycho Brahe was a Danish aristocrat who had his own lab and studied the solar system. Johannes Kepler, a mathematician, was his assistant who worked on this problem for forty years and then published his findings in the form of three profound laws that were well worth the wait. So what if Kepler gave us just three laws after forty years? It is still better than Congress.

(I have to tell all of you who are going into science not to wait forty years to publish. You're not going to get a job; you are not going to get a grant; and if you get a Ph.D, you will be at retirement age when you graduate. In today's climate such long-term projects are at risk. One of the rare counter-examples that comes to my mind is the work by Raymond Davis Jr. on neutrinos, which lasted more than thirty years and ultimately helped resolve an outstanding puzzle connected to them.)

Here are Kepler's three laws, followed by explanations.

- The planets move in elliptical orbits with the sun S at a focal point.
- As the planet moves, its position vector SP in Figure 7.1 sweeps out equal areas in equal times, say as between 1 and 2 and between 3 and 4 .
- The ratio $\frac{T^2}{a^3}$, where T is the time period and a the semi-major axis of the ellipse, is the same for all planets.

Let me remind you that an ellipse is the locus of points the sum of whose distances from the two focal points ($r + r'$ in Figure 7.1) is a constant. To draw an ellipse, you take a string of length $2a$ and you nail its two ends to the *focal points S* and *E* using thumbtacks and pass it over the pencil tip at P as shown in Figure 7.1. Keeping the string taut, you trace a closed loop, and you have the ellipse whose *major axis* is $2a$ and whose *semi-major axis* is a. Thus for the ellipse

$$r + r' = 2a. \tag{7.1}$$

Let us understand why the string length $2a$ is also the major axis, the width of the ellipse in the x-direction. Imagine the pencil at point A. The length

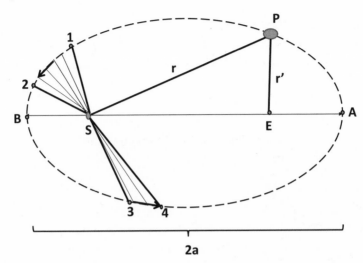

Figure 7.1   The solar system seen from afar, with just one planet P shown. The orbit is an ellipse with the sun S at one focal point, and a major axis $2a$ that obeys $r + r' = 2a$. The position vector sweeps out equal areas in equal times, for example, between 1 and 2 and between 3 and 4.

of the string goes from S to A, and back from A to E. Take the segment AE that is covered twice, and borrow one cover to match the segment BS, and you see that the length of the string spans the major axis $2a$.

If you move the two focal points to a single point, you get the circle with radius $R = a$. (The *minor axis* $2b$ is the width in the perpendicular direction and will not figure in this elementary treatment.) If the sun is at S, what is at the other focal point E? Nothing, as far as anyone can tell, though there are rumors Elvis has been sighted there. The labeling in the figure acknowledges this very credible possibility.

Now for Kepler's second law. Follow any one planet for a fixed period, say one week, as it moves from 1 to 2. Measure the *area swept out*, which is the area between the radius vectors $\mathbf{r}_1$, $\mathbf{r}_2$ and the part of the orbit between 1 and 2. Repeat this for any other one-week period, say between 3 and 4, and you will measure the same area. Indeed, this rate of sweeping area is equal even for infinitesimal times, which allows us to say

$$\frac{dA}{dt} = \text{constant}. \tag{7.2}$$

Kepler's third law states that

$$\frac{T^2}{a^3} = \text{same for all planets in the solar system} \qquad (7.3)$$

where T is the time period and a is the semi-major axis. In other words, if you plug in $T = 365$ days and $a = 93,000,000$ miles for the earth (assuming it is on a circular orbit, which it nearly is), you get some number: roughly $2.96 \cdot 10^{-19} \, s^2 m^{-3}$. Now compute this ratio for Jupiter, and you get $3.01 \cdot 10^{-19} \, s^2 m^{-3}$. Very impressive!

Students often ask if Kepler's laws have corrections. They certainly do, like every other law. First of all, planets are not moving just under the influence of the sun but also other planets, especially Jupiter. Secondly, the Newtonian law of gravitation has been modified by Einstein's general theory of relativity. Both these effects prevent the orbit from being closed. The major axis slowly rotates with time, and this is called *precession of the perihelion*, the effect being most pronounced for Mercury. After all the corrections explicable in Newtonian terms, a tiny amount, 43 degrees of an arc per century, remained unexplained. (A degree of arc is 1/3600 of a degree.) In a remarkable feat of human invention, the general theory of relativity explained that last discrepancy.

Kepler's data were available to Newton. Newton, as you know, was sent home from college because there was a plague in Cambridge. He went and lived in his old village, contemplating gravitation. Newton had already invented $F = ma$ but not the F for the particular case of gravity. He then went on to do just that.

7.2 The law of universal gravity

Newton built on his earlier insight that you have to associate a force with acceleration and not with velocity. If you follow a planet as it moves around, thinking it is force that causes velocity, you won't get any definitive answer as to what is behind the force. On the other hand, if you calculate the acceleration of the planet, you will find that, at every instant, it points toward the sun. (This is obvious for the circular orbit.) If all bodies are accelerating toward the sun, it's fairly clear the reason for the acceleration is the sun. You then postulate a force that's exerted by the sun on the planet that bends it into a circle. Your job is to find that force.

What's the nature of the force? Again, it was Newton who figured out that the force that bends the planets around the sun is the same as the force that bends the moon around the earth or makes the apple fall to the earth. Now, the fact that the moon, orbiting at a radius R_m at a speed v_m, is accelerating toward the earth at a rate

$$a_m = \frac{v_m^2}{R_m} \tag{7.4}$$

is something you have already learned in this course. As for the apple, it also is accelerating toward the earth at a rate

$$a_a = g \simeq 9.8 \ ms^{-2}, \tag{7.5}$$

independent of its mass m_a.

Let us guess the formula for the force on the apple, looking at just the magnitude, the direction being obviously toward the center of the earth. Near the earth, we all know, the acceleration of every body is the same. Therefore from $a = \frac{F}{m}$, we deduce that the force of gravity on a falling body is itself proportional to the mass of the body, so that the m may cancel out. So we can write for the apple

$$F_a = m_a f(M_e, R_a) \tag{7.6}$$

where the unknown function f depends on M_e, the mass of the earth, and its radius R_e, which is also the distance R_a between the apple and the center of the earth.

What else do we need? The third law says that if there's a force on the apple exerted by the earth, the apple must exert an equal and opposite force on the earth. Consider the earth and the apple—but imagine the apple getting bigger and bigger. The formula is not going to change. Wait until the apple is huge compared to the earth. Then you will have to agree that the earth is falling toward the apple rather than the other way around. Under this exchanged role, the force on the earth must then be proportional to M_e. Thus, for any two bodies of mass m and M, we obtain an expression for the force compatible with Newton's third law,

$$F = (Mm)f(R), \tag{7.7}$$

where R is the distance between their centers. We don't know the distance dependence; we don't know the function $f(R)$.

To find it, let us compare the acceleration of the apple and the moon due to the pull of the earth:

$$a_a = \frac{M_e m_a}{m_a} f(R_a) \tag{7.8}$$

$$a_m = \frac{M_e M_m}{M_m} f(R_m) \tag{7.9}$$

$$\frac{a_a}{a_m} = \frac{f(R_a)}{f(R_m)}. \tag{7.10}$$

We know $a_a = 9.8 ms^{-2}$. What about a_m? It is given by kinematics to be $a_m = \frac{v_m^2}{R_m}$. Does anybody know how far the moon is? If you guessed a million miles, that is not bad; the correct answer is 238,000 miles. If you make an estimate that's off by a factor of 4 in astrophysics, it's fine, but if you say $R_m = 1000$ miles, we should have a very long talk. Anyway, let us round it off and say $R_m = 240,000$ miles.

Next, what's the radius of the earth? You have some idea, right? How far is California? Three thousand miles. And how many hours is the time difference? Three hours. So that is one hour per thousand miles. If you go all the way around the earth and come back, the accumulated time difference has to be 24 hours. That means the earth has a circumference of roughly 24,000 miles. Dividing by $2\pi \simeq 6$, we get $R_e \simeq 4000$ miles.

I know that we should work with meters and kilometers but, like the rest of you, once I get on the freeway I'm watching how many miles per hour I'm driving, not how many meters per second. Nonetheless, we Americans do use a lot of British units. If you go buy insulation at Home Depot, it's rated in BTUs per slug per poundal, right? Sometimes you wonder why we fought the War of Independence if we're still using those units. Anyway, my brain is split. When I do physics, I use the metric system. When I shop at Home Depot, I use the Home Depot units.

Back to finding the acceleration of the moon, $a_m = \frac{v_m^2}{R_m}$. We'll assume every orbit is a circle; that assumption turns out to be not so bad, even for planets. We already have R_m, which also gives us the length of the orbit

$2\pi R_m$, which it completes in roughly $T = 28$ days, yielding a velocity $v_m = \frac{2\pi R_m}{T}$ and an acceleration

$$a_m = \frac{(2\pi R_m/T)^2}{R_m}. \tag{7.11}$$

If you plug in the numbers, you find

$$\frac{a_a}{a_m} \simeq 3600 = \frac{f(R_a)}{f(R_m)}. \tag{7.12}$$

Given that $\frac{R_m}{R_a} = \frac{240000}{4000} = 60$, you don't have to be a Newton to figure out that

$$f(R) = \frac{1}{R^2}. \tag{7.13}$$

Combining this with Eqn. 7.7, we find the great law of universal gravity:

$$F = G\frac{Mm}{R^2} \tag{7.14}$$

$$G = 6.67 \cdot 10^{-11} Nm^2 kg^{-2}, \tag{7.15}$$

where G is the *universal gravitational constant* that balances the units in Eqn. 7.14 and ensures that the numerical value we obtain at the surface of the earth reproduces $g = 9.8ms^{-2}$.

In this argument, we are assuming that the distance between the apple and the earth is R_e, the radius of the earth. Why not use the height of the tree from which the apple fell? Because Newton's formula is actually written down for two point-like objects, with an unambiguous distance between them. The correct way to handle the earth is to divide it into many small pieces and find the force on the apple due to each piece and add, or rather, integrate, over their contributions. *The result will be that the earth acts as if all its mass were concentrated at its center.* Newton knew this to be true, but he could not prove it to his satisfaction for many years, which is why he delayed publication. Even today it is a hard problem in integration.

Here is another similar result. Suppose you are *inside* a hollow spherical shell of some mass M. What force will you feel? It is clear that if you are at the center, you will feel no force because for every piece of matter in the shell pulling you one way, there is an identical one pulling you the opposite way. What is not obvious, but true, is that the gravitational force will be zero inside the *entire* shell. Of course, outside the shell the force will be that of a point mass M sitting at the center. In summary, for any spherically symmetric distribution of mass, the force felt by a body at radius r is due to all the mass inside a sphere of radius r, acting as a point mass at the center, while the mass outside contributes nothing.

Equation 7.14 is rightly called the law of universal gravitation. It was a tremendous leap of faith to believe that the laws that are operative near the earth also apply to the moon and beyond. This was the year 1687; people believed in witchcraft and harbored all kinds of superstitions. They were not thinking in modern scientific terms. They had a lot of illusions about what the heavens were made of. To believe they're made of the same stuff, and controlled by the same laws, was far from obvious in those days.

Newton's leap of faith has proven extraordinarily prescient: not only the law of gravitation but all the laws of physics that we deduce near the earth seem to work over the entire universe, not just now, but even in the distant past and, we hope, in the future. Indeed, given the long times light takes to get to us from far away galaxies and quasars, much of what we see in the heavens today happened a long time back, and yet we analyze them using the recently discovered laws. We have sampled a very tiny part of the universe, over a tiny period of time, but we apply the laws we deduce here and now to the far reaches of the universe and all the way back in time to the big bang. We confidently predict the future fate of the universe. It's a great break for us that the laws we find seem to be universal and eternal. It need not have been so.

7.3 Details of the orbits

I will apply the law of gravity to the simple case where a planet of mass m orbits a sun of mass $M >>> m$ so that the latter may be assumed to stay put despite the pull of the planet. The origin of coordinates is chosen to be at the sun. We have the equation we need,

$$m\frac{d^2\mathbf{r}}{dt^2} = -\frac{GMm}{r^2}\mathbf{e}_r, \tag{7.16}$$

where $\mathbf{e}_r = \frac{\mathbf{r}}{r}$ is a unit vector that points from the sun to the planet.

This is now a problem in calculus, and everything Kepler said should come out of its solution. You should find that planets move in elliptical orbits. You should find that equal areas are swept out in equal times. You should find the square of a time period divided by the cube of the major axis is independent of the planet's mass.

Even centuries later, I find it takes the class in advanced mechanics quite an effort to solve this equation. It is one thing to write it down, another to actually solve it and get the elliptical orbits. But Newton did all that hundreds of years ago.

Just imagine that you wrote Eqn. 7.16 but couldn't solve it. You would have found the correct law of gravity, but you could never be sure it was right, or convince others, because you could not find the consequence of your equations. That is not unheard of. Consider the theory of quarks, which we believe to be the constituents of protons, neutrons, and so forth. We think we know the underlying equations of motion and forces between quarks. But we do not yet have a way to show, *analytically*, that the underlying equations imply the phenomenon or particles that we see. However, by solving them approximately on big computers, we are fairly certain, after years of work, that the equations are correct. For a new theory to be accepted, its signature consequences must be worked out exactly or approximately, and these must agree with experiment to persuasive accuracy.

Returning to our problem: although we have the means to prove the orbits are elliptical in general, we are going to specialize to circular orbits. You're always allowed in an equation to make an assumption and plug it in to see if it works. We are going to assume there is a circular orbit of radius r, in which the planet is moving at speed v, and see if it is permitted by the laws of motion and gravity. In the radially *inward* direction, $\mathbf{F} = m\mathbf{a}$ gives us

$$m\frac{v^2}{r} = \frac{GMm}{r^2}. \tag{7.17}$$

The left-hand side is the effect, the right-hand side is the cause. If you're spinning a rock tied to a string in a circle, it has an acceleration toward the center, and the string is providing the requisite force. Here, it's the unseen force of gravity from the sun, reaching out to a planet and pulling it in. So let's cancel the m and one power of r from both sides to get a very useful equation:

$$v^2 r = GM. \tag{7.18}$$

We find we can have a circular orbit of any radius we like, provided the speed satisfies this equation. If you want to launch a satellite at some radius r, launch it at this speed at that radius. And as long as you satisfy the equation, it does not matter if the thing that is orbiting is actually a satellite, a space station, or a potato, because the mass of the object has dropped out, taking with it the identity of the object.

This is all we can get out of Newton's laws. Let's go back to see how much of Kepler follows. We have shown that the circle, which is a special case of the ellipse, is possible. How about equal areas in equal time? It's obviously true in this problem, because by assumption, the planet is going at a constant speed on a circle of fixed radius r. So, the only thing left to knock off is Kepler's third law, relating the time period to the size of the orbit.

Let us plug into Eqn. 7.18 the fact that the velocity of the planet is the circumference divided by the time period

$$v = \frac{2\pi r}{T}, \tag{7.19}$$

to obtain

$$\frac{4\pi^2 r^3}{T^2} = GM \tag{7.20}$$

$$\frac{T^2}{r^3} = \frac{4\pi^2}{GM}, \tag{7.21}$$

which is Kepler's third law, since r is the semi-major axis a for a circle, and the right-hand side does not depend on the planet, just the sun.

What did Newton do that went further than Kepler? Kepler said $\frac{T^2}{a^3}$ was a constant for all planets but did not say what the constant was in terms of anything else. Newton tells us what it is in terms of the mass of the sun, π and G. If we plug in $M = 2 \cdot 10^{30}$ kg for the sun, we find $\frac{4\pi^2}{GM}$ is essentially the same number $\simeq 3 \cdot 10^{-19} s^2 m^{-3}$ that we encountered earlier in connection with the $\frac{T^2}{a^3}$ data for the earth and Jupiter.

A similar thing happened in atomic physics. A schoolteacher called Johann Balmer was analyzing the frequencies of light emitted by atoms like hydrogen, and he observed that all the frequencies were given by the formula

$$f = R\left(\frac{1}{n_1^2} - \frac{1}{n_2^2}\right) \tag{7.22}$$

where R is a constant (for a particular atom) and n_1 and $n_2 > n_1$ are any two positive integers. He knew the value of R from the data but not in terms of anything more fundamental. Bohr then derived a formula of this form from making a quantum postulate, and in the process he obtained an expression for R in terms of fundamental constants like Planck's constant, the mass, and charge of the electron and so forth. Balmer did for Bohr what Kepler did for Newton, which was to condense the complicated data into some simple form so that theorists could have a crack at it. Bohr did for atoms what Newton did for gravity, which was to provide the theory underlying the phenomenologically observed behavior.

Now you can do a variety of problems using $\frac{T^2}{a^3} = \frac{4\pi^2}{GM}$, where M is the mass of the heavy object, the "sun" in the problem. One interesting example is the following, depicted in Figure 7.2. It shows the the earth, as we look down from above the North Pole. I'm at point A and I want to watch a tennis game being played at B. I have access to radio waves, but they cannot go through the earth and they can only travel in straight lines. The solution is to have three satellites forming a nice triangle as shown. Each covers a part of the surface of the earth. If B can send the image to satellite 3, then 3 can send it to 1 along the dotted line, and then 1 can beam it to me since I lie within its broadcast cone. If you have three suitably placed satellites, they can help you connect any point on the earth to any other point in this manner. But the satellites better be where you think they are at all times. If they're constantly moving around, it doesn't work. So, what you really want are *geosynchronous satellites*. If you look down at the North Pole, you see the earth is spinning counterclockwise. To stay on top of a fixed point above the earth, these satellites should be rotating around the earth once every 24 hours. The only question is, at what altitude should I launch them? I put in $T = 24$ hours in

$$\frac{T^2}{a^3} = \frac{4\pi^2}{GM} \tag{7.23}$$

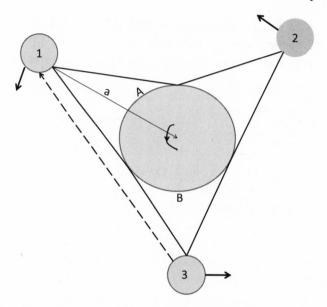

Figure 7.2   The view as we look down at the North Pole shows three
geosynchronous satellites, each of which hovers over a fixed point on the earth,
takes 24 hours per revolution, and covers a part of the earth as shown. By
communicating with them, and allowing them to communicate with each other,
any point on the earth can communicate with any other. If B can send the image
of a game to satellite 3, it can send it to 1 along the dotted line, and that in turn
can beam it to me at A.

and get the radius of $a = 42,200\ km$. Once I have the radius, $v = \frac{2\pi a}{T}$ (Eqn.
7.19) will tell me at what velocity they should be launched into orbit. Naturally, using more than three satellites will provide better television and
cell phone connections.

7.4 Law of conservation of energy far from the earth

What's the potential energy I can associate with the gravitational force? We
have already seen that near the earth

$$\mathbf{F}_g = -\mathbf{j}mg \tag{7.24}$$

is conservative since $\frac{\partial F_x}{\partial y} = \frac{\partial F_y}{\partial x} = 0$. The corresponding potential we have been using is

$$U = mgy, \tag{7.25}$$

where y measured vertically up from the ground. With this choice $U = 0$ on the ground.

Now consider the formula for the force valid for all distances:

$$\mathbf{F}_g = -\mathbf{e}_r \frac{GMm}{r^2} = -\mathbf{r} \frac{GMm}{r^3} = -(\mathbf{i}x + \mathbf{j}y + \mathbf{k}z) \frac{GMm}{r^3}, \tag{7.26}$$

using the fact that $\mathbf{e}_r$, the unit vector in the radial direction, is just $\frac{\mathbf{r}}{r}$.

I claim that this force comes from the following potential upon taking partial derivatives:

$$U(r) = -\frac{GMm}{r}. \tag{7.27}$$

If this were true we should find $F_x = -\frac{\partial U}{\partial x}$:

$$-x\frac{GMm}{r^3} = -\frac{\partial U}{\partial x} \tag{7.28}$$

and likewise for y and z. By symmetry, if it works for x, it will work for y and z. Consider

$$-\frac{\partial U}{\partial x} = GMm \frac{\partial(1/r)}{\partial x} \tag{7.29}$$

$$= -GMm \frac{1}{r^2} \cdot \frac{\partial r}{\partial x} \tag{7.30}$$

$$= -\frac{GMm}{r^2} \cdot \frac{\partial \sqrt{x^2 + y^2 + z^2}}{\partial x} \tag{7.31}$$

$$= -\frac{GMm}{r^2} \cdot \frac{1}{2} \frac{1}{\sqrt{x^2 + y^2 + z^2}} 2x \tag{7.32}$$

$$= -x\frac{GMm}{r^3} = F_x \quad \text{Q.E.D.} \tag{7.33}$$

As in $d = 2$ if $F_x = -\frac{\partial U}{\partial x}$ and similarly for y and z, it follows that

$$\mathbf{F} \cdot d\mathbf{r} = -\left[\frac{\partial U}{\partial x} dx + \frac{\partial U}{\partial y} dy + \frac{\partial U}{\partial z} dz \right] = -dU \qquad (7.34)$$

$$\int_1^2 \mathbf{F} \cdot d\mathbf{r} = -\int_1^2 dU = U(1) - U(2), \qquad (7.35)$$

that is, the line integral of $\mathbf{F}$ is path-independent and U is the corresponding potential energy.

Note that g may be written in terms of G, the mass of the earth M_e and its radius R_e by invoking the law of universal gravity for a body of mass m at the surface of the earth:

$$F = \frac{GM_e m}{R_e^2} \equiv mg \quad \text{which means} \qquad (7.36)$$

$$g = \frac{GM_e}{R_e^2}. \qquad (7.37)$$

If we want to consider motion on a celestial scale we must use the formula for U that is valid for all distances and use for the conserved energy the expression

$$E = \frac{1}{2} mv^2 - \frac{GMm}{r}. \qquad (7.38)$$

7.5 Choosing the constant in U

Since for any r

$$E = \frac{1}{2} mv^2 - \frac{GMm}{r} \qquad (7.39)$$

is the exact conserved energy, we may expect that for an object moving at a modest height y above the earth, this must reduce to

$$E = \frac{1}{2} mv^2 + mgy. \qquad (7.40)$$

The potential energy in the exact result will equal the approximate one valid near the earth if

$$-\frac{GMm}{R_e + y} = mgy = \frac{GMm}{R_e^2}y \tag{7.41}$$

upon using $g = \frac{GM}{R_e^2}$.

But there is no way this equality can be correct, because the two sides have opposite signs!

The resolution is that when you define a potential U, you are free to add a constant. In different schemes you may choose different constants without any contradiction, because only the difference $U_1 - U_2$, in which the constant drops out, is ever invoked in a physical problem. However, if a direct comparison of U itself in two different schemes is made, there need not be any agreement. That is what is happening here. The person working near the earth chooses the potential U_e that vanishes at the surface of the earth, $y = 0$, while a person doing celestial mechanics chooses a potential U_c that vanishes at $r = \infty$. Let c be the difference between the two schemes:

$$U_e(r) = U_c(r) + c. \tag{7.42}$$

To find c, choose a point on the surface of the earth with $r = R_e$ and $y = 0$ where U_e vanishes:

$$mg \cdot 0 = -\frac{GMm}{R_e} + c \tag{7.43}$$

which means

$$c = \frac{GMm}{R_e} \tag{7.44}$$

$$U_e(r) = U_c(r) + \frac{GMm}{R_e}. \tag{7.45}$$

Thus the exact potential, shifted by a suitable constant to make it vanish at $y = 0$, gives for U_e the exact expression

$$U_e(R_e + y) = -\frac{GMm}{R_e + y} + \frac{GMm}{R_e} \tag{7.46}$$

$$= GMm\left(\frac{1}{R_e} - \frac{1}{R_e + y}\right) \tag{7.47}$$

$$= GMm\frac{y}{R_e(R_e + y)} \tag{7.48}$$

$$\simeq m\frac{GM}{R_e^2}y = mgy \tag{7.49}$$

upon ignoring the y compared to R_e in the $R_e + y$ in Eqn. 7.48.

Finally, consider the total energy of a circular orbit. Recall that for the radial direction we have from the second law

$$m\frac{v^2}{r} = \frac{GMm}{r^2} \quad \text{so that} \tag{7.50}$$

$$mv^2 = \frac{GMm}{r}. \tag{7.51}$$

So the kinetic energy is exactly half the magnitude of the potential energy. The total energy is

$$E = \frac{1}{2}mv^2 - \frac{GMm}{r} \tag{7.52}$$

$$= \frac{GMm}{2r} - \frac{GMm}{r} = -\frac{GMm}{2r} = -K. \tag{7.53}$$

So, for a particle in a circular orbit, the total energy is negative and equals half the potential energy or the negative of the kinetic energy. (This assumes we are using the potential U_c appropriate for celestial mechanics, which vanishes at infinity: $U_c(r \to \infty) = 0$.) You will find that even if you solved for the elliptical orbit, the total energy would be negative. This is a general property: an object that is never able to escape the sun's pull, that keeps orbiting it, has negative total energy. Let us see why. A body with negative total energy $E < 0$ can never run off to infinity: at infinity the potential energy is zero and the entire energy is kinetic, and that is supposedly negative, which is impossible. So the assumption that a body of negative energy escaped to infinity is wrong.

So, if you see a comet and want to know if it will come back again, add the kinetic and potential energies. If it's positive, it won't come back; if it's negative, the comet is trapped. Zero is the dividing line, when the comet will collapse at the finish line at infinity. (See if you understand why you do not need the mass of the comet to perform this test.)

Suppose you start at the surface of the earth and start shooting things upward. As you crank up the speed, they will go farther and farther away, and beyond some *escape velocity* v_e they will never come back. What is the minimum speed for escape? You should ensure that it goes to infinity with no kinetic energy to spare. You want it to just manage to get infinity, stagger, and fall down. Well, if it has no kinetic energy at infinity and no potential energy at infinity, its total energy is 0. Therefore, by the law of conservation of energy, the total energy at launch must be zero:

$$0 = \frac{1}{2}mv_e^2 - \frac{GMm}{R_e}, \quad \text{which determines } v_e \text{ to be} \qquad (7.54)$$

$$v_e = \sqrt{\frac{2GM}{R_e}}. \qquad (7.55)$$

You can now go ahead and read some cutting-edge articles. For example, you have heard about dark matter, right? Most of the universe seems to be made out of stuff we cannot see. You, me, we all add up to a very small percentage of the total mass. How do people know there is dark matter if we cannot see it? By the use of $v^2R = GM$, which relates the speed of an orbiting object at radius R to the mass inside the orbit that is pulling it into orbit. As you follow the trajectories of objects orbiting around the galactic center at radius R, the enclosed mass should grow with R, and it should stop growing once the orbit size passes the observed radius of the matter in the galaxy (except for a few odd balls like the one whose orbit we are following). But it keeps growing for a considerable distance beyond, telling us there is a halo of dark matter. Even dark matter cannot hide its gravitational effect. Every galaxy seems to have a dark matter halo that extends beyond the visible part.

CHAPTER 8

Multi-particle Dynamics

8.1 The two-body problem

Next we begin our study of the dynamics of more than one body. You might think we already did this when we studied the solar system, consisting of the sun and all the planets. But we considered only one planet, and the sun just stood there as a source of the gravitational force. That was essentially a one-body problem.

As usual, let me start with the simplest possible case of two bodies moving in one dimension. They have coordinates x_1 and x_2 and masses m_1 and m_2 as shown in Figure 8.1. The first body obeys

$$m_1 \frac{d^2 x_1}{dt^2} = F_1,$$ (8.1)

which I am going to rewrite as

$$m_1 \ddot{x}_1 = F_1$$ (8.2)

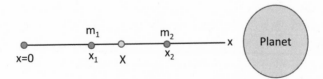

Figure 8.1   A system of two bodies in $d = 1$ at x_1 and x_2 with their CM at X. Far to the right is a planet.

118

where the two dots on $\ddot{x}$ tell you we are taking two time derivatives, a convenient notation if you take only a few time derivatives. Divide the forces on body 1 into two parts:

$$m_1\ddot{x}_1 = F_{12} + F_{1e} \tag{8.3}$$

where F_{12} is the force on 1 due to 2 and F_{1e} the sum of all external forces on 1 due to everything else. Similarly for 2, and in the same notation

$$m_2\ddot{x}_2 = F_{21} + F_{2e}. \tag{8.4}$$

The universe has many bodies, and I have just picked these two as part of my system and lumped the rest under the label "external." For example, a spring could be mediating a force between the masses, which could also be "falling" under the gravitational pull of some large planet far to the right on the x-axis. The spring is just a way of transmitting force from one body to the other; it is the source of the internal forces $F_{12} = -F_{21}$. The external force F_e is the gravitational pull of the planet. Now I squash the spring and let go. The masses will both accelerate to the right under F_e and also oscillate relative to each other under the elastic force due to the spring, which I will refer to as simply spring force.

8.2   The center of mass

We are going to manipulate Eqns. 8.3 and 8.4 to get some interesting results. Let us add their left-hand sides and equate them to the sum of the right-hand sides:

$$m_1\ddot{x}_1 + m_2\ddot{x}_2 = F_{12} + F_{1e} + F_{21} + F_{2e}. \tag{8.5}$$

I now invoke Newton's third law,

$$F_{12} = -F_{21}, \tag{8.6}$$

whatever the underlying force: gravity, spring, electrostatic, and so forth. This whole chapter is about milking this one simple result, this cancellation of F_{12} and F_{21}. Next, I lump all the external forces F_{1e} and F_{2e} into one external force F_e:

$$F_e = F_{1e} + F_{2e}. \tag{8.7}$$

I then have this equation:

$$m_1\ddot{x}_1 + m_2\ddot{x}_2 = F_e. \tag{8.8}$$

I multiply and divide the left-hand side by the total mass

$$M = m_1 + m_2 \tag{8.9}$$

to obtain

$$M\left[\frac{m_1\ddot{x}_1 + m_2\ddot{x}_2}{M}\right] = F_e, \tag{8.10}$$

$$M\frac{d^2X}{dt^2} = F_e, \quad \text{where} \tag{8.11}$$

$$X = \left[\frac{m_1 x_1 + m_2 x_2}{M}\right] \tag{8.12}$$

is called the *center-of-mass coordinate or the CM*.

What I have done is correct, but why did I do that? I have introduced a fictitious entity, the center of mass. The center of mass has a location X, which is a weighted average of x_1 and x_2:

$$X = \frac{m_1 x_1 + m_2 x_2}{M} = \frac{m_1}{m_1 + m_2} x_1 + \frac{m_2}{m_1 + m_2} x_2. \tag{8.13}$$

If $m_1 = m_2 = m$, then $M = 2m$ and

$$X = \frac{x_1 + x_2}{2} \tag{8.14}$$

lies midway between the particles. If $m_1 > m_2$, X will be closer to m_1 and vice versa. The weighted sum gives a certain coordinate, but there is nothing there. All the stuff is either at x_1 or x_2. The center of mass is the location of a mathematical entity. It's not a physical entity. But we care, because it behaves like a body. After all, if you were shown only Eqn. 8.11, you would say, "Well, this guy's talking about a body of mass M undergoing some acceleration due to the force F_e." Thus, the center of mass is a fictitious body, whose mass is the total mass of these two particles, and whose acceleration is controlled by only the total *external* force. This is the key. All the internal forces have canceled out, and what remains is the external force.

If you have three bodies, you can do a similar manipulation with extra forces like $F_{23} = -F_{32}$ and so on, and you will end up with Eqn. 8.11 with one obvious change:

$$X = \frac{m_1 x_1 + m_2 x_2 + m_3 x_3}{M} = \frac{m_1 x_1 + m_2 x_2 + m_3 x_3}{m_1 + m_2 + m_3}. \qquad (8.15)$$

We write such sums in a compact notation:

$$X = \frac{\sum_{i=1}^{3} m_i x_i}{\sum_{i=1}^{3} m_i}, \qquad (8.16)$$

which can be generalized to N particles by replacing the 3 by N.

Once more with feeling: the CM responds only to the total external force; it doesn't care about internal forces. I'll give an example. A couple of samurai are having a fight in an airplane, punching each other and so on. It is a flight and fight situation. The rest of the passengers get fed up and throw them out. So, they're falling down, affecting each other's dynamics. This samurai will feel a force due to that samurai, that samurai will feel a force due to this samurai, but the center of mass is still going to drop like a rock. It's going to feel a force $(m_1 + m_2)g$, and it will have an acceleration g.

Suppose at some point one falling samurai cuts the other into two pieces. So now we have three bodies: the first protagonist and the other two, who used to be one. You can take these three bodies, find their center of mass, and it will be the same story—the center of mass will just keep accelerating at the same g, as if nothing happened.

Indeed, if you were following the CM alone, you would see no sign of all this violence and the involuntary partitioning of the second samurai. The system is becoming more and more complicated, but nothing changes the dynamics of the center of mass, undergoing free fall under gravity.

In summary, the CM can accelerate only due to external forces, like gravity in this example. If there were no external forces, then the center of mass would behave like a free particle. If it was not moving to begin with, it won't move later. If it was moving to begin with, it will maintain the initial velocity.

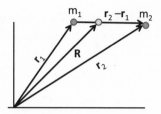

Figure 8.2   The CM vector $\mathbf{R}$ lies on $\mathbf{r}_2 - \mathbf{r}_1$, the line joining the tips of $\mathbf{r}_1$ and $\mathbf{r}_2$.

If you're living in two dimensions, you define a CM vector $\mathbf{R}$ for two particles as follows and as indicated in Figure 8.2:

$$\mathbf{R} = \mathbf{i}X + \mathbf{j}Y = \frac{m_1\mathbf{r}_1 + m_2\mathbf{r}_2}{m_1 + m_2}, \tag{8.17}$$

which is equal to two relations

$$X = \frac{m_1 x_1 + m_2 x_2}{m_1 + m_2} \tag{8.18}$$

$$Y = \frac{m_1 y_1 + m_2 y_2}{m_1 + m_2}. \tag{8.19}$$

I have shown the CM situated on the line connecting the masses. Why? One way to understand this is to choose a new x'-axis that passes through the masses and a new y'-axis perpendicular to it. Since the masses only have an x' coordinate, the weighted average must also have only an x' coordinate.

Here's another example. You take a complicated collection of masses and springs, connected by ropes and chains and whatnot. You throw the whole mess into the air. All the different parts of it are jiggling and doing complicated movements, but if you follow the center of mass, in other words, at every instant you compute

$$\mathbf{R} = \frac{\sum_{i=1}^{N} m_i \mathbf{r}_i}{\sum_{i=1}^{N} m_i}, \tag{8.20}$$

it will simply follow the parabolic path of a body curving under gravity. If at some point this complicated object fragments into two disconnected chunks that fly off on their own paths, the CM will continue as before.

Here is a very useful alternative for finding the CM of a collection of masses.

1. Take a subset of them and replace them with all their mass sitting at their center of mass. Replace the rest by their total mass sitting at their center of mass.
2. Find the center of mass of these two centers of mass, properly weighted.

Let us verify the equivalence of this to the standard definition, for the case of three bodies. The original recipe to compute X was

$$X = \frac{m_1 x_1 + m_2 x_2 + m_3 x_3}{m_1 + m_2 + m_3}. \tag{8.21}$$

Next we follow the new recipe and divide the system into two parts, one made of masses numbered 1 and 2, of total mass $M_{12} = m_1 + m_2$, and the third one by itself. We first compute the CM of $1 + 2$ following the standard definition:

$$X_{12} = \frac{m_1 x_1 + m_2 x_2}{M_{12}} \tag{8.22}$$

and combine it with object 3 and see what happens if we follow the recipe

$$X_{\text{Recipe}} = \frac{m_3 x_3 + M_{12} X_{12}}{m_3 + M_{12}} \tag{8.23}$$

$$= \frac{m_3 x_3 + m_1 x_1 + m_2 x_2}{m_1 + m_2 + m_3} \tag{8.24}$$

$$= X. \tag{8.25}$$

Along the way, I have used Eqn. 8.22, which implies that $M_{12} X_{12} = m_1 x_1 + m_2 x_2$.

Clearly, in a general case, we can subdivide the masses into more than two subsets and follow the same procedure.

As long as you are dealing with a countable number of masses, finding the CM is just plug and chug, in one or higher dimensions. Things become more interesting if I give you an extended body, like a rod of mass M and length L, shown in the upper part of Figure 8.3. Where is the center of mass? We have to adapt the previous definition for discrete masses. The

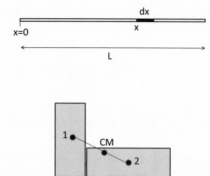

Figure 8.3   (Top) A rod of length L and mass M. (Bottom) An L-shaped object made of two rectangles. The points 1 and 2 are the CMs where all their mass may be assumed to be concentrated. The point marked CM is the weighted sum of 1 and 2.

trick is to partition the rod into tiny pieces of length dx located a distance x from the origin, chosen to be at the left end of the rod. The mass of this segment is $\frac{M}{L}dx$, the product of the mass per unit length and the length of the sliver. Its location is x.

Now, you might object that the segment extends from x to $x + dx$ and doesn't have a definite coordinate. This objection is valid for any finite dx, but in the end, we will let $dx \to 0$ and the objection will disappear in that limit. The CM is now given by the ratio of *integrals*, rather than *sums* as in Eqn. 8.20:

$$X = \frac{\int_0^L \frac{M}{L} x\, dx}{\int_0^L \frac{M}{L} dx} \tag{8.26}$$

$$= \frac{1}{L} \left. \frac{x^2}{2} \right|_0^L \tag{8.27}$$

$$= \frac{L}{2}. \tag{8.28}$$

So the center of mass of this rod, to nobody's surprise, is right at the midpoint. We knew this before doing the integral. What was behind that intuition? If you take the origin at the midpoint of the rod, you can argue that for every sliver with coordinate x, you have another equally massive one at $-x$, and that the weighted average of these two points is zero.

The CM, which is the weighted average of all these zeros, also lies at the geometric center.

This symmetry argument applies even for a two-dimensional body, like a rectangle. We can argue that its CM is at its geometric center. This is because for each tiny square segment of size $dx \cdot dy$ at (x, y) (measured from the geometric center), we can find an equal one at $(-x, -y)$, and the weighted average of the two will be at the origin $(0, 0)$. The weighted average of all these zeros will also be $(0, 0)$.

Now consider a rod whose linear mass density, or mass per unit length, varies as some function $\rho(x)$. Then by the usual argument

$$X = \frac{\int_0^L \rho(x)x\,dx}{\int_0^L \rho(x)\,dx}. \tag{8.29}$$

For example, if $\rho(x) = Ax$, where A is some constant

$$X = \frac{\int_0^L Axx\,dx}{\int_0^L Ax\,dx} \tag{8.30}$$

$$= \frac{L^3/3}{L^2/2} = \frac{2L}{3}, \tag{8.31}$$

which is biased to the right as you would expect.

Consider next an L-shaped object shown in the lower part of Figure 8.3. Where is its CM? We resort to the trick of first reducing each rectangle to its CM and then doing the weighted sum of the two CMs. By the symmetry argument, the CM of each rectangle is at its geometric center, the points numbered 1 and 2. We may imagine their total masses M_1 and M_2 to be concentrated at these points. The CM is the weighted sum of these two point masses, and it lies along the line joining them. Its precise location is easily found if the masses and dimensions of the rectangles are known.

Now for one final CM calculation, the most difficult one you are supposed to know. The object is a triangle of mass M, base $2w$, and height h, as shown in Figure 8.4. It has an area

$$A = \frac{1}{2}2wh = wh \tag{8.32}$$

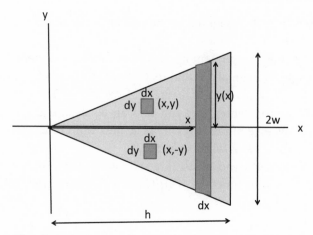

Figure 8.4   The CM calculation of a triangle of base $2w$ and height h. It is viewed as a weighted sum over rods of width dx and height $2y(x)$.

and a mass per unit area or areal density

$$\rho = \frac{M}{A} = \frac{M}{wh}. \qquad (8.33)$$

Where is the center of mass of this object? Again, by symmetry, you can tell that Y, the y coordinate of the center of mass, must be zero. For every tiny square $dxdy$ with some coordinate (x, y), there is a matching one with coordinate $(x, -y)$. For X, you have to do some honest work. We will divide and conquer.

Let us imagine the triangle as composed of thin rectangles of width dx and height $2y(x)$, as indicated. (Each strip is not quite a rectangle, because the edges are slightly tapered, but when $dx \to 0$, they will reduce to rectangles.) The mass dm of the rectangle at a given x is

$$dm = \frac{M}{A} 2y(x)dx = \frac{M}{wh} 2y(x)dx, \qquad (8.34)$$

which is just the product of the mass per unit area $\frac{M}{A}$ and the area of the strip $2y(x)dx$. We find $y(x)$ using similar triangles:

$$\frac{y(x)}{w} = \frac{x}{h} \quad \text{which means } y(x) = \frac{wx}{h}. \qquad (8.35)$$

The weighted average of x is then

$$X = \frac{1}{M} \int_{x=0}^{h} \frac{M}{wh} 2y(x)x\,dx \tag{8.36}$$

$$= \frac{1}{wh} \int_{0}^{h} 2\frac{wx}{h}x\,dx \tag{8.37}$$

$$= \frac{2}{h^2} \int_{0}^{h} x^2 dx \tag{8.38}$$

$$= \frac{2}{3}h. \tag{8.39}$$

We could have anticipated that X would be skewed to the right, and this formula quantifies that intuition. Note in Eqn. 8.37 that this two-dimensional problem maps onto a one-dimensional one, with a linear density proportional to x, that is, $\rho(x) \propto x$. This is because each vertical strip may be replaced by a point mass on the x-axis proportional to $y(x)$, which in turn grows linearly with x.

To summarize, when we work with extended bodies or more than one body, we can replace the entire body by a single point for certain purposes. The single point is called a center of mass or CM. The CM is fictitious. It has a mass equal to the total mass. It has a location $\mathbf{R}$ that moves in response to the total external force:

$$M\frac{d^2\mathbf{R}}{dt^2} = \mathbf{F}_e. \tag{8.40}$$

The center of mass is not aware of internal forces, and that's what we want to exploit.

One class of problems has a net external force $\mathbf{F}_e$, and there we know that the CM responds as a point to $\mathbf{F}_e$, regardless of its constituents. For example, a jumbled mass of constituents tossed in the air follows the parabolic trajectory of a point mass, in response to gravity. This is just a one-body problem, which we have studied extensively. So we move on.

8.3 Law of conservation of momentum

Now consider the case $\mathbf{F}_e = 0$. That means

$$\frac{d^2\mathbf{R}}{dt^2} = 0 \tag{8.41}$$

$$\frac{d\mathbf{R}}{dt} = \text{some constant.} \tag{8.42}$$

Let us multiply both sides by the total mass M, which is itself a constant, to obtain

$$M\frac{d\mathbf{R}}{dt} = \text{some other constant.} \tag{8.43}$$

What is this constant? This question requires a digression into the concept of momentum. The *momentum* $\mathbf{p}$ of a single particle is given by

$$\mathbf{p} = m\mathbf{v}. \tag{8.44}$$

We may rewrite $\mathbf{F} = m\mathbf{a}$ as

$$\mathbf{F} = m\frac{d^2\mathbf{r}}{dt^2} = m\frac{d\mathbf{v}}{dt} = \frac{dm\mathbf{v}}{dt} = \frac{d\mathbf{p}}{dt}, \tag{8.45}$$

that is, the *force is the rate of change of momentum.*

The CM, though fictitious, is endowed with a well-defined position

$$\mathbf{R} = \frac{m_1\mathbf{r}_1 + m_2\mathbf{r}_2}{M} \tag{8.46}$$

and a velocity, given by differentiation of both sides

$$\frac{d\mathbf{R}}{dt} = \frac{m_1\frac{d\mathbf{r}_1}{dt} + m_2\frac{d\mathbf{r}_2}{dt}}{M}. \tag{8.47}$$

Let us define its momentum as we did for a real particle

$$\mathbf{P} = M\frac{d\mathbf{R}}{dt}. \tag{8.48}$$

Now we find, using Eqn. 8.47, that

$$\mathbf{P} = M\frac{d\mathbf{R}}{dt} = \left[m_1\frac{d\mathbf{r}_1}{dt} + m_2\frac{d\mathbf{r}_2}{dt}\right] \qquad (8.49)$$

$$= \mathbf{p}_1 + \mathbf{p}_2. \qquad (8.50)$$

In summary, *the CM momentum is the sum of the momenta of all the particles and only external forces can change it*:

$$M\frac{d^2\mathbf{R}}{dt^2} = \frac{d\mathbf{P}}{dt} = \mathbf{F}_e. \qquad (8.51)$$

If $\mathbf{F}_e = 0$, the CM momentum $\mathbf{P}$ is conserved. This means the sum over all the individual momenta is conserved, as long as the particles interact only with each other, and nothing external.

A classic example is two people standing on ice. Their total initial momentum is $\mathbf{0}$. The ice is going to support them vertically against gravity, but if it's frictionless, it cannot apply any force parallel to the plane of the ice. For example, if you and I are standing on ice, and we push against each other and fly apart, my momentum vector has to be exactly the opposite of your momentum vector.

Next consider a mass m_1 that collides with a mass m_2 in the absence of external forces. Then $\mathbf{P}$, the total initial momentum, equals $\mathbf{P}'$, the total final momentum:

$$\mathbf{P} = m_1\mathbf{v}_1 + m_2\mathbf{v}_2 = m_1\mathbf{v}_1' + m_2\mathbf{v}_2' = \mathbf{P}'. \qquad (8.52)$$

I use no primes and primes on the initial and final velocities rather than the labels 1 and 2, which are now used to distinguish the two particles. During the collision, one mass exerts a force on the other mass, but the other mass exerts an opposite force on the first. So the rate of change of momentum of one is equal and opposite at every instant to the rate of change of momentum of the other. Hence the total momentum is unaffected, even if the individual momenta change.

This is called the *law of conservation of momentum*. In terms of $\mathbf{p}$,

$$\mathbf{p}_1 + \mathbf{p}_2 = \mathbf{p}_1' + \mathbf{p}_2'. \qquad (8.53)$$

Let us be sure we get it. Take a collection of bodies. At a given instant every one has its own velocity and its momentum. Add up all the momenta. If you are in one dimension, just add the numbers; if in two dimensions, add the vectors to obtain the total momentum. That total momentum does not change, if there are no outside forces acting.

The law of conservation of momentum survived the revolutions of relativity and quantum mechanics, although the explicit formula $\mathbf{p} = m\mathbf{v}$ did not.

So far, we have considered the following two cases:

- The CM is subject to an external force $\mathbf{F}_e \neq 0$, and it moves in response like a single particle of mass M, as illustrated by the falling samurai.
- The external force $\mathbf{F}_e = 0$, and the CM has a non-zero velocity or momentum that is conserved.

Now I finally consider a problem where there are no external forces, and the CM is initially at rest and stays put.

Recall the dynamics of the sun and the earth. My earlier description of the sun sitting still and the earth moving around it is not acceptable for the following reason. The momentum of the sun, $\mathbf{P}_s$, is not changing, and it equals $\mathbf{0}$. The momentum of the earth, $\mathbf{P}_e$, is changing as it orbits the sun. Their total momentum is therefore changing, which is not allowed, when the only force is their mutual attraction. Stated differently, what is wrong with our earlier description is that the center of mass of the sun and earth does not move uniformly in a straight line, or remain at rest, as it should. Instead, as the earth orbits the fixed sun, the CM, which lies on the line joining the sun and the earth, goes around the sun as well. The CM of the earth and sun should not accelerate. If it was initially at rest, it should remain at rest. If it had an initial velocity, it should maintain that velocity, and we can choose to view it from a frame in which it is at rest. (Going to a frame at a fixed relative velocity to our initial one will not alter our status as inertial observers.) Figure 8.5 portrays the correct description. I'm considering a solar system where the sun is not as big as ours relative to the planets, so I can show the center of mass distinctly. (Our sun is so much more massive than the average planet that the CM typically lies inside the sun.) When the two bodies are at 1 and 2, the CM is where it is as shown, and it will be there later when they are at 3 and 4. So both the bodies will revolve around their common CM, which stays put.

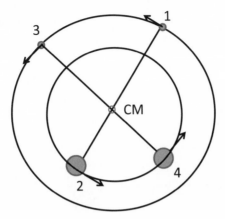

Figure 8.5   The sun (big circle) and a planet (small circle) orbiting around their CM in some solar system where the sun is not quite as massive as ours. The arrows give the instantaneous momenta at points 1 through 4. Our sun is so massive compared to most planets that the CM typically lies inside the body of the sun.

Our earlier description with the fixed sun, while not strictly correct, is an excellent approximation to the truth for our solar system, given how much more massive our sun is than the planets (except for Jupiter). In the limit $\frac{m}{M} \to 0$, the CM falls right on top of the sun, and neither moves. This was the limit we were tacitly assuming earlier.

You have to be careful when you apply $\mathbf{F} = m\mathbf{a}$ in the radial direction, for, say, the planet. You should write

$$\frac{GMm}{r_{12}^2} = \frac{mv^2}{r_1} \tag{8.54}$$

where v is its velocity, r_{12} is the distance between the sun and the planet, and r_1 is the distance between the planet and the CM, the point about which it goes in its circular orbit. In other words, the force of gravity on the planet or the sun is a function of the distance between the planet and the sun, not the distance to the CM, whereas the centripetal force depends on the radius of the orbit, which is the distance to the CM.

Next, a few more examples of a CM coordinate $\mathbf{R}$ that is free from forces and immobile.

Consider a closed railway carriage of length L and mass M that contains a point horse of mass m. The horse is at the left end of the carriage, which we choose to be $x=0$. The CM of the carriage is at $x=\frac{L}{2}$. The initial CM of the horse and carriage is at

$$X = \frac{m \cdot 0 + M\frac{1}{2}L}{m + M}. \tag{8.55}$$

Assuming the track is frictionless, the CM cannot move.

Now the horse decides to walk over to the right end of the carriage. You will know something is going on without looking in, because when the horse moves to the right, the carriage will move to the left, to keep the CM fixed. Let us find out by how much the midpoint of the carriage moves. Say it goes to $x=\frac{1}{2}L - \delta$. The horse then has a coordinate $\frac{1}{2}L - \delta + \frac{1}{2}L$, because it is now $\frac{1}{2}L$ to the right of the center of the carriage. Equating X before and after, we find

$$\frac{m \cdot 0 + M\frac{1}{2}L}{m + M} = \frac{m \cdot (L - \delta) + M(\frac{1}{2}L - \delta)}{m + M} \tag{8.56}$$

$$M\frac{L}{2} = m \cdot (L - \delta) + M\left(\frac{L}{2} - \delta\right) \quad \text{with a solution} \tag{8.57}$$

$$\delta = \frac{mL}{m + M}. \tag{8.58}$$

Once again, if $\frac{m}{M} \to 0$, the carriage will not move when the horse does.

Yet another problem is depicted in Figure 8.6. There is a boat of length L and mass M, whose left-most point is located d meters from the shore. You have a mass m, and you are x meters from the left end. You want to jump ship. Rather than leap from where you are, you decide to walk to the left-most part of the boat and then jump, because you would rather jump d meters than $d+x$ meters. Again, assuming the water exerts no horizontal force on the boat, you will find you have to jump more than d. If you move to the left, the boat will move to the right, to keep the CM, shown by a big cross, at a fixed distance X from the shore. When you arrive at the left edge of the boat, you will be more than d meters from shore. Let us find out how far you have to jump.

Let D be the final distance from the shore to the left end of the boat, which is also how much you need to jump. The CM of the boat has a coordinate $D+\frac{L}{2}$. With the origin at the shore, let us equate the X before

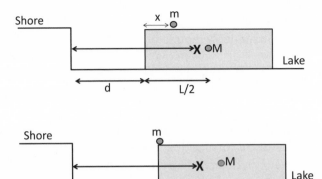

Figure 8.6 (Top) You (tiny circle of mass m) and the boat of mass M, before. (Bottom) You at the left end of the boat, which has shifted slightly to the right keeping the CM fixed at X.

and after, after first canceling the total mass from the denominator in both sides:

$$m(d+x) + M\left(d+\frac{L}{2}\right) = mD + M\left(D+\frac{L}{2}\right) \qquad (8.59)$$

$$D = \frac{m(d+x)+Md}{M+m} = d + \frac{mx}{m+M}, \qquad (8.60)$$

which is more than d and less than $d+x$. Thus it helps to walk to the edge of the boat, but not by as much as you would naively expect.

Let's ask what happens when you leap to the shore and are airborne. The CM cannot move, so if you zoom to the left, the boat will move to the right. Equivalently, the total momentum cannot change. Originally, the momentum was **0**; nobody was moving. Suddenly you're moving to the left, and the boat has to move the other way. Of course, its velocity is not the opposite of yours; its momentum is. So, the big mass of the boat times the small velocity of the boat will be equal and opposite to your small mass times your big velocity.

Now you have landed on the shore. Your momentum is **0**. What's happening to the boat? Is it going to stop now? No, the boat will not stop just because you hit the shore. The boat will keep moving because there's no force on the boat; it will keep moving. The question you may have is, "How come there is suddenly momentum in the system when it had none before?" The answer is that an external force has now come into play: the force of friction between you and the earth. Previously, it was just you and the boat, and you couldn't change your total momentum. But the ground is now obviously pushing you to the right because you, who were initially flying to the left, have stopped. So the combined system—you and the boat—have a rightward force acting during the time it took to stop you; it's that momentum that's carried by the boat. A better way to say this is as follows. You and the boat exchanged momenta: you push the boat to the right, you move to the left. Then your momentum is reduced to zero by the shore, and the earth as a whole takes up the missing momentum. The boat has no reason to change its momentum, and it keeps going. Can you calculate the speed of the boat? No. I only told you that you jumped and landed on the shore, and that's not enough to predict how fast the boat will be moving. But if I told you your velocity when you leaped off the boat, then of course you know your momentum and can deduce that of the boat.

8.4 Rocket science

There is no use struggling through a physics course if at the end you cannot claim to do rocket science. So we are going to derive the rocket equation. Everyone knows that if you blow up a balloon and let it go, the balloon goes one way, and the air goes the other way. Action and reaction are equal; even lay people know that. I don't want to go into the rocket problem in any great detail, just enough to get you familiarized with how the key equations are derived.

Figure 8.7 part A shows a rocket whose mass at time t is M and whose velocity is v. The rocket emits gases, and the gases have a certain *exhaust velocity* of magnitude v_0, pointing away from the rocket. *The value of v_0 is fixed relative to the rocket*, not relative to the ground. If you are riding inside the rocket and you look at the fumes coming out of the back, they will be leaving you at that speed v_0. Their velocity relative to the ground will be $v - v_0$: the velocity $-v_0$ relative to the rocket plus the velocity v of the rocket relative to the ground.

A short time dt later, the rocket has a mass $M - \delta$ because it has lost some of its own body mass δ in the form of exhaust fumes. The rocket's

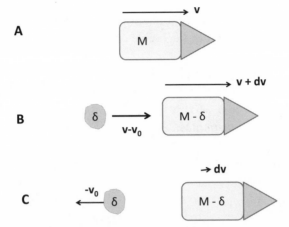

Figure 8.7   (A) The rocket at time t as seen from the ground. (B) The rocket at time $t + dt$ as seen from the ground. It has lost mass δ in the form of emitted gases and gained speed dv. The fumes of mass δ (the blob) are moving at a speed $(v - v_0)$ relative to the ground. (C) The rocket and fumes at time $t + dt$ in the frame moving at velocity $v(t)$. The fumes have an exhaust velocity $-\mathbf{v}_0$ in the rocket frame.

velocity is now $v + dv$. Let us balance the momentum before and after:

$$Mv = (M - \delta)(v + dv) + (v - v_0)\delta. \tag{8.61}$$

The left-hand side is clearly just the momentum of the rocket before this short interval. The right has first the new mass of the rocket times its new velocity. Next is the momentum of the blob of fumes emitted: its mass δ times its velocity relative to the ground of $v - v_0$. I write δ to the right of $(v - v_0)$ so that when you open the brackets you do not make the mistake of thinking δv is a change in v. Let us open up the brackets and manipulate as follows (keeping δ to the right):

$$Mv = Mv + Mdv - v \cdot \delta - dv \cdot \delta + v \cdot \delta - v_0 \cdot \delta \tag{8.62}$$

which simplifies to

$$v_0\delta = Mdv \quad \text{or} \tag{8.63}$$

$$\frac{\delta}{M} = \frac{dv}{v_0}. \tag{8.64}$$

Along the way, I have dropped a quantity $dv \cdot \delta$ because it is quadratic in the infinitesimals, and hence negligible compared to the infinitesimals dv and δ.

The next thing is to integrate these equations. The rocket emitted gas of mass δ in time dt. The mass of the rocket then was reduced to $M - \delta$. In the language of calculus, $M(t)$ is the mass of the rocket at time t, and dM is the change in the *function* M:

$$dM = M(t + \Delta t) - M(t). \tag{8.65}$$

We have seen that

$$dM = (M - \delta) - M = -\delta, \tag{8.66}$$

so that Eqn. 8.64 becomes

$$-\frac{dM}{M} = \frac{dv}{v_0} \quad \text{which integrates to} \tag{8.67}$$

$$\ln \frac{M_0}{M} = \frac{v}{v_0} \tag{8.68}$$

$$v(t) = v_0 \ln \frac{M_0}{M(t)}, \tag{8.69}$$

assuming the rocket had an initial velocity of 0 and mass M_0. Remember that at large times, $M(t)$ cannot fall below the mass of the empty rocket.

8.5   Elastic and inelastic collisions

A body of mass m_1 and velocity v_1 collides with a body of mass m_2 and velocity v_2, with all velocities shown as positive in Figure 8.8. Our goal is to find the final velocities v_1' and v_2'. We need two conditions to find two unknowns, right? The conservation of momentum is always good for one equation as long as there are no external forces like friction:

$$m_1 v_1 + m_2 v_2 = m_1 v_1' + m_2 v_2'. \tag{8.70}$$

You need a second equation to solve for the two unknowns, and that's where there are two extreme cases.

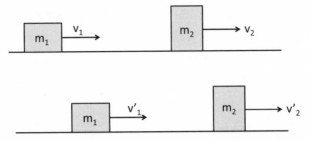

Figure 8.8   (Top) A body of mass m_1 and velocity v_1 collides with a body of mass m_2 and velocity v_2, with all velocities shown as positive. (Bottom) The two bodies with final velocities v_1' and v_2'.

One is called the *totally inelastic collision* in which the two masses stick together and move at a common velocity $v_1' = v_2' = v'$. That means there is just a single unknown v', which we find by going back to Eqn. 8.70

$$m_1 v_1 + m_2 v_2 = (m_1 + m_2)v' \quad \text{with a solution} \tag{8.71}$$

$$v' = \frac{m_1 v_1 + m_2 v_2}{m_1 + m_2}. \tag{8.72}$$

The other kind of collision is called *totally elastic*. Here the kinetic energy is conserved:

$$\frac{1}{2}m_1 v_1^2 + \frac{1}{2}m_2 v_2^2 = \frac{1}{2}m_1(v_1')^2 + \frac{1}{2}m_2(v_2')^2. \tag{8.73}$$

You can juggle Eqns. 8.70 and 8.73 and solve for v_1' and v_2'. The answer is

$$v_1' = \left[\frac{m_1 - m_2}{m_1 + m_2}\right]v_1 + \left[\frac{2m_2}{m_1 + m_2}\right]v_2 \tag{8.74}$$

$$v_2' = \left[\frac{m_2 - m_1}{m_1 + m_2}\right]v_2 + \left[\frac{2m_1}{m_1 + m_2}\right]v_1. \tag{8.75}$$

Because one of these two simultaneous equations (Eqn. 8.73) is quadratic in the velocities and will not yield to the familiar trick for linear ones, here is some help for those who want to derive Eqns. 8.74 and 8.75. Rewrite Eqn. 8.73 as

$$\frac{1}{2}m_1 v_1^2 - \frac{1}{2}m_1(v_1')^2 = \frac{1}{2}m_2(v_2')^2 - \frac{1}{2}m_2 v_2^2 \qquad (8.76)$$

$$m_1(v_1 - v_1')(v_1 + v_1') = m_2(v_2' - v_2)(v_2 + v_2') \qquad (8.77)$$

while Eqn. 8.70 says

$$m_1(v_1 - v_1') = m_2(v_2' - v_2). \qquad (8.78)$$

Upon dividing Eqn. 8.77 by this we get

$$(v_1 + v_1') = (v_2 + v_2'). \qquad (8.79)$$

The last two *linear* equations 8.78 (which is just a rearrangement of Eqn. 8.70) and 8.79 may be readily solved for v_1' and v_2' to yield Eqns. 8.74 and 8.75.

   You have to be very careful in using the conservation laws. You cannot use the law of conservation of energy in an inelastic collision. For example, if two identical bodies with opposite velocities collide and stick together to form a single mass at rest, they have lost their initial kinetic energy.

   Here is an example that illustrates the proper use of conservation laws. You have a pistol and you want to know with what speed the bullet comes out. In the old days, the following technique was used. Figure 8.9 shows a ballistic pendulum. You hang a chunk of wood of mass M from

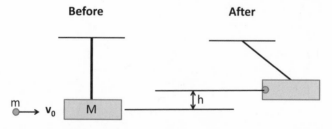

Figure 8.9   A bullet of mass m and velocity v_0 plows into a suspended block of wood of mass M, and the two of them rise as a pendulum to height h.

the ceiling. Then you fire the bullet of known mass m and unknown speed v_0. The bullet rams into this chunk of wood, and the whole thing is set in motion. The combination is like a pendulum, which will rise up to a certain maximum height h that you can easily measure. From that maximum height, you can calculate the speed of the bullet.

You might be naive and decide not to care about intermediate details, and find v_0 by equating the initial kinetic energy of the bullet $\frac{1}{2}mv_0^2$ to the final potential energy $(M+m)gh$ of the block and bullet. That is

$$(M+m)gh = \frac{1}{2}mv_0^2 \tag{8.80}$$

where the only unknown is v_0. That's wrong because you cannot use the law of conservation of energy in the totally inelastic collision between the bullet and the block. Some energy will go into heating up the block. But you can use the law of conservation of horizontal momentum during the collision, since gravity cannot change the total horizontal momentum during the practically instantaneous collision. So in the first totally inelastic collision you may assert that

$$mv_0 = (M+m)V, \tag{8.81}$$

which determines the velocity V with which the block and bullet begin to swing as a pendulum. The pendulum then climbs to a height h *with no further loss of energy*, allowing you to write

$$\frac{1}{2}(M+m)V^2 = (M+m)gh. \tag{8.82}$$

Combining Eqns. 8.81 and 8.82 we find

$$(M+m)gh = \frac{1}{2}(M+m)V^2 \tag{8.83}$$

$$= \frac{1}{2}(M+m)\left[\frac{mv_0}{M+m}\right]^2 = \frac{m^2v_0^2}{2(M+m)}. \tag{8.84}$$

The initial or muzzle velocity of the bullet is then

$$v_0 = \left[1+\frac{M}{m}\right]\sqrt{2gh}. \tag{8.85}$$

Notice that after the initial inelastic collision the energy of the block and bullet is

$$\frac{m^2 v_0^2}{2(M+m)} = \frac{1}{2} m v_0^2 \left[\frac{m}{m+M} \right]$$ (8.86)

$$< \frac{1}{2} m v_0^2,$$ (8.87)

which reflects the energy loss in the inelastic collision.

8.6   Scattering in higher dimensions

The problems we did on scattering explain how the basic conservation laws of energy and momentum are to be applied. But they fail to indicate the great importance of scattering experiments in our understanding of fundamental physics. For example, we learned only from scattering that the atom has a hard nucleus. It was known, in the early twentieth century, that the negative charge of the point-like electrons was balanced by a compensating positive charge. In one popular model, called the "plum-pudding model," the atom was a positive spherical charge, with the electrons embedded in it. The scattering experiments of Ernest Rutherford changed everything. By shooting a beam of alpha particles (helium nuclei with two protons and two neutrons) at a gold foil and detecting the scattered particles in various directions, he deduced the structure of the nucleus. In particular, he was stunned to notice that some of the alpha particles came right back, turned around by $180°$. No soft mush of positive charge (as in the plum-pudding model) could have done that. He then assumed the atomic nucleus was hard and point-like, contained all the mass and charge, and exerted a $1/r^2$ electric repulsion on the alpha particles. He computed what fraction of incident particles would suffer a deflection by an angle θ from the incident direction. The scattering data were in perfect agreement with his prediction and provided a brilliant confirmation of the point-nucleus, the cornerstone of Bohr's model of the atom. (Luckily for Rutherford, his treatment of the scattering as a Kepler problem with a *repulsive* $1/r^2$ force gave answers that coincided with that of the quantum treatment, which came many years later.)

During the years 1966 to 1978 Jerome Friedman, Henry Kendall, and Richard Taylor of the Stanford Linear Accelerator Center, shot high-energy electrons at nucleons (a common name for protons and neutrons).

The incident electron emitted photons of extremely short wave length, and these were exquisite probes of nucleon structure at very short length-scales. They revealed that the proton was not a point charge but was made up of two "up" quarks, which carried a fraction $\frac{2}{3}$ of the proton charge, and one "down" quark, which carried $-\frac{1}{3}$ of the proton charge. Likewise, the neutron was made up of two down quarks and one up quark. These charges were in agreement with the assignments made by the inventors of the quark model, Murray Gell-Mann and George Zweig. (Besides the three quarks, the nucleons also contain a cloud of short-lived quark–antiquark pairs and "gluons," which bind the quarks. Scattering can also probe this cloud.)

In three dimensions, the law of conservation of momentum for a two-particle collision takes the form of a vector equation

$$\mathbf{p}_1 + \mathbf{p}_2 = \mathbf{p}'_1 + \mathbf{p}'_2. \tag{8.88}$$

If the collision is totally inelastic, the final blob will have mass $m_1 + m_2$ and move at a velocity $\mathbf{V}$ that satisfies momentum conservation:

$$\mathbf{V} = \frac{\mathbf{p}_1 + \mathbf{p}_2}{m_1 + m_2}. \tag{8.89}$$

If the collision is elastic we may assume exactly as in $d = 1$,

$$\frac{1}{2}m_1 v_1^2 + \frac{1}{2}m_2 v_2^2 = \frac{1}{2}m_1(v'_1)^2 + \frac{1}{2}m_2(v'_2)^2, \tag{8.90}$$

which may be rewritten in terms of momenta as

$$\frac{p_1^2}{2m_1} + \frac{p_2^2}{2m_2} = \frac{(p'_1)^2}{2m_1} + \frac{(p'_2)^2}{2m_2} \tag{8.91}$$

since

$$\frac{p^2}{2m} = \frac{m^2 v^2}{2m} = \frac{1}{2}mv^2. \tag{8.92}$$

In $d = 2$, Eqns. 8.88 and 8.90 furnish three equations, not enough to determine the four numbers $\mathbf{p}'_1$ and $\mathbf{p}'_2$. This is because, unlike in $d = 1$, a collision is not fully specified by the incoming momenta. We need to

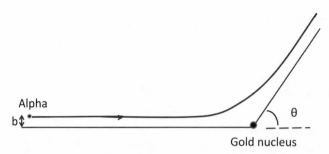

Figure 8.10   An alpha particle approaching a gold nucleus with impact
parameter b. If you assume the nucleus is immobile and exerts a Coulomb force
$1/r^2$, you can calculate the direction θ in which the alpha particle will emerge, as
a function of b and the energy. If a beam of particles with a distribution of b's is
given, one can predict the particle flux in any direction θ.

know the *impact parameter b* shown in Figure 8.10, and the nature of the
force between them (e.g., $1/r^2$) if they are point-like, or their radii, if they
are impenetrable spheres that exert a force on each other upon contact.

There is one problem where we can make a definite prediction with-
out knowing b. Let $m_1 = m_2 = m$, and let us assume initially $\mathbf{p}_2 = 0$. Upon
equating the lengths squared of both sides of

$$\mathbf{p}_1 + \mathbf{p}_2 = \mathbf{p}_1' + \mathbf{p}_2' \tag{8.93}$$

we find for this case when $\mathbf{p}_2 = 0$,

$$(\mathbf{p}_1 + \mathbf{p}_2) \cdot (\mathbf{p}_1 + \mathbf{p}_2) = (\mathbf{p}_1' + \mathbf{p}_2') \cdot (\mathbf{p}_1' + \mathbf{p}_2') \tag{8.94}$$

$$p_1^2 = (p_1')^2 + (p_2')^2 + 2\mathbf{p}_1' \cdot \mathbf{p}_2' \tag{8.95}$$

while Eqn. 8.91 says

$$p_1^2 = (p_1')^2 + (p_2')^2, \quad \text{which means} \tag{8.96}$$

$$\mathbf{p}_1' \cdot \mathbf{p}_2' = 0. \tag{8.97}$$

Because $\mathbf{p}_1' \cdot \mathbf{p}_2' = p_1' p_2' \cos\theta_{12}$ this means either the angle θ_{12} between the
final particles will be $90°$ or $p_1' = 0$, in which case the outgoing particle 2
and the incoming particle 1 have exchanged momenta, clearly conserving
energy and momentum.

Observe that we do not have enough information to figure out the
individual angles, just the relative angle θ_{12}.

CHAPTER 9

Rotational Dynamics I

9.1 Introduction to rigid bodies

In this chapter we graduate to objects like potatoes that are not point-like. For such extended objects, it is not simple to say where "it" is. We can pick a point on the object, like the CM, and locate it, but still we do not have the whole story. We need to say which way the potato is pointing, an issue we did not have with point particles. We could go all the way and consider a body like a snake, which is not only extended but also capable of changing its shape. That is too hard, so I will focus on extended but *rigid* bodies, whose shape is fixed—like a dead snake. By definition, if you take any two points on a rigid body, the distance between them will not change with time. No body is exactly rigid, though Al Gore during debates comes close.

The dynamics of rigid bodies in three dimensions is fairly complicated, so we will begin with two dimensions. Our usual ploy of starting with one dimension will not work here, because no rotations are possible in $d = 1$.

Consider a planar object—say a thin sheet of metal cut into some shape—that moves in the plane of the page, as shown in Figure 9.1. As you can see, it does something fairly complicated. Regardless of how it actually got from state 1 to state 2, we can attain the final state 2 from the initial state 1 by (i) moving the body until any one point F, which could be, but need not be, the CM, has reached its final location and (ii) following this with a rotation around an axis passing through F. The translation brings the body to the dotted configuration. The rotation axis passes through F, so as to attain the final state 2 without disturbing F, already in its final

143

position. There is a nonobvious generalization to three dimensions: we can go from the initial to the final configuration of a rigid body by a translation, followed by a rotation by a suitable angle, about a suitable axis. Let us return to the planar body $d = 2$ for now.

To specify it completely at any time, I have to tell you the coordinates of F and the rotation. To define the rotation, we draw an imaginary line on the body from F to any other fixed point, say the pointy edge, as shown in the inset of Figure 9.1. Then we say what angle θ this line makes with a fixed direction, usually the x-axis. If I give you the location of F and θ, you can reconstruct the configuration of the body shown in Figure 9.1.

Translations are what we have been studying all the time. We are experts on that subject. So we want to focus initially on a body that is only rotating but not translating. Once we have sharpened our skills, we will introduce the translations. To this end, imagine that I have driven a nail through one point F. The nail keeps it from translating but not from rotating about the nail axis. This rotation is fully specified by θ, which is the rotational analog of x.

We are going to set up an analogy between one dimensional translation described by $x(t)$ and two dimensional rotations described by $\theta(t)$.

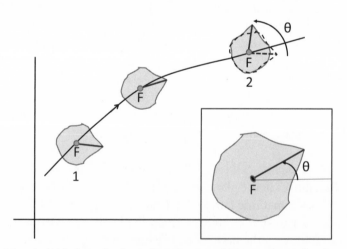

Figure 9.1   A rigid planar body undergoing translation and rotation in a plane as it goes from state 1 to state 2. The dot represents the fixed point F. The solid line from F to the pointy edge of the object allows us to follow its rotation. In state 2 the original orientation is shown in dotted lines and the curved arrow shows the rotation θ needed to bring it to the final orientation.

You'll find the analogy very helpful. The number of things you have to remember will be reduced if you learn to map the problem of rotations to the problem of translation in one dimension.

9.2 Angle of rotation, the radian

Even though the body is in two dimensions, you need just this one angle θ to specify its orientation. How do we want to measure θ? The standard preference in daily life is to measure it in degrees. When the body completes one full revolution, we say it has turned by an angle of 360°. We're going to use something else: *radians*. Why would anybody think of a radian? What's wrong with degrees? All the novels say, "I was going down the wrong path and then I did a 180."

Here is how a radian arises and why we like it. Consider a point on the body, at a distance r from the center at angle θ from the x-axis as shown in Figure 9.2. You agree that if the rigid body is rotating, this point will be traversing a circle of radius r. How long is the arc s that it traces out from the time it was at $\theta = 0$? Here is one way to calculate it. If the point traversed a full circle, we know $s = 2\pi r$. If it has rotated by $\theta°$, the arc length, by simple proportionality, is

$$s = \frac{\theta°}{360}(2\pi r). \tag{9.1}$$

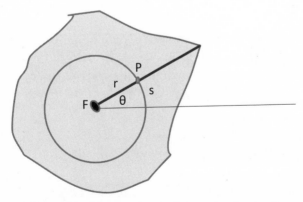

Figure 9.2   As the rigid body rotates around some fixed point F, a typical point P at a distance r from F moves on a circle of radius r.

Someone then decided to make life easy by calling θ^o in degrees times $\frac{2\pi}{360}$ as the same angle measured in radians. The angle in radians will simply be denoted by θ with no superscript. That is

$$\theta = \frac{2\pi\theta^o}{360}. \tag{9.2}$$

In terms of θ you find a very neat formula

$$s = r\theta. \tag{9.3}$$

The linear distance traveled, s, is simply the angle traversed in radians times the distance r to the axis of rotation. That a full circle is worth 2π radians follows from Eqn. 9.2. Equation 9.3 agrees with the familiar expression $C = 2\pi r$ for the circumference.

If $360^o = 2\pi$ radians, a radian is roughly 57.4^o. It looks like an odd thing to pick, but it's odd only if you start with 360 degrees. People from an alien culture might not use 360 at all. On the other hand, I believe 2π would be discovered in any advanced civilization.

You have to know a few popular angles in radians. For instance, a quarter circle is $90^\circ = \frac{\pi}{2}$, while $60^\circ = \frac{\pi}{3}$, $45^\circ = \frac{\pi}{4}$, and $30^\circ = \frac{\pi}{6}$.

If we take the time derivative of Eqn. 9.3, $s = r\theta$, we find

$$\frac{ds}{dt} = r\frac{d\theta}{dt} \tag{9.4}$$

because r is a constant by the rigid body condition. Clearly $\frac{ds}{dt}$ is the actual tangential speed v_T, or the magnitude of the velocity in the tangential direction. If at this instant this point in the body separated and flew off, it would emerge with that speed tangent to the circle. Thus

$$v_T = r\omega \quad \text{where} \tag{9.5}$$

$$\omega = \frac{d\theta}{dt} \tag{9.6}$$

is called the *angular velocity* and is measured in radians per second. It is positive for counterclockwise rotation.

What is the angular velocity of the moon as it goes around the earth? In roughly 28 days it covers 2π radians so

$$\omega \simeq \frac{2\pi}{28 * 24 * 60 * 60} = 2.6 \cdot 10^{-6} rad \cdot s^{-1}. \qquad (9.7)$$

Remember, the rigid body as a whole has a single angular velocity, but the tangential velocity of a point a distance r from the axis of rotation varies as ωr.

9.3 Rotation at constant angular acceleration

Let's take a problem where the angular velocity ω is itself changing. We define the *angular acceleration* α by

$$\alpha = \frac{d\omega}{dt} = \frac{d^2\theta}{dt^2} \qquad (9.8)$$

measured in $rad \cdot s^{-2}$. Consider a rotating circular saw where its ω itself is changing. Any point on it, at a distance r from the center, has two components of the usual (linear) acceleration. First, even if ω is a constant, it has the centripetal acceleration directed toward the center:

$$a_r = \frac{v_T^2}{r} = \frac{\omega^2 r^2}{r} = \omega^2 r. \qquad (9.9)$$

If in addition ω itself is changing, the tangential velocity $v_T = \omega r$ will also change, and there will be a tangential acceleration

$$a_T = \frac{dv_T}{dt} = r\frac{d\omega}{dt} = r\alpha. \qquad (9.10)$$

If you are driving around a circular racetrack in a car, a non-zero a_T will result when you step on the gas and see the speedometer needle move up. In addition, even at constant speed, the seat belt will remind you of the centripetal acceleration a_r.

In summary here are the relations between angular and tangential quantities:

$$s = r\theta \tag{9.11}$$

$$v_T = r\omega \tag{9.12}$$

$$a_T = r\alpha. \tag{9.13}$$

The bottom two equations follow upon differentiating the top one, bearing in mind that in a rigid body r, the distance of any point from the axis of rotation, will not change with time.

Suppose you are standing on a rotating platform that is undergoing angular acceleration, and you want to know what frictional force you need to stay fixed on the platform. The answer is

$$F_T = ma_T = mr\alpha \tag{9.14}$$

$$F_r = -ma_r = -m\omega^2 r \tag{9.15}$$

in the tangential and radial directions.

We can borrow many results from linear motion described by x by making the obvious substitutions. For a body that has a constant angular acceleration α, in obvious notation,

$$\theta = \theta_0 + \omega_0 t + \frac{1}{2}\alpha t^2 \tag{9.16}$$

$$\omega = \omega_0 + \alpha t \tag{9.17}$$

$$\omega^2 = \omega_0^2 + 2\alpha(\theta - \theta_0). \tag{9.18}$$

Suppose a chain-saw blade spinning at ω_0 is brought to rest after n revolutions. What is α, assuming it was constant? We simply set $\omega = 0$ and $\theta - \theta_0 = 2\pi n$ radians in Eqn. 9.18 and solve for α.

9.4 Rotational inertia, momentum, and energy

Now let's find the kinetic energy of a rotating rigid body. It has a mass, and if it's spinning, all the little particles making up the body are moving, and they have their own $K = \frac{1}{2}mv^2$. We want to compute the total K summed

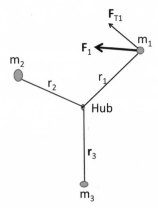

Figure 9.3   A simple rigid body made of three point masses attached to a fixed hub by three massless rigid rods. When a force $\mathbf{F}_1$ acts on 1, only the tangential component F_{T1} produces angular acceleration.

over all the particles. For that purpose, it's convenient to begin with the following simpler rigid body, shown in Figure 9.3. It is made up of masses m_1, m_2, and m_3 attached by rigid massless rods of length r_1, r_2, and r_3 to a fixed massless point-like hub about which the body can rotate. Say it is rotating with a common angular velocity ω. The kinetic energy of this object is the kinetic energy of each mass summed over all the masses. Now, what's the velocity of each mass? If you consider m_1, its velocity is necessarily perpendicular to the line joining it to the point of rotation. It has no radial velocity because r_1 cannot change. The magnitude of its velocity, entirely tangential, is $v_1 = \omega r_1$. In general for mass i

$$v_i = \omega r_i \tag{9.19}$$

and the kinetic energy of all of them is

$$K = \frac{1}{2}m_1 v_1^2 + \frac{1}{2}m_2 v_2^2 + \frac{1}{2}m_3 v_3^2 \tag{9.20}$$

$$= \sum_{i=1}^{3} \frac{1}{2}m_i v_i^2 \tag{9.21}$$

$$= \sum_{i=1}^{3} \frac{1}{2}m_i \omega^2 r_i^2. \tag{9.22}$$

In my example, the summation goes from $i=1$ to $i=3$; you can make up a body where the sum goes from 1 to 30,000. Often I will not indicate the range of values of i in the sum.

Notice that in Eqn. 9.22, unlike m_i and r_i, ω does not have a subscript i because it is the same for all parts of the rigid body. So, you can pull it out of the sum and write K as follows:

$$K = \frac{1}{2}\omega^2 \sum_{i=1}^{3} m_i r_i^2 \tag{9.23}$$

$$\equiv \frac{1}{2}I\omega^2 \quad \text{where} \tag{9.24}$$

$$I = \sum_{i=1}^{3} m_i r_i^2 \tag{9.25}$$

is called the *moment of inertia.*

The moment of inertia is determined not only by the masses that make up the body but also by how far they are from the axis of rotation. If all the masses just fell on top of the axis, the body would have no moment of inertia, though it would have a mass. Likewise, if the same mass is spread out more from the axis of rotation, the moment of inertia will be more. It requires a calculation to find the moment of inertia.

If someone says, "Here are the masses; please find me the moment of inertia," you should say, "I cannot do it until you tell me the point around which you plan to rotate the body." The moment of inertia is defined with respect to a point. The mass is just the mass, but the moment of inertia depends on the point with respect to which you're computing it. For example, in Figure 9.3 we are assuming the rotation is around the hub. If, however, it is around some other point, we would need the distances to that point to find the relevant moment of inertia.

When we compare $K = \frac{1}{2}mv^2$ to $K = \frac{1}{2}I\omega^2$, we see that in the world of rotations, ω plays the role of v and I plays the role of m. It is then natural to construct the rotational analog of the momentum $p = mv$, called the *angular momentum*:

$$L = I\omega. \tag{9.26}$$

Next we seek the rotational analog of $F = ma = \frac{dp}{dt}$. That's going to be equal to some mystery object, which we will call the *torque* τ. What does it equal?

$$\tau = \frac{dL}{dt} \tag{9.27}$$

$$= \frac{d[I\omega]}{dt} = I\frac{d\omega}{dt} \tag{9.28}$$

$$= I\alpha \tag{9.29}$$

where we can pull the I out of the derivative because it is not changing: even as the body rotates and the masses move, r_i, the distance of mass i from the axis of rotation, does not change. We will now relate the torque to the external forces acting on the masses by the following manipulations:

$$\tau = I\alpha \tag{9.30}$$

$$= \sum_i m_i\, r_i^2\, \alpha \tag{9.31}$$

$$= \sum_i m_i\, r_i\, \alpha\, r_i \tag{9.32}$$

$$= \sum_i m_i\, a_{Ti}\, r_i \tag{9.33}$$

$$= \sum_i F_{Ti}\, r_i \tag{9.34}$$

where in arriving at the last equation I have used the fact that mass m_i times the tangential acceleration $a_T = r_i\alpha$ is F_{Ti}, the tangential component of the force on mass i. Note that the tangential component of the force is positive if it is in the direction of increasing angle, that is, counterclockwise.

In summary, the torque is a sum of torques on each mass, and the torque on each mass is the product of the tangential component of the external force acting on it and the distance r_i to the axis of rotation. (I say "external" because there are internal forces that keep the body rigid and that do not figure in this.) Figure 9.3 shows an example with just one force acting on mass 1.

We will follow the convention that a torque is positive if it will cause positive acceleration, that is, in the direction of increasing θ.

Here are two illustrations, shown in Figure 9.4. The left half shows a merry-go-round that you want to accelerate. You should apply a force tangent to the wheel for maximum effect. If you apply a force in a general direction as shown by the bold arrow, the tangential part will contribute to τ and produce some α, while the radial part, which tries to change r, will be balanced by internal forces from the rigid body.

The right half shows a door. People from some civilization have just invented the door; they have thought about hinges but have not quite fig-

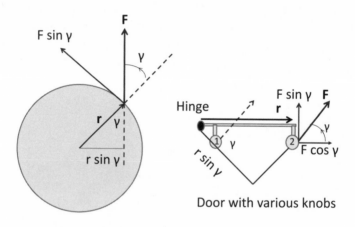

Door with various knobs

Merry-go-round

Figure 9.4   (Left) A merry-go-round being pushed by a force $\mathbf{F}$. Only the tangential part $F \sin \gamma$ produces torque and angular acceleration. The angle γ is measured from (the continuation of) $\mathbf{r}$ to $\mathbf{F}$. If we reverse the direction of $\mathbf{F}$ in the figure, γ will exceed π and the torque will become negative and lead to clockwise acceleration. The radial part $F \cos \gamma$ is canceled by internal forces in the rigid body. (Right) A view from above of a door on hinges in the early days, with two possible placement of knobs. The figure shows a poorly placed door knob numbered 1 and applied force (dotted line), and a better choice (number 2 and solid line). We can read $\tau = Fr \sin \gamma$ in two equivalent ways: r times $F \sin \gamma$, the component of the force perpendicular to the separation from the axis $\mathbf{r}$, or F times $r \sin \gamma$, the component of $\mathbf{r}$ perpendicular to the line of action of the applied force $\mathbf{F}$. For computing torque, the door knob may be taken to have zero size, so that $\mathbf{F}$ acts where $\mathbf{r}$ ends.

ured out where to put the doorknob. They say, "As the proud inventors of the door, we have complete freedom. Maybe we will place the knob right next to the hinges." Then they realize that something is wrong—they are applying a lot of force but not getting anywhere. They finally get it: whereas force was everything in linear motion, for rotations, the placement is relevant. If you want to get your money's worth, you have to take the doorknob as far as you can from the hinges and put it near the other end. But they are not done yet. They apply a force on the knob as far as possible from the hinges but along the line joining the knob to the hinge, the horizontal direction in the figure, and they end up ripping the door off the hinges. Eventually it dawns on them: if you want to get some serious rotation going, you should really go as far as possible from the pivot point and apply the force in the direction perpendicular to the vector joining the pivot point and the point of application of the force. All this is encoded in the definition of the torque as

$$\tau = Fr \sin \gamma \qquad\qquad\qquad (9.35)$$

where γ is the angle measured from $\mathbf{r}$, vector from the axis to the point of application of the force, to the force vector $\mathbf{F}$. In both the merry-go-round and the door depictions, the torque is positive because $\gamma < \pi$. If we reverse $\mathbf{F}$, the torque will change sign because $\sin \gamma$ will be negative for $\gamma > \pi$.

We can read $\tau = Fr \sin \gamma$ in two equivalent ways: r times $F \sin \gamma$, the component of the force perpendicular to $\mathbf{r}$, the separation from the axis, or F times $r \sin \gamma$, the component of $\mathbf{r}$ perpendicular to $\mathbf{F}$, as indicated in the figure.

Now we turn to the concept of work. Once again, everything can be related to what we know for translations, simply reexpressed in terms of quantities more pertinent to rotations. Consider a force that acts on one of the masses. Its radial part will do no work, because it is trying to change r, which is impossible if the body is indeed rigid. Let its tangential part rotate the body by an angle $d\theta$. The arc length traversed is $rd\theta$ and the work done is

$$dW = F_T ds = F_T r d\theta \qquad\qquad (9.36)$$

$$= \tau d\theta \qquad\qquad\qquad (9.37)$$

in accord with our expectations that the product of force and linear displacement should be replaced by the product of torque and angular displacement. Because K has the same meaning, the work-energy theorem becomes

$$dK = dW = \tau\, d\theta. \tag{9.38}$$

9.5 Torque and the work-energy theorem

Here is an example of

$$dW = \tau\, d\theta. \tag{9.39}$$

Consider a pendulum that is simply a massless string of length l attached at one end to the ceiling and at the other to a bob of mass m, as shown in Figure 9.5. The angle θ between the string and the vertical is initially 0. I want to bring it to an angle θ_0. I want to know how much work is required. Consider some intermediate angle θ. The force $\mathbf{T}$ supplied by the string necessarily points along the string. The gravitational force acting on the bob may be resolved into the tangential and radial parts as shown. The radial part $mg\cos\theta$ will be balanced by T. The tangential part $mg\sin\theta$ has to be balanced by a force $\mathbf{F}_{me}$ that I must apply to keep it from slipping back to the vertical. (I displace the bob without giving it any kinetic energy.) If

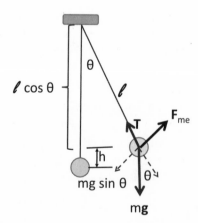

Figure 9.5 A pendulum made up of a massless string length l and a bob of mass m. The force $m\mathbf{g}$ is resolved in the radial and tangential directions. The bob has climbed up a height $h = l - l\cos\theta$.

I increase θ by $d\theta$, this force moves a distance $ds = ld\theta$ along an arc. As $d\theta \to 0$, this arc becomes a linear segment of the same length, and the work done, which is force times distance, becomes

$$dW = mg \sin \theta (ld\theta), \tag{9.40}$$

which may equally well be written as the torque times angular displacement:

$$dW = (mgl \sin \theta)d\theta = \tau \, d\theta. \tag{9.41}$$

The work done by me to bring it up to angle θ_0 is

$$W = \int_0^{\theta_0} mgl \sin \theta \, d\theta = mgl(1 - \cos \theta_0). \tag{9.42}$$

Here is a cross-check. From the figure it is clear the bob has climbed up a height $h = l - l \cos \theta_0$ and its potential energy is $U = mgl(1 - \cos \theta_0)$, which in turn equals the work done. The kinetic energy was always 0.

Table 9.1 contains a list of correspondences between angular and linear dynamics.

To this list we just need to add that the tangential displacement s, velocity v_T, and acceleration a_T are simply r times the angular counterparts, θ, ω, and α.

Note that all the previously described rotational dynamics simply follow from Newton's laws.

Table 9.1   Linear and angular quantities

Entity	Linear	Angular
Displacement	x	θ
Velocity	v	ω
Acceleration	a	α
Inertia	m	$I = \sum mr^2$
Momentum	$p = mv$	$L = I\omega$
Rate of change of momentum	$F = ma = \frac{dp}{dt}$	$\tau = I\alpha = \frac{dL}{dt} = Fr \sin \theta$
Kinetic energy	$\frac{1}{2}mv^2$	$\frac{1}{2}I\omega^2$
Work	$dW = Fdx$	$dW = \tau \, d\theta$

You are now ready to do a variety of simple problems. For example, if a force $\mathbf{F}_1$ is applied to mass 1 in Figure 9.3, what will be the angular acceleration? It is τ/I. We know how to find I by summing mr^2 over the three masses, and τ is just $F_{T1}r_1$. Once you have α, if it is constant, you can compute ω at later times and so on. There is really just one technical obstacle you have to overcome if you want to do rigid body dynamics, and that is to know how to compute the moment of inertia for all kinds of objects. If I give you 37 masses m_i, each with a distance r_i from the point of rotation, it's a trivial thing. But often you are not given a discrete set of masses, but a continuous blob. Just as you did in determining the center of mass, you have to replace the sums by integrals.

9.6 Calculating the moment of inertia

Let's take some rigid bodies and try to find their moments of inertia. First consider one-dimensional objects, a rod of length l and mass M shown in Figure 9.6. Taking the origin at the CM of the rod, we divide it into tiny pieces of length dx centered at some x, just as we did for the CM

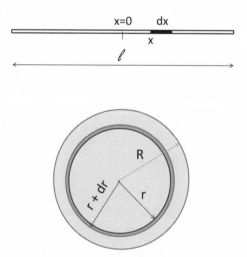

Figure 9.6   (Top) The moment of inertia of a rod found as the integral over tiny segments of width dx. (Bottom) The moment of inertia of a disk viewed as a sum over contributions from annuli of radius r and thickness dr.

calculation. The only difference is that now we consider the mass of the tiny segment times the *square of x* rather than *x*:

$$I_{CM} = \int_{-l/2}^{l/2} \frac{M}{l} x^2 dx \tag{9.43}$$

$$= \frac{M}{l} \frac{x^3}{3} \Big|_{-l/2}^{l/2} \tag{9.44}$$

$$= \frac{Ml^2}{12}. \tag{9.45}$$

The simplest two-dimensional object to consider is a ring of mass M and radius R. To find I_{CM}, its moment of inertia around the CM, which is the center of the circle, imagine dividing it into tiny pieces. Every one of the pieces is at the same distance R from the center. So the sum $m_i r_i^2$, with every $r_i = R$, is just MR^2. The mass was spread out, but luckily spread out in such a way that every part of it was the same distance R from the center.

More challenging is the disk of mass M and radius R. To find I_{CM}, we have to organize our thinking. We must think of the disk as made up of concentric rings of radius r and thickness dr, one of which is shown. The mass of this annulus may be found as follows. The area of the annulus is

$$dA = \pi \left((r+dr)^2 - r^2 \right) = 2\pi r dr + \mathcal{O}(dr^2). \tag{9.46}$$

Its mass is the mass per unit area times the area:

$$dM = \frac{M}{\pi R^2} 2\pi r dr \tag{9.47}$$

and its contribution to I_{CM} is just this mass dM times its r^2:

$$dI_{CM} = \frac{M}{\pi R^2} 2\pi r \cdot r^2 dr \quad \text{which integrates to} \tag{9.48}$$

$$I_{CM} = \int_0^R \frac{M}{\pi R^2} 2\pi r \cdot r^2 dr \tag{9.49}$$

$$= \frac{MR^2}{2}. \tag{9.50}$$

Suppose I made an error somewhere and obtained $I_{CM} = MR^2$. You should know it is wrong, because $I_{CM} = MR^2$ is possible only if all the mass is at a distance R from the center, but we know some of the mass is a lot closer. So the moment of inertia for a disk has to be less than MR^2.

We saw that

$$I_{CM} = \frac{Ml^2}{12} \tag{9.51}$$

for a rod. If you plan to nail it at its CM and spin it, this is the rotational inertia you will be up against. But suppose you want to spin it around the left end? We repeat the old calculation of I_{CM} but we measure x from the left end and integrate from 0 to l:

$$I_{End} = \int_0^l \frac{M}{l} x^2 dx \tag{9.52}$$

$$= \frac{M}{l} \frac{x^3}{3} \Big|_0^l \tag{9.53}$$

$$= \frac{Ml^2}{3}. \tag{9.54}$$

Notice that $I_{End} > I_{CM} = \frac{Ml^2}{12}$. More precisely

$$I_{End} = I_{CM} + M \left[\frac{l}{2}\right]^2. \tag{9.55}$$

In fact I will be smallest about the CM. We will prove in the next chapter that, in general, the moment of inertia I of a planar object around any perpendicular axis that is a distance d away from the CM is $I_{CM} + Md^2$.

CHAPTER 10

Rotational Dynamics II

10.1 The parallel axis theorem

Let us recall what we have learned about rigid bodies that are confined to lie and rotate in a plane, such as a rod or a sheet of some metal cut out into some arbitrary shape. The body has a mass M. It can translate and rotate, but for now we nail a point on it to the plane and let it rotate about that axis, with plans to bring in translations later on. A single angle θ, measured in radians, suffices to tell us what it is doing, because all it can do is rotate about the fixed point. This angle θ is to rotations what x was to translations. There is a corresponding angular velocity $\omega = \frac{d\theta}{dt}$ and acceleration $\alpha = \frac{d\omega}{dt} = \frac{d^2\theta}{dt^2}$. The novel attribute, owing to the object being extended and not point-like, is its moment of inertia I. If we imagine it as being made of point masses m_i sitting at a distance r_i from the axis, we find

$$I = \sum_i m_i r_i^2. \tag{10.1}$$

Note that I is a quantity that depends on how the mass is distributed relative to the axis about which it is rotating. It plays the role that mass did in translational motion. If the body is continuous, the sum is replaced by the corresponding integral. The angular momentum L

$$L = I\omega \tag{10.2}$$

159

plays the role of momentum $p = mv$, and $F = ma = \frac{dp}{dt}$ is replaced by

$$\tau = I\alpha = \frac{dL}{dt} \tag{10.3}$$

where the torque is defined by

$$\tau = \sum_i r_i F_i \sin \gamma_i \tag{10.4}$$

where γ_i is the angle measured from $\mathbf{r}_i$, the vector from the point of rotation to the point of application of the force, and force vector $\mathbf{F}_i$. We may rewrite the expression for torque in two other useful ways:

$$\tau = F_\perp r = F r_\perp \tag{10.5}$$

where $F_\perp$ is the component of the force perpendicular to $\mathbf{r}$ and $r_\perp$ is the component of $\mathbf{r}$ perpendicular to $\mathbf{F}$.

In our convention a torque was positive (negative) if it implied counterclockwise (clockwise) acceleration.

The kinetic energy is

$$K = \frac{1}{2}I\omega^2. \tag{10.6}$$

All the rotational equations, say the one for K, are simply the results from linear motion as applied to the constituents of the rigid body.

Finally, at each point in the body, the tangential displacement, velocity, and acceleration are just r times the angular ones

$$s = r\theta \quad v_T = r\omega \quad a_T = r\alpha. \tag{10.7}$$

We practiced computing moments of inertia of various objects (rod and disk) and concluded with the observation that for a rod

$$I_{end} = I_{CM} + M\left[\frac{L}{2}\right]^2. \tag{10.8}$$

We will now see that this is an example of a more general result:

The parallel axis theorem: The moment of inertia for a planar body about any axis is

$$I = I_{CM} + Md^2 \qquad (10.9)$$

where d is the distance from the CM to the new axis. (In the rod example $d = \frac{L}{2}$.)

Here is the proof. As a warm-up, we will do it in $d = 1$ for a rod of a given linear mass density $\rho(x)$, that is, where $\rho(x)dx$ is the mass of an infinitesimal segment of width dx at the point x. Let us choose as our origin the CM, and measure x from it. Let the new axis pass through the point $x = d$, as shown in Figure 10.1. The moment of inertia about the new axis is

$$I = \int_{Rod} \rho(x)(x - d)^2 dx \qquad (10.10)$$

$$= \int_{Rod} \rho(x)(x^2 + d^2 - 2xd)dx \qquad (10.11)$$

$$= \int_{Rod} \rho(x)x^2 dx + d^2 \int_{Rod} \rho(x)dx - 2d \int_{Rod} \rho(x)xdx \qquad (10.12)$$

$$= I_{CM} + Md^2 - 2d \int_{Rod} \rho(x)xdx. \qquad (10.13)$$

The first two terms are welcome: they are required in the theorem. The cross term is like a leftover part after you have assembled your bookshelf from Ikea. We need to get rid of it. Luckily it is zero, for the following reason. By definition

$$\int_{Rod} \rho(x)xdx = M\frac{\int_{Rod} \rho(x)xdx}{M} = MX \qquad (10.14)$$

where X is the CM. What is X? Our origin is at the CM itself and with respect to that origin, the CM has zero coordinate, $X = 0$.

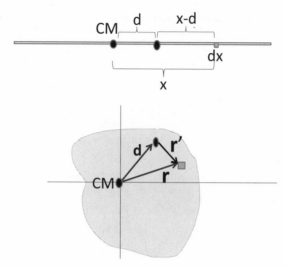

Figure 10.1   (Top) The moment of inertia calculation about an axis a distance d from the CM, which is not necessarily at the midpoint of the rod if the density is not uniform. The rod is made up of segments of width dx, one of which is shown at a distance x from the CM and $x - d$ from the new axis. (Bottom) The parallel axis now passes through a point with a vector separation $\mathbf{d}$ from the CM. The body is made up of tiny areas of size $dx\,dy$, one of which is shown, separated by $\mathbf{r}$ from the CM and $\mathbf{r}'$ from the new parallel axis.

Thus, the moment of inertia is the smallest with respect to the CM; any other axis adds an Md^2.

Now let us do it in $d = 2$. I remind you that the length squared of $\mathbf{A} + \mathbf{B}$ is

$$|\mathbf{A} + \mathbf{B}|^2 = (\mathbf{A} + \mathbf{B}) \cdot (\mathbf{A} + \mathbf{B})$$

$$= \mathbf{A} \cdot \mathbf{A} + \mathbf{B} \cdot \mathbf{B} + 2\mathbf{A} \cdot \mathbf{B} = A^2 + B^2 + 2\mathbf{A} \cdot \mathbf{B}. \quad (10.15)$$

Let $\rho(x, y)$ be the mass per unit area, let $\mathbf{r}$ be the position vector of a point with the origin at the CM, and let $\mathbf{d}$ be the location of the new axis. Clearly the position vector of a point relative to the new axis is

$$\mathbf{r}' = \mathbf{r} - \mathbf{d}. \quad (10.16)$$

Now we repeat the earlier proof with vectors galore:

$$I = \int_{body} \rho(x,y)|\mathbf{r} - \mathbf{d}|^2 dxdy \tag{10.17}$$

$$= \int_{body} \rho(x,y)(r^2 + d^2 - 2\mathbf{d} \cdot \mathbf{r})dxdy \tag{10.18}$$

$$= \int_{body} \rho(x,y)r^2 dxdy + d^2 \int_{body} \rho(x,y)dxdy \tag{10.19}$$

$$- 2\mathbf{d} \cdot \int_{body} \rho(x,y)\mathbf{r}dxdy$$

$$= I_{CM} + Md^2 - 2M\mathbf{d} \cdot \mathbf{R} \tag{10.20}$$

where the last term vanishes for the same reason as in $d = 1$: $\mathbf{R}$ is the CM position in a coordinate system with the CM itself as the origin.

Here is an illustration of the power of this result. Suppose that instead of rotating a disk about its center you wanted to hold fixed a point at its circumference. We cannot view the disk as a union of concentric annuli centered around the new axis as we did for I_{CM}. Only a part of every annulus would fit into the disk, and we would need to figure out how much, if we want to compute I directly. Of course, we will do no such thing: we will invoke the parallel axis theorem to say, for a disk of radius R

$$I_{edge} = I_{CM} + MR^2 = \frac{3MR^2}{2}. \tag{10.21}$$

Consider for example a coin that is standing on its rim on some surface. It has just one point of contact with the surface. If we demand that there be no slipping, the point of contact cannot move relative to the surface as it rolls. So the coin will simply rotate around this point (at this instant) and the relevant I will be $\frac{3MR^2}{2}$.

10.2 Kinetic energy for a general N-body system

Now let's turn to a result involving kinetic energy, whose derivation has a flavor similar to the previous one, which is why I want to present it side-by-side. Imagine a whole collection of masses m_i, *not necessarily forming a rigid body.* For example, they could be stars forming a galaxy. The formula for the kinetic energy of this collection is

$$K = \sum_i \frac{1}{2} m_i v_i^2, \tag{10.22}$$

where v_i are the magnitudes of the velocities $\mathbf{v}_i$ as seen by some generic observer. These velocities will of course depend on the frame of reference of the observer. Let $\mathbf{V}_{CM}$ be the velocity of the CM as seen by this observer. Consider a special observer who is co-moving with the CM. *To her the CM is at rest.* If $\mathbf{v}_i^{CM}$ is the velocity of the object as seen by this co-moving observer, then by the law of composition of velocities derived in Chapter 2 (Eqn. 2.44),

$$\mathbf{v}_i = \mathbf{V}_{CM} + \mathbf{v}_i^{CM}. \tag{10.23}$$

In this notation, the superscript CM means "in the frame moving with the CM." No superscript means in the original generic frame, and the subscript CM or i means "of the CM" or "of the particle i." Thus $\mathbf{v}_i^{CM}$ is the velocity of particle i in the CM frame and v_i is the speed of the particle i in the original generic frame. We now manipulate Eqn. 10.22 as follows:

$$K = \frac{1}{2} \sum_i m_i v_i^2 \tag{10.24}$$

$$= \frac{1}{2} \sum_i m_i |\mathbf{V}_{CM} + \mathbf{v}_i^{CM}|^2 \tag{10.25}$$

$$= \frac{1}{2} \sum_i m_i (V_{CM}^2 + |v_i^{CM}|^2) + \mathbf{V}_{CM} \cdot \sum_i m_i \mathbf{v}_i^{CM} \tag{10.26}$$

$$= \frac{1}{2} M V_{CM}^2 + K^{CM} + 0. \tag{10.27}$$

The first term is the kinetic energy of the center of mass, a point containing the entire mass of the object and moving with a velocity $\mathbf{V}_{CM}$. The second term is the K as measured by the co-moving observer who uses the velocities v_i^{CM} for particle i. If you were really traveling with the center of mass, and you made a measurement of how fast every particle in it was moving, this would be the kinetic energy you would attribute to these particles. The third term in Eqn. 10.26 vanishes because

$$\sum_i m_i \mathbf{v}_i^{CM} = \mathbf{0}. \tag{10.28}$$

Here is the reason. In any generic frame, by definition,

$$M\mathbf{R} = \sum_i m_i \mathbf{r}_i. \tag{10.29}$$

Taking the time-derivative of both sides we find, also in any frame,

$$M\mathbf{V}_{CM} = \sum_i m_i \mathbf{v}_i. \tag{10.30}$$

Now apply this to the observer co-moving with the CM:

$$M\mathbf{V}_{CM}^{CM} = \sum_i m_i \mathbf{v}_i^{CM}. \tag{10.31}$$

The left-hand side vanishes because $\mathbf{V}_{CM}^{CM}$ is the *velocity of the CM as seen by a person riding with the CM*. This means the right-hand side, which appears in the last term in Eqn. 10.26, also vanishes.

Equation 10.27 is valid for any collection of objects, whether they be parts of a rigid body or stars in a galaxy. It is going to be very useful to us when we study rigid bodies that rotate and translate. In that context, it means that as far as the kinetic energy goes, we may simply add the K of the translational motion of the CM to the K of rotations *about the CM*.

$$K = K_{CM} + K_{rot} \tag{10.32}$$

10.3 Simultaneous translations and rotations

Now consider a disk, say a tire in a car. There are different things you could imagine. Lift the car off the ground, and let the tires spin; that has rotational energy $K = K_{rot} = \frac{1}{2}I_{CM}\omega^2$. That's the pure rotation we have been studying so far. Next, let the tire hit the ground and set the car in motion. Suddenly slam the brakes and prevent the tire from turning. The skidding tire still moves with a certain velocity, which is due to just the translational motion of the center of mass, with $K = K_{CM} = \frac{1}{2}mV_{CM}^2$. This too is familiar, just like that of a point particle.

In general, the tire will rotate around the center and also translate. The linear velocity of the center of mass and the angular velocity around the center don't have to be connected in any way. Suppose you start your car on an icy road. The tire is spinning, but the car is not moving. So, that's a case of $V_{CM} = 0$, $\omega \neq 0$. In the case of a skid, we have $V_{CM} \neq 0$, $\omega = 0$. We are interested in the case where there is no slipping relative to the ground; the wheels are not burning rubber. Now there is a correlation between the angular velocity and the linear velocity, and the motion is called "rolling without slipping." When the tire rolls without slipping, by the time it finishes one full revolution, the car will have moved the distance equal to the circumference of the tire. So every part of the tire touches the ground once, while the car moves a distance equal to the circumference.

Let us calculate the velocity when the tire rolls without slipping. If it made f revolutions per second and in each revolution the car moves $2\pi R$ meters

$$V_{CM} = 2\pi R f = \omega R \tag{10.33}$$

where I have used the fact that $\omega = 2\pi f$. Now we have

$$K = \frac{1}{2}MV_{CM}^2 + \frac{1}{2}I_{CM}\omega^2 \quad \text{because of Eqn. 10.27} \tag{10.34}$$

$$= \frac{1}{2}M\omega^2 R^2 + \frac{1}{2}\frac{MR^2}{2}\omega^2 = \frac{1}{2}\frac{3MR^2}{2}\omega^2 \tag{10.35}$$

$$= \frac{1}{2}I_{edge}\omega^2 \tag{10.36}$$

because of Eqn. 10.21, the parallel axis theorem.

This very interesting result tells you that you can view the entire energy of the tire as due to pure rotation with angular velocity ω around this point of contact P in Figure 10.2. If the tire is indeed rotating around that point, it means that point cannot be moving. And I will now try to

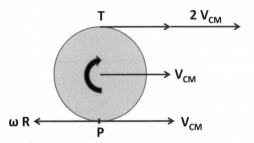

Figure 10.2   A tire of radius R that is rolling clockwise at ω without slipping. At the contact point P, the CM velocity $\mathbf{V}_{CM}$ cancels the tangential velocity ωR, while at the top, T, it doubles it. The point P is instantaneously at rest with respect to the road.

convince you. How fast are the different parts of the tire moving? The car as a whole is moving at $\mathbf{V}_{CM}$. In addition, the tire is spinning. To find the velocity of any point on the circumference, you should add the tangential velocity of magnitude ωR to the CM velocity. At P, the point of contact, the two cancel, while at the top T, they add to give $2\mathbf{V}_{CM}$. Thus, in a car going past you at 200 miles per hour, there is a part with zero velocity and a part with a velocity 400 miles per hour relative to the ground. It's not obvious that a zooming car has one part of it that's not moving at all. But at every instant, the part of the tire that touches the ground has zero instantaneous velocity.

10.4   Conservation of energy

Suppose I release an object of mass M from the top of a hill of height h. If it is a point mass, I know its speed when it comes down:

$$Mgh = \frac{1}{2}Mv^2 \tag{10.37}$$

$$v = \sqrt{2gh}. \tag{10.38}$$

But if it is a coin of radius R that is rolling without slipping, when it reaches the bottom the center of mass cannot move at $V_{CM} = \sqrt{2gh}$ because it has to be rotating about its axis to avoid slipping. Some rotational energy is mandated. So now we write

$$Mgh = \frac{1}{2}MV_{CM}^2 + \frac{1}{2}I_{CM}\omega^2 \qquad (10.39)$$

$$= \frac{1}{2}\frac{3MR^2}{2}\omega^2 = \frac{1}{2}\frac{3M}{2}(\omega R)^2 = \frac{3M}{4}V_{CM}^2 \qquad (10.40)$$

$$V_{CM} = \sqrt{\frac{4gh}{3}} < \sqrt{2gh}. \qquad (10.41)$$

This formula for the disk (of zero thickness) is also valid for a cylinder rolling about its symmetry axis, because the cylinder may be viewed as a coaxial stack of disks whose moments of inertia add.

For a solid sphere, we can do this again using $I_{CM} = \frac{2}{5}MR^2$ and the parallel axis theorem. How about a hollow sphere of the same mass? Argue that it will have a greater moment of inertia. The exact answer is $\frac{2}{3}MR^2$. If I roll a hollow sphere and a solid sphere of the same mass and radius, the latter will come down faster.

You can readily imagine all kinds of new problems. You remember that in the loop-the-loop, the velocity squared of a point mass at the top of the loop has to be bigger than or equal to Rg if it is to follow the track. Suppose the object in question is a cylinder or sphere. The condition on the velocity at the top is still the same, but the height from which you release it has to be greater, because if it rolls without slipping, it also has to have a rotational energy correlated to its translational energy.

To summarize, when rigid bodies move, they have translational and rotational energy. In general, they are independent numbers, but when you have rolling without slipping, the angular velocity and the linear velocity are connected by $V_{CM} = \omega R$. So, it's not surprising that the total energy has a contribution from both, which you can write either in terms of the angular velocity or the linear velocity. In other words, if I know how fast the wheel is spinning, I can tell you how fast the car is moving. If I tell you how fast the car is moving, I know the rate at which the wheels are spinning.

10.5   Rotational dynamics using $\tau = \frac{dL}{dt}$

Let me take a simple example. A group of rogue toddlers decided their parents were not pushing their merry-go-round of mass M fast enough, and

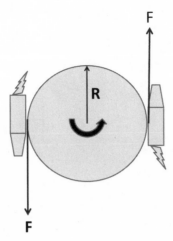

Figure 10.3   A rocket-powered merry-go-round of mass M. Each engine has a thrust F, mass m, and contributes mR^2 to the moment of inertia.

so they decided to use rocket propulsion. Initially they had only one rocket of mass m, but it nearly blew the axle holding the platform. While the rocket did provide a torque, it also provided a force that would have caused linear acceleration of the entire system, but for the axle, which applied a counterforce. To avoid this, they got a second identical rocket and installed it diametrically opposite to the first, with its thrust oriented to assist the first as in Figure 10.3. The engines have a thrust F each and together they apply a torque $\tau = 2FR$. This will produce an angular acceleration

$$\alpha = \frac{\tau}{I} = \frac{2FR}{\frac{1}{2}MR^2 + 2mR^2}. \tag{10.42}$$

Given this, we can manufacture any number of trivial problems. For example, how many revolutions would the merry-go-round have completed after time t if it started from rest? It would have rotated by

$$\theta(t) = \frac{1}{2}\alpha t^2 \text{ radians.} \tag{10.43}$$

10.6   Advanced rotations

Now we consider slightly more complicated problems. A block of mass m is supported by a massless rope that wraps around a pulley of radius R and

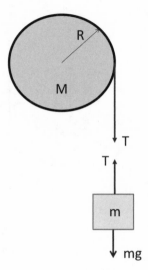

Figure 10.4   Free-body diagram of a block of mass m supported by a massless rope that wraps around a pulley of radius R and mass M.

mass M as shown in Figure 10.4. We want to find out with what acceleration the block descends. Let T be the tension in the rope, as shown in the free-body diagram. We have the following equations of motion for the block ($F = ma$) and the pulley ($\tau = I\alpha$) and the no-slip condition relating α and a:

$$mg - T = ma \tag{10.44}$$

$$TR = I\alpha = \frac{1}{2}MR^2\alpha \tag{10.45}$$

$$a = R\alpha. \tag{10.46}$$

I have chosen the down direction as positive for the block and clockwise as positive for the torque, in contrast to the usual. This should prepare you for dealing with unconventional choices.

I have not shown the force exerted by the pivot supporting the pulley because this force does not contribute to the torque about the pivot: $r_\perp = 0$. In Eqn. 10.46 I have related the angular acceleration of the pulley to the linear acceleration of the block. The mass can go down one inch only if the pulley turns by a corresponding amount to release one more inch of rope. By taking derivatives, we equate the tangential acceleration αR to a, the linear acceleration of the mass.

In Eqn. 10.45, I have used $I_{CM} = \frac{1}{2}MR^2$. Canceling one power of R in that equation and adding it to Eqn. 10.44 we find

$$mg = ma + \frac{1}{2}MR\alpha = ma + \frac{1}{2}Ma \qquad (10.47)$$

$$a = g\frac{m}{m + \frac{1}{2}M}. \qquad (10.48)$$

We can check our result as follows. If the block drops a height h, then its final velocity is, from the relation $v^2 = v_0^2 + 2ad$,

$$v^2 = 2ah = 2g\frac{m}{m + \frac{1}{2}M}h, \qquad (10.49)$$

which can be rearranged to give

$$\frac{1}{2}mv^2 + \frac{1}{2}\frac{Mv^2}{2} = mgh \qquad (10.50)$$

$$\frac{1}{2}mv^2 + \frac{1}{2}\frac{MR^2}{2}\omega^2 = mgh, \qquad (10.51)$$

which respects energy conservation.

Again we can add more complications. We could go back to the problem of one block sliding down the incline and couple it to another block moving vertically. We could decide to assign a non-zero moment of inertia to the pulley.

10.7 Conservation of angular momentum

So far, we have looked at examples of $\tau = I\alpha$ with $\tau \neq 0$. Now I'm going to consider problems where the total torque on a collection of bodies is zero. Then because $\tau = \frac{dL}{dt}$, the total angular momentum will be conserved.

Here's one example. There is a disk spinning around a spindle passing through a hole at its center. The disk has some moment of inertia I_1, and angular velocity ω_1. Infinitesimally above this disk, and on the same spindle, is a second non-rotating disk with moment of inertia I_2. If the upper disk falls on the lower one and the two rotate as one unit, what will be the final ω? Angular momentum conservation tells us

$$I_1\omega_1 + I_2 \cdot 0 = (I_1 + I_2)\omega \tag{10.52}$$

$$\omega = \frac{I_1}{I_1 + I_2}\omega_1. \tag{10.53}$$

This should remind you of the problem in translational motion where m_1 slams into m_2, which is at rest, and the two move together as a unit following the totally inelastic collision.

Here is a similar problem. A merry-go-round of radius R is spinning at ω_1 and has angular momentum $L_1 = I_1\omega_1$ where $I_1 = \frac{1}{2}MR^2$. I drop a point kid of mass m on its outer rim. The kid and the merry-go-round now rotate together. What will be the final ω? The kid has to be brought from rest to the tangential speed of the rim. The merry-go-round will apply a force on the kid to speed up the kid, and the kid will exert an equal and opposite force that will slow the merry-go-round. The torques will be equal and opposite. So we may use the conservation of angular momentum for the entire system, to find ω:

$$I_1\omega_1 = (I_1 + mR^2)\omega \tag{10.54}$$

$$\omega = \frac{I_1\omega_1}{I_1 + mR^2} \tag{10.55}$$

where mR^2 is the moment of inertia of the kid sitting at the edge.

10.8   Angular momentum of the figure skater

Consider a single particle with momentum $p = mv$. If there are no forces, the momentum mv cannot change, and that translates into a constant v. If the body could reduce its mass in half, it could double its velocity keeping the momentum constant. But bodies have no way of changing their mass. But *if they are not rigid*, they *can* change their moment of inertia, by rearranging their mass distribution. This way they can change their ω, keeping the product $L = I\omega$ constant.

Consider a figure skater with her hands extended outward, spinning at some rate, as shown in the left half of Figure 10.5. She is a rigid body with some I_1 and ω_1. Then the skater briefly becomes non-rigid as she raises her

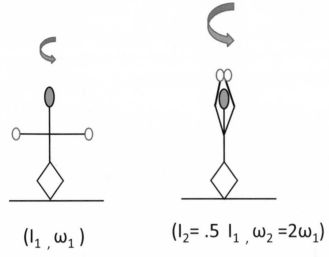

(I_1, ω_1) $(I_2 = .5\ I_1, \omega_2 = 2\omega_1)$

Figure 10.5   Left: A spinning figure skater with her hands extended outward. Right: Raising her arms over her head, the skater has reduced her moment of inertia and increased her angular velocity, without changing their product, her conserved angular momentum.

hands over her head and reduces her moment of inertia to I_2 and angular frequency to ω_2 such that

$$I_1\omega_1 = I_2\omega_2. \tag{10.56}$$

If she extends her arms out again, she slows down. Though she is non-rigid during some of this exercise, her L is conserved because there are no external torques.

However, her kinetic energy changes in this process. For example, let her initial moment of inertia with arms extended be double the final one with arms drawn over her head: $I_1 = 2I_2$. This means

$$\omega_2 = 2\omega_1 \tag{10.57}$$

$$K_2 = \frac{1}{2}I_2\omega_2^2 = \frac{1}{2}\left(\frac{1}{2}I_1\right)(2\omega_1)^2 = 2K_1. \tag{10.58}$$

The final kinetic energy is double the initial one. While the halving of the I compensates the doubling of the ω where angular momentum is concerned, this is not so for K, which is quadratic in ω. Where does the extra energy come from? It comes from her muscles that apply the centripetal force as she pulls in her arms. The centripetal force is always there to keep her extended arms moving in a circle, but it does work only when her arms move radially, for that is when $dW = \mathbf{F} \cdot d\mathbf{r} \neq 0$. (We ignore the gravitational work needed to raise her arms.)

CHAPTER 11

Rotational Dynamics III

11.1 Static equilibrium

I'm going to consider cases where there is no external torque. If there's no torque, we know the angular velocity is constant. But I'm going to take a case where this constant value of angular velocity is itself zero. There is no motion, there is no torque. So you might say, "What's there to study?" Well, sometimes it's of great interest to us to know that the object has no angular velocity, for example, if the object is a ladder we have climbed. The ladder better not have any angular acceleration either. What does it take to keep the ladder from falling over? That's the kind of problem we're going to discuss.

Obviously the forces on the ladder should add up to zero; otherwise $\mathbf{F} = m\mathbf{a}$ will ensure the CM accelerates. But $\mathbf{F} = 0$ is not good enough, as is clear from the the rocket-propelled merry-go-round: the thrusts are equal and opposite and cancel as forces, but the torques they produce add.

Because the cause of rotation is the torque, we want the total torque to vanish as well. So the condition for static equilibrium of a rigid body is

$$\sum_i F_{xi} = 0 \qquad\qquad (11.1)$$

$$\sum_i F_{yi} = 0 \qquad\qquad (11.2)$$

$$\sum_i \tau_i = 0 \qquad\qquad (11.3)$$

where the index i in the sums runs over all the applied forces in the x and y directions and all torques.

11.2 The seesaw

Look at Figure 11.1, which shows a seesaw with a kid of mass m_1 sitting to the left of the supporting pivot P and similarly a kid numbered 2 sitting to the right. Let these kids have coordinates x_1 and x_2 with respect to the origin P. First we write down all the forces on the seesaw. We see $\mathbf{F}_1 = m_1\mathbf{g}$, $\mathbf{F}_2 = m_2\mathbf{g}$ (where $\mathbf{g} = -9.8\mathbf{j}m/s^2$) due to the weight of the kids. Of course, that cannot be the whole story, because the seesaw is not falling to the ground. There is the pivot P, which exerts an upward normal force N.

The force equations are

$$0 = 0 \quad \text{along the } x \text{ direction} \tag{11.4}$$

$$N + F_1 + F_2 = 0 \quad \text{along the } y \text{ direction,} \tag{11.5}$$

where F_1 and F_2, are the y-components of $\mathbf{F}_1$ and $\mathbf{F}_2$. Assume that F_1 is known and that we are trying to solve for F_2. Clearly we cannot solve for F_2 and N using just one useful equation. This is where the torque equation comes in.

Now, if a body is rotating about an axis, we are used to equating the torque around that axis to the rate of change angular momentum about that axis or point. But what if it is not rotating? About what point should we compute the torque? You can say it's not rotating through one axis,

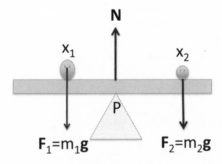

Figure 11.1   Two kids on a seesaw. The supporting pivot provides a force $\mathbf{N}$ upward. The total torque due to the kids' weights $\mathbf{F}_1 = m_1\mathbf{g}$, $\mathbf{F}_2 = m_2\mathbf{g}$ and the pivot force $\mathbf{N}$ is zero.

compute the total torque with respect to that axis, and equate it to zero to get the third equation. But then someone else can say it is not rotating about some other point and get yet another equation. So we can write down an infinite number of equations, one for each choice of pivot point or axis. But we have only two unknowns, F_2 and N, and they can only satisfy two equations. So, we have to hope that the extra equations we get by varying the pivot point all say the same thing. And that's what I will show you.

First consider just this rod in equilibrium. It has forces $\mathbf{F}_1, \ldots \mathbf{F}_N$ acting on it, and these add up to zero. They act at positions $x_1, \ldots x_N$, measured from some point on the rod I have chosen for torque computation. That it is in equilibrium according to me implies

$$\sum_i \mathbf{F}_i = 0 \tag{11.6}$$

$$\tau = \sum_i \tau_i = \sum_i F_{i\perp} x_i = 0. \tag{11.7}$$

If you choose a different point along the rod for torque computation, you will replace every x by $x - d$ if your axis is to the right of mine by d. The torque you get will be

$$\tau' = \sum_i F_{i\perp}(x_i - d) = \sum_i F_{i\perp} x_i - d \sum_i F_{i\perp} = 0 + 0, \tag{11.8}$$

where the first zero is there because it is just the total τ according to me, and the second zero arises because the total force $\mathbf{F}_T = \sum_i \mathbf{F}_i$ is zero and so its perpendicular component $F_{T\perp}$.

The proof gets tricky if the rotation axis is chosen off the rod or off the x-axis. We will address it later.

Anyway, when you compute the torque, you may pick any point you like provided the total force is zero. However, some choices are better suited than others for some purposes. For example, if you were asked simply to find F_2 but not the normal force N, there's a particular choice of pivot point that is optimal. That choice is the one in which this normal force doesn't get to enter the torque equation and corresponds to computing the torque about the pivot point P, where the normal force N acts. Now N will drop out of the torque equation, because $r_\perp = 0$. In general, always compute the torque relative to a point where an unknown force is acting, because then the unknown force cannot contribute to the torque equation.

Following that rule we get

$$F_1 x_1 + F_2 x_2 = 0 \qquad\qquad (11.9)$$

where x_1 is negative and x_2 is positive. Because F_1 and F_2 are negative, $F_1 x_1$ will be positive (counterclockwise) and $F_2 x_2$ will be negative (clockwise). We can readily solve Eqn. 11.9 for F_2. Once we have it, we can go back to Eqn. 11.5 and find the normal force N. Suppose $F_1 = -100$ Newtons, $x_1 = -4$ m and $x_2 = 5$ m, then $F_2 = -80$ Newtons and it follows from Eqn. 11.5 that $N = 180$ Newtons.

11.3 A hanging sign

Now, Figure 11.2 depicts a more complicated problem. There is a rod of length L and mass m, supported by some kind of pivot on the wall at one end and a wire anchored to the wall at the other end. The rod will eventually support a sign, which was not ready in time for this problem. My concern is the tension on the wire, which I do not want to exceed some

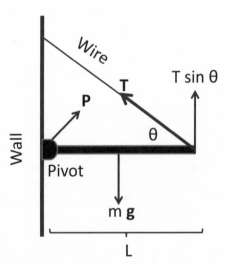

Figure 11.2   A rod of length L supported by a wire with some tension **T** and a pivot on the wall that applies a force **P**.

limit. The straightforward way to find T is to write the free-body diagram for the rod and demand that the net force and torque vanish:

$$P_x - T\cos\theta = 0 \tag{11.10}$$

$$T\sin\theta + P_y - mg = 0 \tag{11.11}$$

$$\tau_P + \tau_W + \tau_m = 0 \tag{11.12}$$

where the three torques are due to the pivot, the tension on the wire, and the weight mg about *any* point.

We have three equations for three unknowns: P_x, P_y, T. But if all I want is the tension T, I can get it from just the torque equation if I choose as the point for its computation the pivot itself. Then **P** drops out and we find

$$T\sin\theta \cdot L - mg \cdot \frac{L}{2} = 0 \quad \text{with a solution} \tag{11.13}$$

$$T = \frac{mg}{2\sin\theta}. \tag{11.14}$$

Observe that I am following the standard convention that a counterclockwise torque is positive.

In a practical problem, you may have a rod of weight $mg = 2000\ N$ and a wire that can only withstand a tension T_{max}. Then the computed T must obey

$$T = \frac{mg}{2\sin\theta} < T_{max} \tag{11.15}$$

$$\sin\theta > \frac{mg}{2T_{max}}. \tag{11.16}$$

If $T_{max} = 4000\ N$, this gives us

$$\sin\theta > \frac{1}{4}. \tag{11.17}$$

Now, if I am curious, I can go back and find out P_x and P_y from the force equations:

$$P_x = T \cos\theta = \frac{mg \cos\theta}{2 \sin\theta} \tag{11.18}$$

$$P_y = mg - T \sin\theta = \frac{mg}{2}. \tag{11.19}$$

Interestingly, the wire and the pivot each support half the weight of the rod no matter what θ is.

Notice that I computed the torque due to gravity as if all the mass was concentrated at the CM. This is true but not obvious, because the definition of the CM as the weighted sum of the masses in the body does not necessarily mean the CM can also stand in for the whole body in computing τ. You should always go back to first principles and check. The only thing we know is $\mathbf{F} = m\mathbf{a}$ and the force of gravity. Let us next imagine dividing the rod into tiny pieces, each of which is small enough to say it has a definite location x measured from the pivot and a mass $\frac{m}{L}dx$, exactly as in the CM calculation. The total torque about the pivot is then the integral

$$\tau = \int_0^L x \frac{mg}{L} dx = g \int_0^L x \frac{m}{L} dx = gmX. \tag{11.20}$$

Except for the factor g, the second integral is familiar from the CM calculation as mX, where $X = \frac{L}{2}$ is the CM. There the x entered as the coordinate to be weighted; here it is the $r_\perp$ for torque calculation. This $\tau = mgX$ is indeed what we would find if we lumped all the mass at the CM. This result is valid even if the mass density $\rho(x)$ is not a constant; the integral would still be mX.

11.4 The leaning ladder

The last of the equilibrium problems is the one I mentioned in the beginning of this chapter. I have a ladder of mass m and length L leaning on a wall at some angle θ, as shown in Figure 11.3. The wall is frictionless and the floor has a static coefficient of friction μ_s, which you may assume is given to you. I have not climbed the ladder yet; I just want to rest it against the wall and make sure it does not slide down. Can you see intuitively that θ has to exceed some lower limit for the ladder not to slide? Let's just write down all the equations for equilibrium to find this limit.

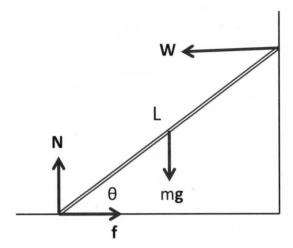

Figure 11.3   A ladder of mass m and length L is leaning against a frictionless wall and standing on a floor with some coefficient of friction μ_s. The normal and frictional forces due to the floor are shown separately.

We know mg can be assumed to be acting at the the center of the rod. The frictionless wall can only exert a force perpendicular to itself, called **W** for wall. The floor is exerting a normal force **N** and a frictional force **f** directed inward to balance **W**. That's it. The force equations along r x and y are:

$$f - W = 0 \tag{11.21}$$

$$N - mg = 0. \tag{11.22}$$

All that is left is the torque, and that's where we have the choice. We can take the torque around any point we like. If we want to punish ourselves, we can take a torque around some crazy point and then **N** and **f** will all come into the problem, and we will have to work that much harder. The trick, again, is to take the torque where the ladder touches the floor. Then **N** and **f** are gone, and what do we have? We have mg trying to rotate it clockwise and **W** trying to rotate counterclockwise. Let us use the first of the two definitions of the torque

$$\tau = Fr_\perp = F_\perp r. \tag{11.23}$$

For the weight $r_\perp = \frac{L}{2}\cos\theta$ and for the wall it is $L\sin\theta$. So we have

$$mg\frac{L}{2}\cos\theta = WL\sin\theta \quad \text{with the solution} \tag{11.24}$$

$$W = \frac{mg}{2}\cot\theta. \tag{11.25}$$

But now we have a restriction: the frictional force must obey $f \leq \mu_s N$. Consider the two equations:

$$f < \mu_s N = \mu_s mg \tag{11.26}$$

$$f = W = \frac{mg}{2}\cot\theta, \tag{11.27}$$

which lead to the inequality

$$\frac{mg}{2}\cot\theta < \mu_s mg \tag{11.28}$$

$$\tan\theta > \frac{1}{2\mu_s}. \tag{11.29}$$

That means θ has to be bigger than some number, because $\tan\theta$ always increases with θ for the angles under consideration. For example, if $\mu_s = .5$, we need $\theta >= 45^\circ = \frac{\pi}{4}$. If I want to climb up the ladder, I need to add my weight to the force and torque equations. The torque will depend on where I am on the ladder, and I will need to make sure I can climb all the way up.

11.5 Rigid-body dynamics in 3d

Rigid-body dynamics is quite easy as long as everything lies in the plane: all the mass is concentrated on the plane, all the forces are in the plane; all the vectors separating the point of application of the force to the point of rotation are in the plane. As we look down the plane, the only possible rotation is clockwise (negative) or counterclockwise (positive) about an axis perpendicular to the plane. The torque is $\tau = F_\perp r = r_\perp F$ where r is the distance to the axis from the point of application of the force.

That was then. But now we have to deal with the fact that in real life objects like tops and potatoes are not just planar and move in three dimensions. This is one case where not all the essential ideas can be conveyed in anything fewer than three dimensions.

So we will go to $d = 3$ but stay with a single point mass for a while. More complicated bodies can be built out of point masses. What is the analog of $\tau = I\alpha$ when the mass is running around in $d = 3$? How do you define torque and angular momentum in $d = 3$?

Suppose I have a point mass m in $d = 3$ at a location $\mathbf{r}$ with respect to some origin, as shown in the left half of Figure 11.4. Let $\mathbf{F}$ be the force on it. As in $d = 2$, we want to combine the two vectors $\mathbf{r}$ and $\mathbf{F}$ and get a torque out of it. We want the magnitude of the torque to be $rF \sin \gamma$ as in $d = 2$ *because the two vectors still lie in one plane, which could have been our old $d = 2$ plane.* But now we need to specify the orientation of this common plane in $d = 3$ to specify the direction of the torque. A natural choice is the normal to the plane. But there are two possible orientations of the normal! We break the tie by defining the torque to point in the direction in which a screw will advance if turned from $\mathbf{r}$ to $\mathbf{F}$. This completely defines the torque vector $\boldsymbol{\tau}$. What we have arrived at is called the *cross product* of $\mathbf{r}$ and $\mathbf{F}$, written as

$$\boldsymbol{\tau} = \mathbf{r} \times \mathbf{F}. \tag{11.30}$$

In general $\mathbf{C}$, the cross product of two vectors $\mathbf{A}$ and $\mathbf{B}$,

$$\mathbf{C} = \mathbf{A} \times \mathbf{B}, \tag{11.31}$$

points in the direction in which a screw would advance if you turned it from $\mathbf{A}$ to $\mathbf{B}$, and it has a magnitude

$$C = AB \sin \gamma, \tag{11.32}$$

where γ is the angle measured from $\mathbf{A}$ to $\mathbf{B}$ in their common plane, as shown in the right half of Figure 11.4.

This rule for fixing the direction of the cross product is called the *right-hand rule* for the following reason. If you grab the z-axis with your right hand and your fingers curl around it from $\mathbf{A}$ to $\mathbf{B}$ as shown by the big arrow, the torque points along the thumb, or the positive z-axis. This is

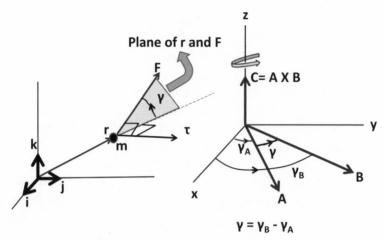

Figure 11.4   Left: The unit vectors **i**, **j**, and **k** shown for later use. We consider the torque τ due to force **F** with respect to a point (the origin) from which it is separated by **r**. The vectors **F** and (the continuation of) **r** (dotted line) define the shaded plane in which angle γ is measured from **r** to **F**. The magnitude of τ is assigned as in $d = 2$: $\tau = Fr\sin\gamma$. The vector $\tau = \mathbf{r} \times \mathbf{F}$, called the cross product of **r** and **F**, is oriented perpendicular to the plane in the sense dictated by the right-hand rule. Right: The cross product of **A** and **B** (whose common plane is chosen to be the $x - y$ plane for convenience) points in the direction a typical screw would advance if turned from **A** to **B**, as indicated by the curved arrow wrapping around the z-axis. It has a magnitude $AB\sin\gamma$, where γ is the angle measured from **A** to **B**. Instead of the screw rule, one can invoke the right-hand rule: if the fingers of the right hand curl from **A** to **B**, the thumb points along the cross product **C**.

why we homo sapiens can teach our children the cross product and lower primates can't.

Let us briefly digress to gain familiarity with the cross product.

Whereas the dot product of two vectors that yields a scalar is defined in any number of dimensions, the cross product of two vectors that yields a vector makes sense only in $d = 3$. This trick of using two vectors to define a unique direction perpendicular to the plane they define works only in $d = 3$. In $d = 4$, there will be *two* independent directions perpendicular to the plane defined by any two vectors.

Back to the cross product. If we reverse the order of the factors in the cross product it changes sign:

$$\mathbf{B} \times \mathbf{A} = -\mathbf{A} \times \mathbf{B} \tag{11.33}$$

because the screw will advance the opposite way in $\mathbf{B} \times \mathbf{A}$, or equivalently, γ will change sign. If we set $\mathbf{A} = \mathbf{B}$, then we find

$$\mathbf{A} \times \mathbf{A} = -\mathbf{A} \times \mathbf{A} = 0 \tag{11.34}$$

in agreement with the result $C = AB \sin \gamma$, which also yields 0 when $\gamma = 0$. Indeed, the cross product vanishes even if $\mathbf{B}$ is any scalar η times $\mathbf{A}$: *the cross product of two parallel vectors is zero.* This is good because the parallel vectors do not define a plane.

Let us compute the cross product for the basis vectors depicted in the left half of Figure 11.4. Consider $\mathbf{i} \times \mathbf{j}$. It points along $\mathbf{k}$ by the right-hand rule and has unit length because $|\mathbf{i}| \cdot |\mathbf{j}| \sin \frac{\pi}{2} = 1$. Here is a table of the nine possible cross products of the unit vectors:

$$\mathbf{i} \times \mathbf{i} = \mathbf{j} \times \mathbf{j} = \mathbf{k} \times \mathbf{k} = 0 \tag{11.35}$$

$$\mathbf{i} \times \mathbf{j} = \mathbf{k} = -\mathbf{j} \times \mathbf{i} \tag{11.36}$$

$$\mathbf{j} \times \mathbf{k} = \mathbf{i} = -\mathbf{k} \times \mathbf{j} \tag{11.37}$$

$$\mathbf{k} \times \mathbf{i} = \mathbf{j} = -\mathbf{i} \times \mathbf{k}. \tag{11.38}$$

Recall that we wrote the dot product in two equivalent ways:

$$\mathbf{A} \cdot \mathbf{B} = AB \cos \theta = A_x B_x + A_y B_y + A_z B_z. \tag{11.39}$$

We now want to write the cross product in terms of components instead of lengths and angles. We can guess the formula for $\mathbf{A} \times \mathbf{B}$ in terms of components by placing them both in the $x - y$ plane or choosing the $x - y$ plane so the two of them lie on it. If the vectors make angles γ_A and γ_B

with the x-axis as shown in Figure 11.4, the cross product $\mathbf{C}$ has only a z-component given by

$$C_z = AB\sin(\gamma_B - \gamma_A) \tag{11.40}$$

$$= AB(\sin\gamma_B \cos\gamma_A - \cos\gamma_B \sin\gamma_A) \tag{11.41}$$

$$= A\cos\gamma_A\, B\sin\gamma_B - A\sin\gamma_A\, B\cos\gamma_B \tag{11.42}$$

$$= A_x B_y - A_y B_x. \tag{11.43}$$

It is not hard to guess and to verify that if the vectors had been in some arbitrary plane so that $\mathbf{C}$ had all three components, the complete answer would have been

$$C_x = A_y B_z - A_z B_y \tag{11.44}$$

$$C_y = A_z B_x - A_x B_z \tag{11.45}$$

$$C_z = A_x B_y - A_y B_x. \tag{11.46}$$

Once you have one component, say C_z, you get the next by making the cyclic change $x \to y, y \to z, z \to x$ everywhere.

In this form it is clear that

$$(\mathbf{A} + \mathbf{B}) \times \mathbf{C} = \mathbf{A} \times \mathbf{C} + \mathbf{B} \times \mathbf{C}. \tag{11.47}$$

For example, if $\mathbf{F} = (\mathbf{A} + \mathbf{B}) \times \mathbf{C}$ then for the x-component, following Eqn. 11.44,

$$F_x = (A_y + B_y)C_z - (A_z + B_z)C_y \tag{11.48}$$

$$= A_y C_z + B_y C_z - A_z C_y - B_z C_y \tag{11.49}$$

$$= (A_y C_z - A_z C_y) + (B_y C_z - B_z C_y), \quad \text{which means} \tag{11.50}$$

$$\mathbf{F} = \mathbf{A} \times \mathbf{C} + \mathbf{B} \times \mathbf{C}. \tag{11.51}$$

Thus the cross product is *distributive*: the brackets can be opened out as with ordinary products of numbers. The reverse is also true: like terms can be grouped into brackets.

Now back to torque. Here are two applications of Eqn. 11.47.

First we will show that if the total force $\mathbf{F}_T$ on a body vanishes, and the total torque $\boldsymbol{\tau}_T$ is zero about one point, it will be zero about any other. Suppose relative to one point we have

$$\boldsymbol{\tau}_T = \sum_i \mathbf{r}_i \times \mathbf{F}_i = 0. \tag{11.52}$$

(Notice that I have used 0 instead of $\mathbf{0}$ as is commonly done.) Switching to another point located at $\mathbf{r}_0$ changes every $\mathbf{r}_i$ into $\mathbf{r}_i - \mathbf{r}_0$:

$$\boldsymbol{\tau}'_T = \sum_i (\mathbf{r}_i - \mathbf{r}_0) \times \mathbf{F}_i = \boldsymbol{\tau}_T - \mathbf{r}_0 \times \sum_i \mathbf{F}_i = 0 + 0 \tag{11.53}$$

where we have pulled out $\mathbf{r}_0$ from the sum over i using the distributive nature of the cross product.

Next consider the torque on a body in a uniform *gravitational field* $\mathbf{g}$, which means simply that every mass m will experience a force $m\mathbf{g}$. Near the earth, $\mathbf{g} = -9.8\ \mathbf{k}\ ms^{-2}$, if the vertical direction is along the z-axis. Assume the body is made of countable masses m_i at positions $\mathbf{r}_i$. The torque is

$$\boldsymbol{\tau} = \sum_i \mathbf{r}_i \times m_i \mathbf{g} \tag{11.54}$$

$$= \left[\sum_i m_i \mathbf{r}_i \right] \times \mathbf{g} \tag{11.55}$$

$$= M\mathbf{R} \times \mathbf{g} \tag{11.56}$$

where $\mathbf{R}$ is the CM. Thus the torque behaves as if all the mass M were concentrated at the CM position $\mathbf{R}$.

Now for the definition of angular momentum. Unlike

$$\tau = I\alpha = I\frac{d\omega}{dt} = \frac{dL}{dt} \tag{11.57}$$

in the planar case, we will now have a vector equation:

$$\boldsymbol{\tau} = \frac{d\mathbf{L}}{dt}. \tag{11.58}$$

Evidently the angular momentum $\mathbf{L}$ has to be a vector because $\boldsymbol{\tau}$ is a vector. In the plane we saw that

$$L = I\omega = mr^2\omega = m(r\omega)r = mv_\perp r \tag{11.59}$$

for an object that is part of a rigid body and hence forced to go around in a circle of radius r, with its velocity vector always perpendicular to the position vector. What if the object in question is not part of a rigid body? Then it could have a velocity that is not necessarily perpendicular to the position vector. The rigid body case suggests we keep only the part of its velocity perpendicular to $\mathbf{r}$ in the magnitude of $\mathbf{L}$:

$$L = mv_\perp r = mvr\sin\gamma \tag{11.60}$$

where γ is the angle between $\mathbf{v}$ and $\mathbf{r}$. The $\sin\gamma$ gives us the clue for the formula for the vector $\mathbf{L}$ in $d = 3$:

$$\mathbf{L} = \mathbf{r} \times \mathbf{p}. \tag{11.61}$$

This cross product would also arise naturally if we were guided by mathematics and sought a way to combine $\mathbf{r}$ and $\mathbf{p}$ to form a vector. Of course (-19) times this answer would also be a cross product, but the above-mentioned choice of $\mathbf{L}$ has the essential feature that its time derivative is the torque $\boldsymbol{\tau}$:

$$\frac{d\mathbf{L}}{dt} = \frac{d\left[\mathbf{r} \times \mathbf{p}\right]}{dt} \tag{11.62}$$

$$= \frac{d\mathbf{r}}{dt} \times \mathbf{p} + \mathbf{r} \times \frac{d\mathbf{p}}{dt} \tag{11.63}$$

$$= \frac{\mathbf{p}}{m} \times \mathbf{p} + \boldsymbol{\tau} \tag{11.64}$$

$$= 0 + \boldsymbol{\tau}. \tag{11.65}$$

These steps call for some explanation. By working with components you may satisfy yourself that the derivative acts on the cross product $\mathbf{r} \times \mathbf{p}$ just as it does on the ordinary product of two functions, *provided we keep the order of the two factors* $\mathbf{r}$ *and* $\mathbf{p}$ *the same*. Next I have used the fact that because the velocity and momentum are parallel (related by m), their cross product vanishes. The bottom line is that

$$\tau = \frac{d\mathbf{L}}{dt}. \tag{11.66}$$

In the case of a planar body that is made up of many point masses in the plane indexed by i, each one of them has its own angular momentum $\mathbf{L}_i$, each of which must be pointing out of the plane or into the plane: because $\mathbf{r}_i$ and $\mathbf{p}_i$ both lie in the plane, their cross product points perpendicular to that plane. That is why both torque and angular momentum were treated as scalars and not as vectors. Once a vector is constrained to lie up or down one axis, we can forget about the fact that it is a vector. We just say it's plus if it's up and minus if it's down.

In the case of a point mass that belonged to a rigid body, angular momentum had to do with actual rotations around some axis. But the definition $\mathbf{L} = \mathbf{r} \times \mathbf{p}$ holds for any particle with linear momentum, even if it is not going around in a circle, even if it is going in a straight line. Is this idea consistent with the older one? Consider Figure 11.5. Suppose a piece from the edge of a counterclockwise-rotating disk of radius r flies off from the 3 o'clock position and moves linearly along the tangential or y-direction. Before the fragmentation it had

$$L_1 = mv_1 r_{1\perp} = p_1 r_1 = p_1 r. \tag{11.67}$$

We believe it will maintain this $\mathbf{L}$ because there are no forces or torques acting on it. Indeed it does: $p_1 = p_2$, because there are no forces and even though $r_2 > r_1$, $r_{2\perp} = r_{1\perp} = r$, the radius of the disk.

So, every moving particle will have an angular momentum about any given point, unless the line of its momentum vector goes through that point. Thus, even if the particle had never been part of the disk but simply traveled along the dotted line with a momentum $\mathbf{p} = \mathbf{p}_1 = \mathbf{p}_2$, it would have had the constant angular momentum $\mathbf{L} = \mathbf{r} \times \mathbf{p}$ with respect to the origin.

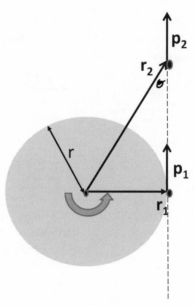

Figure 11.5   Angular momentum is defined by $\mathbf{L} = \mathbf{r} \times \mathbf{p}$ whether or not the
particle is actually going around a point. The figure shows a point at the 3 o'clock
position at the rim of a rotating disk. It has an angular momentum $\mathbf{L}_1 = \mathbf{r}_1 \times \mathbf{p}_1$.
Now it flies off along the tangent to the position $\mathbf{r}_2$ and has the angular momen-
tum $\mathbf{L}_2 = \mathbf{r}_2 \times \mathbf{p}_2$. Because there are no forces or torques on it when it flies off, we
expect that $\mathbf{L}_1 = \mathbf{L}_2$. This is true: even though $r_2 > r_1$ in magnitude, the
component perpendicular to the momentum is unchanged and equal to the
radius r of the disk, while $p_1 = p_2$ for a free particle. $r \sin \theta$

$L_1 = r_1 \times r_1$

$L_2 = r_2 \sin \theta \times r_2$ $L_1 = L_2$

Consider now a collection of bodies. Each one obeys

$$\frac{d\mathbf{L}_i}{dt} = \boldsymbol{\tau}_i = \boldsymbol{\tau}_{ie} + \sum_j \boldsymbol{\tau}_{ij} \qquad (11.68)$$

where $\boldsymbol{\tau}_{ie}$ is the torque on i due to external forces and $\boldsymbol{\tau}_{ij}$ the torque on it
due to force $\mathbf{F}_{ij}$ exerted on it by particle j. If we sum the terms over i, we
get for the total angular momentum

$$\frac{d\mathbf{L}}{dt} = \boldsymbol{\tau}_e + \sum_{i,j} \mathbf{r}_i \times \mathbf{F}_{ij}. \qquad (11.69)$$

Suppose the total external torque $\tau_e = 0$. Then is $\frac{d\mathbf{L}}{dt} = 0$? We can still use $\mathbf{F}_{ij} = -\mathbf{F}_{ji}$ but we have *different* factors of $\mathbf{r}_i$ and $\mathbf{r}_j$ that multiply the forces. Let us first consider just two particles $i = 1$ and $j = 2$ and see why $\mathbf{L}$ is conserved. (The argument applies unchanged with more particles.) Consider the internal torques and use $\mathbf{F}_{21} = -\mathbf{F}_{12}$:

$$\mathbf{r}_1 \times \mathbf{F}_{12} + \mathbf{r}_2 \times \mathbf{F}_{21} = (\mathbf{r}_1 - \mathbf{r}_2) \times \mathbf{F}_{12}. \tag{11.70}$$

It will vanish provided the cross product vanishes and that in turn is assured if $\mathbf{r}_1 - \mathbf{r}_2$ and $\mathbf{F}_{12}$ are parallel, that is, *the force between the bodies points along the vector separating them.* This is true for the case of gravitation, as we have seen, and true for the electrostatic force as well. One could argue on philosophical grounds that if i and j were the only two particles in the universe, the only possible direction for their mutual force to point along is their separation vector, *unless the universe has some other intrinsically preferred direction.* We believe it does not, and the conservation of $\mathbf{L}$ follows from this isotropy of space. Of course, once there are other bodies there is no reason the mutual force between any two could not be affected by the others. But it seems that is not the case, at least in classical physics. In other words, once we find how a pair of particles interact in isolation, we do not need to modify that interaction in the presence of additional particles: each pair continues to interact as before. This is another lucky break for us.

11.6 The gyroscope

The gyro is one problem where the math rather than your intuition will serve you better. Here is the gyro in Figure 11.6. Focus on the side view of the gyro. The gyro is a massless rod, supported by the tower at one end and attached to a cylinder of mass m at the other. The cylinder is free to spin about its axis, but first assume that it is not spinning and that I am propping up the cylinder with my finger. Now I let go. Everybody should know what follows: the rod and gyro will start rotating clockwise in the vertical plane about an axis perpendicular to the page. You don't need to take Physics 200 to know that. Let us make sure that agrees with $\tau = \frac{d\mathbf{L}}{dt}$. The torque $\tau = \mathbf{r} \times m\mathbf{g}$ has a magnitude $\tau = mgl$ and is directed into the page by the right-hand rule. We should be very clear that when angular momentum points into the paper, it doesn't mean the gyro goes into the

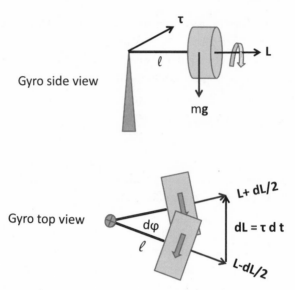

Figure 11.6   Top: The gyro resting on a tower, spinning with angular momentum **L** as seen from the side at $t = 0$. The rotating cylinder has all the mass and moment of inertia. Gravity exerts a downward force and a corresponding torque going into the page. Bottom: The view from the top at times $\mp\frac{1}{2}dt$. The bold arrows indicate the direction of rotation of the cylinder as seen from the top. The gyro has acquired a $d\mathbf{L} = \boldsymbol{\tau}\,dt$ perpendicular to **L**. The horizontal vector $\mathbf{L}(0) = \mathbf{L}$ is not shown to prevent cluttering. The angles are exaggerated in the figure for clarity. To first order in the infinitesimals $|\mathbf{L} \pm \frac{1}{2}d\mathbf{L}|^2 = L^2$ because $\mathbf{L} \cdot d\mathbf{L} = 0$. Thus **L** simply gets rotated by an angle $d\phi = \frac{dL}{L}$ in time dt.

paper. It just means the axis around which the gyro rotates is pointing into the paper.

This is it, if we have a gyro and it does not occur to us to spin the cylinder.

So let us spin it. Now it's a different ballgame. The gyro has an initial angular momentum. Which way is it pointing? That's the first thing you have to understand. Every part of the cylinder is spinning, and the angular momentum of every part—if you do the $\mathbf{r} \times \mathbf{p}$—will point radially out, in the direction of the total **L** as shown.

Let us look at the impact of the gravitational torque over a time dt symmetrically chosen around $t = 0$. The top view shows the gyro at $t = \mp\frac{1}{2}dt$. The angular momentum change $d\mathbf{L} = \boldsymbol{\tau}\,dt$ is going into the paper in the side view and pointing straight up in the top view. The angular momenta at $t = -\frac{1}{2}dt, 0,$ and $\frac{1}{2}dt$ all have the same length *to first order in the infinitesimals*:

$$|\mathbf{L} \pm \frac{1}{2}d\mathbf{L}|^2 = L^2 \pm \mathbf{L} \cdot d\mathbf{L} + \frac{1}{4}|dL|^2 \simeq L^2 \qquad (11.71)$$

because $\mathbf{L} \cdot d\mathbf{L} = 0$. As usual we drop the last term $|dL|^2 \propto dt^2$ because it is a quadratic infinitesimal.

That's why the gyro doesn't fall even though it acquires a $d\mathbf{L}$. For a static rod, getting an angular momentum going into the paper really means swinging down. But if a rotating gyro has an angular momentum to begin with, and we add on a perpendicular change, the new angular momentum is just a rotated version of the old angular momentum. So, when seen from the top, the gyro will be executing what's called *precession*. The gyro itself will just slowly go round and round with one end resting on top of the tower.

Does this remind you of the satellite going around the earth versus the dropped apple? Both accelerate toward the earth. In the case of the apple, the gravitational acceleration produces a downward velocity that adds to zero initial downward velocity, and the apple simply picks up speed toward the earth. A satellite, on the other hand, has a large velocity in the tangential direction to begin with, and adding a tiny radial bit only changes its direction as it moves to a different part of its circular orbit a little bit later.

Same thing with the gyro. It gets the same change in angular momentum $d\mathbf{L}$ whether or not the cylinder is spinning. But in one case, you add $d\mathbf{L}$ to zero and conclude it's going to swing down; in the other case, you add $d\mathbf{L}$ to a non-zero $\mathbf{L}$ and conclude that $\mathbf{L}$ will rotate.

Let us calculate Ω_p, the frequency of precession. Look at the lower half of the figure, the view from the top. In a time dt, the $d\mathbf{L}$ vector grows to a length τdt and $\mathbf{L}$ itself rotates by $d\phi$. Applying the formula for the arc length $ds = rd\phi$ to the triangle made of $\mathbf{L} - d\mathbf{L}/2$, $d\mathbf{L}$ and $\mathbf{L} + d\mathbf{L}/2$,

$$\tau dt = L d\phi, \quad \text{which means} \qquad (11.72)$$

$$\Omega_p = \frac{d\phi}{dt} = \frac{\tau}{L} = \frac{mgl}{I_{CM}\omega} \qquad (11.73)$$

where $I_{CM} = \frac{1}{2}mR^2$ for the rotating cylinder. (You may worry that $d\mathbf{L}$ is a chord and not an arc, but this difference vanishes when $d\mathbf{L} \to 0$.)

CHAPTER 12

Special Relativity I:

The Lorentz Transformation

Although the general public associates the theory of relativity with Einstein's monumental work of 1905, it is actually a lot older, going back to Galileo and Newton. According to the relativity principle, two observers in uniform relative motion will deduce the same laws of physics. That view of relativity has remained unchanged even after Einstein. However, in the Galilean version, the laws considered were those of mechanics, which was pretty much everything in those days. In the nineteenth century, it began to look as if the laws of electromagnetism and light did not respect the relativity principle. Einstein then rescued the principle, but he threw many cherished Newtonian ideas of space and time under the bus in the bargain. His 1905 work is referred to as the special theory of relativity, in contrast to his general theory, which came out in 1915. It was a theory of gravitation, and it is universally considered one of the greatest feats of human imagination and invention.

We will limit ourselves to special relativity, once again working with the minimum number of spatial dimensions, which happens to be just one. Time, which was once viewed as an absolute parameter, will turn out to be an additional dimension, in a sense to be made precise later.

12.1 Galilean and Newtonian relativity

The standard pedagogical technique for explaining relativity is in terms of some high-speed trains. Imagine two such (infinite) trains parked in parallel tracks in the station, along the x-axis. You board one train and see the other at rest. All the blinds are now closed; you are not to look outside yet. You settle down and explore the world around you. You pour yourself a drink, you play pool, you juggle some tennis balls, you play with your mass-spring system, and so on, and you develop a certain awareness and understanding of the mechanical world. Then you go to sleep. While you are sleeping, one of two things happens: either your train is left alone or some unseen hand gives your train a velocity of 200 miles per hour. The question is, when you wake up, can you tell which of the two things has happened without looking outside? Will anything inside this train betray that velocity? Now, you might say, "I know I am not moving because the sign up there says Amtrak." That is a kind of sociological reason, not a physics reason based on experiments. The claim of relativity is that you just will not know if you are moving or not. However, if the train picks up speed or slows down, you will know right away. If it picks up speed, you find yourself pushed against the back of the seat; if the driver slams on the brake, you will slam into the seat in front of you. So accelerated motion can be detected in a closed train without looking outside. The question is whether uniform velocity can produce detectable changes inside the closed train. The answer is no, according to relativity.

There is an equivalent way to say that everything looks the same when you wake up: Newton's laws continue to be valid in the moving train. If Newton's laws are valid, everything mechanical will look the same. Our expectations of what happens when, say, two billiard balls collide, or a mass and spring system oscillates, are all based on Newton's laws. So the claim is that Newton's laws will be unchanged when this uniform velocity is imparted to you.

Recall that if Newton's laws worked for you, you are called an inertial observer, and your frame of reference is called an inertial frame. Not all observers or frames are inertial. For example, $F = ma$ won't be true in an accelerating train: You leave things on the floor and they will slide backward. So with no apparent force acting on them, things will begin to accelerate. That's why an accelerating train is a non-inertial frame. We are not interested in that situation. So assume you started out in the station as an observer for whom the laws of Newton hold. The claim is they will hold when you wake up, even if in the meantime you have been given a uniform velocity.

Remember the train standing next to yours in the beginning. Suppose when you wake up, you open the blinds and see the other train moving at 200 miles per hour. That there is motion between you and the other train is an undeniable, experimental fact. The question is, can you tell if your train is responsible for this relative motion? Maybe nothing happened to you and the other train is moving the opposite way at 200 miles per hour? The relativity principle states that you really cannot tell. You can only tell that there is *relative motion* between the two trains that was not there before, but not "who is really moving."

This means that if you observe relative motion between you and the other inertial observer, you have every right to insist that you are not moving and the other observer is moving the opposite way and vice versa. But I repeat: You can make this argument only for uniform relative motion. If your train is accelerating, you cannot say "I'm not accelerating; the other train is accelerating in the other direction," because you are the one who is slamming your head on the seat in front of you, and nothing is happening to the other person. Or if you are in a rocket that is taking off and the G forces are enormous, you are in danger, not everyone on earth.

Now, back to the train. Suppose you look outside the window on the other side, and you see some cows going 200 miles per hour. Now what? I know you are thinking it must be the train traveling at 200 miles per hour. That conclusion is again based on non-physics notions. You are forgetting that it is logically possible that somebody put the whole landscape on wheels when you went to sleep, making the cows outside the train appear to move the opposite way. It is not likely that someone would go to such lengths just to fool you, *but if they did, you would not know.*

That's why in these thought experiments we don't like to open the window and look at the landscape, because then we have a bias. We will just look at the other train. Then, when you detect relative motion, you really cannot tell who is actually moving. This is clearer if, instead of two trains, you have two spaceships in outer space in relative motion and no other background.

12.2 Proof of Galilean relativity

I will now show that the laws of mechanics that you will deduce after waking up will be the same, namely Newtonian, even if you have acquired a uniform velocity while you slept. This will explain once and for all why everything mechanical will look the same when you wake up. First we need to do some groundwork.

We begin with the notion of an *event*. An event is something that happens at a certain place at a certain time. For example, if a firecracker goes off somewhere at some time, the x is where it happened and the t is when it happened. So (x, t) are coordinates in *spacetime*. Once again spacetime does not require Einstein's involvement at all. Even cave dwellers organizing a party knew that you have to say where (next to the dead T-rex) and you have to say when (at sunset). The fact that you need x and t, or, if you're living in three spatial dimensions, the fact that you need x, y, z, and t is not new. What is new will be clear soon.

Figure 12.1 shows two frames of reference, with their x-axes aligned. All the action is along the x-axis. We draw a perpendicular y-axis for pedagogical reasons, but we rarely deal with the y coordinate. I am in frame S and you are in frame S'.

Actually S is not just me sitting at the origin, $x = 0$: I am part of a huge team of people who are at rest relative to me. I have my agents all over the x-axis; they are my eyes and ears. If there's a firecracker exploding to the right, my guys will tell me, even if I am not personally there. So when I say *I* see something, I really mean my buddies and I, all traveling in the same train at the same speed, spread all over space, taking notes on what's happening. We may have to pool our information later, but we will know

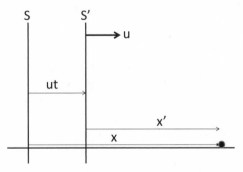

Figure 12.1   The same event (solid dot) is assigned coordinates (x, t) by me (S) and (x', t') by you (S'). In the pre-Einstein days $t = t'$ not only initially (when our origins crossed and clocks were synchronized) but always. At time t, your origin is to the right of mine by an amount ut. So the firecracker going off at location x at time t according to me happens at $x' = x - ut$ according to you.

this explosion took place at (x, t). In short, it takes a village to measure the spacetime coordinates of events. And you, sitting at your origin $x' = 0$, also have your own village of informers in S'.

Now, imagine you are sliding to the right according to me at a uniform velocity u. You start out to my left; you pass me; and then a little later you are somewhere to the right. We arrange it so that when you pass me, that is, when our origins cross, you set your clock to zero and I set my clock to zero. So here's an event: You and I crossed. What are the coordinates for that event? According to me, that event occurred at my origin $x = 0$, and the time was chosen to be $t = 0$. According to you, because your origin was also on top of my origin, $x' = 0$ as well. In Newtonian mechanics there is one time for all of us, and it was chosen to be 0 when our origins crossed. We made the coordinate of the event, our crossing, $(0, 0)$ for both you and me.

Now for a second event: a firecracker going off, shown by a black dot in Figure 12.1. I assign to it coordinates (x, t).

What do I expect you to say? You measure the distance from your origin, and you call this point x'; the time is still $t' = t$. What's the relation of x' to x? Well, this event took place at time t, so I know that your origin is off to the right by u times t. So the distance from your origin for this event, I maintain, is $x' = x - ut$. In summary we have

$$x' = x - ut \qquad\qquad (12.1)$$
$$t' = t. \qquad\qquad (12.2)$$

Henceforth we will denote the common time by t. This is called the *Galilean transformation* of coordinates in honor of the man who played a major role in articulating the principle of relativity.

Eqn. 12.1 may be inverted to express my x in terms of your x':

$$x = x' + ut. \qquad\qquad (12.3)$$

So far x and t have been unrelated, and they could be chosen independently. At any time t, the coordinates of an event x can have any value. Let us now imagine that I am using (x, t) to describe a moving particle, say a bullet, and that $x(t)$ is its location at time t according to me. Likewise $x'(t)$ is its location according to you at the same common time $t' = t$. By the previous logic, the functions $x(t)$ and $x'(t)$ also differ by ut:

$$x'(t) = x(t) - ut. \tag{12.4}$$

Let $v = \frac{dx}{dt}$ be its velocity according to me. Then w, the velocity according to you, follows from differentiating Eqn. 12.1 with respect to t and bearing in mind that u is a constant:

$$w = \frac{dx'}{dt} = \frac{dx}{dt} - u = v - u. \tag{12.5}$$

This is how velocities transform in Galilean relativity. This agrees with common sense and our day-to-day experience. For example, if I say the bullet is going at $v = 600$ mph, and you are going to the right in your train at $u = 200$ mph, you should measure the bullet speed to be reduced by 200 mph, and you should find that $w = 600 - 200 = 400$ mph. Taking one more derivative of Eqn. 12.1, we find for the acceleration of the moving object

$$a'(t) = \frac{d^2x'}{dt^2} = \frac{d^2x}{dt^2} = a(t) \tag{12.6}$$

because u is a constant. That means you and I agree on the acceleration of the body. We disagree on where the body is. We disagree on how fast the body is moving. But we agree on the acceleration of the body, because in going from me to you all we do is add a constant to all velocities. Therefore, if according to you the velocity of the body is constant, according to me the velocity of the body is also a constant, *but a different constant*. Or, if the body has an acceleration, we'll both get the same acceleration. This fact is key to determining the fate of Newton's law when we change frames.

Imagine that I study, in my inertial frame, two bodies with coordinates x_1 and x_2, which exert a force that depends on the distance between them, say $F(x_1 - x_2) = A/|x_1 - x_2|$, where A is some constant with units Newton-meters. The masses obey

$$m_1 \frac{d^2x_1}{dt^2} = \frac{A}{|x_1 - x_2|} \tag{12.7}$$

$$m_2 \frac{d^2x_2}{dt^2} = -\frac{A}{|x_1 - x_2|}, \tag{12.8}$$

where $F_{12} = -F_{21}$ in accordance with the third law. We have already seen that we can replace my accelerations on the left-hand side by your primed ones. But we can also replace my coordinate differences on the right by your primed ones because this is implied by $x'_1 = x_1 - ut$ and $x'_2 = x_2 - ut$:

$$|x'_1 - x'_2| = |(x_1 - ut) - (x_2 - ut)| = |x_1 - x_2|. \tag{12.9}$$

Thus my Eqns. 12.7 and 12.8 imply identical equations for you, in S':

$$m_1 \frac{d^2 x'_1}{dt^2} = \frac{A}{|x'_1 - x'_2|} \tag{12.10}$$

$$m_2 \frac{d^2 x'_2}{dt^2} = -\frac{A}{|x'_1 - x'_2|}. \tag{12.11}$$

It should be clear that this argument is quite general. You should try to apply it to a mass-spring system seen in the two frames, using the fact that the extension or compression of the spring is the same for both observers.

So this is the trick. Newton's laws work in my inertial frame, S; my unprimed coordinates obey them. I can then use the Galilean transformation to show that you, in S', will find that your primed ones will obey them as well.

We can also say it differently. If you wake up from your nap in a moving train and examine the world around you, you are going to get the same Newtonian laws as before. This can be proven by a person on the ground using the previous argument with you as the observer S'.

This is the way one can *prove* the principle of relativity in Newtonian mechanics.

Finally, suppose that the relative velocity u is not constant. Then the accelerations will not agree because $a' = a - \frac{du}{dt}$ and $F = ma$ will fail in the accelerating frame. The person will see objects accelerating without any force acting on them and know she is non-inertial.

12.3 Enter Einstein

We now fast forward about three hundred years. By this time André-Marie Ampère, Michael Faraday, Carl Gauss, and others had discovered the basic equations of electricity and magnetism. James Clerk Maxwell condensed these findings into his compact "Maxwell's equations" and, by modifying one of them to achieve mathematical and physical consistency, predicted electromagnetic waves. The wave velocity that Maxwell calculated, $3 \cdot 10^8 m/s$, was widely known at that time to be the velocity of light, c. It was then correctly concluded that light must just be an electromagnetic wave. And the question was, for which observer is this $c = 3 \cdot 10^8 m/s$ going to be the velocity of light?

Let me elaborate. Later we will calculate the velocity of waves on a string clamped at two ends as a function of the tension on the string and the mass density. That's the velocity *as seen by a person in the lab, one for whom the string is at rest except for its tiny transverse vibrations.* Likewise, if you calculate the speed of sound in a room, that speed is with respect to the air in that room, because sound waves travel in air. In general, the calculated velocity of a wave will be with respect to the medium that supports it. An observer moving relative to the medium will measure a different velocity.

So it was naturally assumed that the speed c calculated by Maxwell was for an observer at rest with respect to the *ether*, the medium that was postulated to carry electromagnetic waves, the way air carries sound waves. People wanted to know more about this ether. First of all, the ether must be everywhere because we can see the sun, and we can see the stars. Then you can ask, how dense is the ether? Usually the denser the medium, the more rapidly signals travel. For example, sound travels much faster in a very dense material like iron than in air. So the ether would have to be very, very dense. And yet it allows celestial bodies to move through it for years and years without slowing down. It's a very peculiar medium. And the question is, how fast are we moving relative to this medium?

In 1887 Albert Michelson and Edward Morley did the experiment to find out. Suppose their lab was moving a speed v relative to ether at the instant of measurement. They should find the speed of light to be $c - v$. They got a speed of exactly c! Perhaps at the very moment Michelson and Morley performed their experiment, their lab just happened to be at rest with respect to the omnipresent ether. Fine. But we know that this has to change in 12 hours, when the lab's velocity due to the earth's rotation is reversed, and in 6 months, when the velocity around the sun is reversed. You cannot be at rest with respect to the ether at all these times, and you cannot get the same answer of c all the time. Yet that is exactly what happens when you perform the experiment!

So people tried other solutions. Consider the speed of sound, 760 miles per hour. Why doesn't that change from day to day as the earth rotates around its own axis and around the sun? How can you and I talk to each other, oblivious of these high speeds? The reason, of course, is that the earth carries the air with it as it spins around itself and around the sun. So if you can carry the medium with you, then it doesn't matter how fast you're moving. When we fly, we carry the air with us in the plane, and sound travels at the same speed inside the moving plane as in a static one. When I ask for a pillow, it takes the flight attendant the same duration of time to pretend he did not hear me, independent of the speed of the plane. So, people tried to argue that the earth carries the ether with it, the same way it carries the air with it. But it's very easy to show, by looking at distant stars, that this cannot be true. I don't have room here to tell you more. (Read up on *aberration of starlight* if you want.) You cannot take the ether with you, and you cannot leave it behind; that was the impasse.

The velocity of light posed quite a problem. Let us pause to absorb that. Imagine a bullet going to the right at a certain speed v. You move to the right at some speed, say $\frac{1}{2}v$. I expect you to measure the speed of the bullet to be $\frac{1}{2}v$. But if this were a light beam and not a bullet, you are supposed to get the value $v = c$ no matter what your speed was. You go at a speed $\frac{3c}{4}$ along the beam, and you still measure the value of c. None of the models of ether or materials could explain this.

At this point Einstein entered the scene, and he explained the baffling behavior of light as follows. Recall that in Galilean relativity, an inertial observer who goes to sleep in a train parked at the station cannot tell, upon waking up, if the train has acquired a uniform velocity. But if the velocity of light depended on how fast the train was moving, then by comparing the velocity of light before and after the nap, the observer could deduce the speed of the train without looking outside. Because motion at uniform velocity produced detectable changes, without any reference to the outside world, it was not relative but absolute. Even though mechanical laws remained the same in a moving train, laws of electricity and magnetism would betray its velocity.

But, of course, this is not what happens, and the speed of light does not change with the speed of the train. This means electric and magnetic phenomena are part of the natural conspiracy to hide our uniform velocity. Just as mechanical phenomena won't tell us how fast we're moving without looking outside, neither will electromagnetic phenomena.

To Einstein it was obvious that nature would not design a system in which mechanical laws are the same in a moving frame, but laws of electromagnetism are different. So he postulated that *all phenomena*, whatever their nature, will be unaffected by going to a frame at constant velocity relative to the initial one. That was a very bold postulate, because it even applies to biological phenomena, which were not his specialty. But he believed that either all natural phenomena would follow the principle of relativity, or none would.

Einstein had a faith that underlying laws of nature would have a certain uniformity across all natural phenomena. It's not a religious issue—otherwise I wouldn't bring it up in the classroom—but it is certainly the credo of all scientists, at least all physicists, that there is some elegance and consistency in the laws of nature. Even though they may not believe in design by any personal god, they do believe in this underlying, rational system that we are trying to uncover. This belief has been reinforced over and over again.

12.4   The postulates

Here are the two great postulates of Einstein's special relativity.

Postulate I. All inertial observers are equivalent.
Postulate II. The velocity of light is independent of the state of motion of
  the source and the observer.

In postulate I, "equivalent" means each inertial observer is as privileged as any other to discover the laws of nature. If I find some laws, and you're moving relative to me at uniform velocity, you'll find the same laws. And if you and I find each other in relative motion, you have as much right to claim you are at rest and I am moving as I have to claim that I am at rest and you are moving. There is complete symmetry between observers in uniform relative motion. There is no symmetry between people in non-uniform motion. As I said, non-uniform motion creates effects that can destroy me and not destroy you. So, no one's trying to talk his way out of acceleration, whereas you can talk your way out of uniform velocity. That's the first postulate. This was understood even in the time of Newton. What is new now is that all inertial observers are equivalent with respect to *all* natural phenomena, including the electromagnetic, and not just mechanical phenomena.

As per postulate II, if a light beam is emitted by a moving rocket, it doesn't matter: it travels at c. If a light beam is seen by a moving rocket, it doesn't matter: it will measure c. All people will get the same answer for the velocity of light. It is a postulate because it cannot be deduced from anything else.

It looks as if Einstein has solved the problem by saying that light behaves this way because it is part of the big conspiracy to hide uniform motion. But you will see that he has made an expensive bargain trying to save relativity. His postulates will be seen to force him and us to give up many cherished notions of Newtonian physics. Think about why. Say a car is going 200 miles per hour according to me. You get into your own car and follow that car at 50 miles per hour. I expect you to measure its speed to be 150 miles per hour. But what if you also got 200 miles per hour? Now, this is not what happens for cars traveling at the speeds I mentioned, but that is exactly what happens if the car is replaced by a pulse of light. You must agree that is really incompatible with our daily notions and the formula $w = v - u$. When you put $v = c$, somehow w has got to come out to be c and not $c - u$. Because that does not happen in the Galilean transformation, it has to go.

12.5   The Lorentz transformation

We are led to seek a new rule or transformation connecting (x, t) and (x', t'), such that when the velocity of light is computed, we get the same answer of c in both frames. Here is a clue to how things will work out. Suppose I send a pulse to the right at speed c, and you are going to the right at $\frac{3c}{4}$. My Newtonian expectation is that you should measure the speed of the pulse as $\frac{c}{4}$. But you insist it is c. What will I say to you? Because you are finding velocity of the pulse as the distance traveled divided by the time taken and you are getting four times the answer I expected, I will say your "meter" sticks are somehow shorter than a meter. Specifically, I would say they have shrunk to one-fourth their original size (when we were at relative rest), resulting in your measurement of a velocity four times what I expected.

But there's another possibility. Your clocks, which used to be identical to mine when we were at relative rest, may be running slow. So you let the light travel for four seconds and thought it was only one second. That's why you measured a velocity four times bigger than what I expect. Or maybe both your meter sticks and clocks are messed up. But something has to give, and so in searching for the replacement to the Galilean transformation of coordinates Eqns. 12.1 and 12.2, *we will no longer assume length and time intervals are the same for both observers.*

Consequently, the spatial coordinate of the event I labeled (x, t) will not be $x' = x - ut$ according to you, but

$$x' = \gamma(x - ut) \tag{12.12}$$

where γ is the fudge factor to convert lengths from me to you. Similarly, I modify your expectation that $x = x' + ut'$ (we admit the possibility that $t' \neq t$) to read

$$x = \gamma(x' + ut') \tag{12.13}$$

where the fudge factor γ is the same from me to you and vice versa, because otherwise we would not be equivalent, and one of us would be holding the shortened "meter" stick. Since $u \rightarrow -u$ when we change from me to you, we expect γ to be a function of u^2. We proceed to nail down γ as follows.

Suppose we sent off a light pulse when our origins coincided, and this pulse set off the firecracker at (x, t) according me and (x', t') according to you. Because the light pulse took t seconds to travel x meters according to me and took t' seconds to go x' meters according to you, and we both agree on the value of c, it must be true for *this particular event* that

$$x = ct \quad \text{and} \quad x' = ct'. \tag{12.14}$$

Let us multiply the left-hand side of Eqn. 12.12 by the left-hand side of 12.13 and equate the result to the product of the right-hand sides to obtain

$$xx' = \gamma^2(xx' + xut' - x'ut - u^2tt'). \tag{12.15}$$

Upon setting $x = ct$, $x' = ct'$ we find

$$c^2tt' = \gamma^2(c^2tt' + uctt' - uct't - u^2tt') \tag{12.16}$$

$$1 = \gamma^2\left(1 - \frac{u^2}{c^2}\right) \tag{12.17}$$

$$\gamma = \frac{1}{\sqrt{1 - \frac{u^2}{c^2}}}. \tag{12.18}$$

Note that once we have found the length conversion factor γ it does not matter that we deduced it from this specific event involving a light pulse. It can be applied to Eqns. 12.12 and 12.13 valid for a generic event. Putting γ back into Eqn. 12.12 we obtain

$$x' = \frac{x - ut}{\sqrt{1 - \frac{u^2}{c^2}}}. \tag{12.19}$$

We now go to $x = \gamma(x' + ut')$, Eqn. 12.13, isolate t' and express it entirely in terms of x and t as follows:

$$\begin{aligned} t' &= \frac{1}{u}\left(\frac{x}{\gamma} - x'\right) \\ &= \frac{1}{u}\left(\frac{x}{\gamma} - \gamma(x - ut)\right) \\ &= \frac{\gamma}{u}\left(\frac{x}{\gamma^2} - (x - ut)\right) \\ &= \frac{\gamma}{u}\left(x(1 - \frac{u^2}{c^2}) - (x - ut)\right) \\ &= \frac{t - \frac{ux}{c^2}}{\sqrt{1 - u^2/c^2}}. \end{aligned} \tag{12.20}$$

To summarize, imagine I am (in frame) S and you are (in frame) S', and you are moving to the right (increasing x direction) at speed u. Let my coordinates for an event be (x, t) and let your coordinates for the same event be (x', t'). The *Lorentz transformation* tells us that

$$x' = \frac{x - ut}{\sqrt{1 - u^2/c^2}} \tag{12.21}$$

$$t' = \frac{t - \frac{u}{c^2}x}{\sqrt{1 - u^2/c^2}}. \tag{12.22}$$

Observe that the Lorentz transformation reduces to the Galilean transformation if the velocity between you and me is much smaller than the velocity of light, that is, $\frac{u}{c} << 1$. So the relativistic formula really kicks in only for velocities comparable to the velocity of light.

You must clearly understand what the formula is connecting. Things are happening in space and in time, right? Say something happens here. That something has a spatial coordinate and a time coordinate, according to two observers who originally had their origins and their clocks coincide when they passed. And one is moving to the right at speed u. Then the claim is that if the event had coordinates (x, t) for one person, for the other person moving to the right at speed u, the same event would have coordinates (x', t'), related to (x, t) as above.

Now we can see why Einstein gets the credit for making the world four-dimensional. After all, the four coordinates x, y, z, and t were present before he came along. But t' was always equal to t in the old days, no matter how you moved. In Einstein's theory, x, y, z, and t get scrambled into primed variables. Here is an analogy. Imagine creatures restricted to the $x - y$ plane. When they rotate axes, x and y get scrambled into each other. Now, there is a third dimension perpendicular to these two, labeled by z, but if all their rotations are limited to the $x - y$ plane, z will never mix with x and y, and $z' = z$ before and after rotation. This is what makes their world two-dimensional. If you now permit these creatures to rotate out of the $x - y$ plane, then z will indeed begin to mix with x and y, by tiny imperceptible amounts for small tilts (and they may not realize that z is just another coordinate) and sizable ones for larger tilts where z gets seriously mixed up with x and y. In our problem $\frac{u}{c}$ is the tilt in spacetime. When our experiments were limited to $\frac{u}{c} \ll 1$, we thought t was an invariant, the same for all observers. Einstein then showed us the complete picture.

You can already see that the theory will not admit velocities bigger than the velocity of light because the square root $\sqrt{1 - u^2/c^2}$ then becomes imaginary. So the one single velocity that we wanted to be the same for everybody is also the greatest possible velocity, and no observer can move with respect to another at a speed that is equal to or in excess of the speed of light.

The fact that an event has different coordinates in different systems doesn't mean the laws deduced are different. For example, suppose I'm on the ground and I throw a piece of chalk; it goes straight up and comes straight down. If you see me from a moving train, you would think it went up and down but along a parabola. No one says the chalk will also go up and down for you, only that its motion will still obey Newton's laws. That's all you really mean by saying things look the same in both inertial frames. Even though Einstein's theory upheld the principle of relativity, it discarded Newton's laws. They had to be modified at velocities comparable to c. However, relativity assures us that the *same* modified laws will be deduced by two observers in uniform relative motion.

CHAPTER 13

Special Relativity II:

Some Consequences

13.1 Summary of the Lorentz transformation

Let us begin with the Lorentz transformation, which relates the coordinates of an event in two different frames of reference, with the primed one moving at a velocity u relative to the unprimed one:

$$x' = \frac{x - ut}{\sqrt{1 - u^2/c^2}} \tag{13.1}$$

$$t' = \frac{t - \frac{u}{c^2}x}{\sqrt{1 - u^2/c^2}}. \tag{13.2}$$

The Lorentz transformation is the cornerstone of relativity; all the funny stuff you hear about—the shrinking rods, the twin paradox—everything comes from these simple equations, derived without even calculus. If you consider the stresses and strains on a loaded steel beam, the mathematics involved is a whole lot more difficult. That is the remarkable thing about relativity. In extracting the bizarre consequences of these equations, Einstein showed as much courage as brilliance. If you derived these equations, you own them and have no choice but to deduce and defend the consequences.

Now, some of you might be rattled by Eqns. 13.1 and 13.2; it's not clear what is being stated. Let me explain. The velocity u is a fixed number.

That's your speed relative to mine. I see some event happening, and I give it a pair of numbers, (x, t). You give the same event another pair of numbers, (x', t'), and they are related by the Lorentz transformation.

Here is the analogy. Consider the $x - y$ plane with a point P as in Figure 13.1. I assign to it a pair of numbers, (x, y). Now, you have a different coordinate system, rotated relative to mine by some angle θ. Our coordinates are related as follows

$$x' = x\cos\theta + y\sin\theta \qquad\qquad (13.3)$$

$$y' = -x\sin\theta + y\cos\theta. \qquad\qquad (13.4)$$

The angle θ is the analog of the velocity u. If $\theta = 0$, you and I agree completely. In general, you plug in my (x, y) into these equations and get your (x', y'). Let's take $\theta = \frac{\pi}{4}$. In that case,

$$x' = \frac{(x+y)}{\sqrt{2}} \quad\text{and}\quad y' = \frac{(-x+y)}{\sqrt{2}}. \qquad\qquad (13.5)$$

So, for every angle, the cosine and sine will reduce to some numbers; both happen to be $1/\sqrt{2}$ here. For example, if (x, y) is $(1,1)$, and your axis is obtained from mine by a counterclockwise rotation of $45°$, my $(1, 1)$ lies right on your x' axis. So, I expect you to say $y' = 0$, and this is indeed so. And how about x'? It becomes $(1+1)/\sqrt{2} = \sqrt{2}$, the length of the position vector, which is entirely along the x' direction.

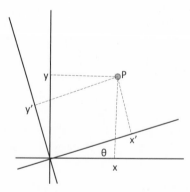

Figure 13.1   The same point P (solid dot) is assigned coordinates (x, y) by me and (x', y') by you. The pairs of coordinates are linearly related. The dotted lines indicate the sizes of the coordinates in the two frames.

The Lorentz transformation is very similar. For example, the first equation can be written as

$$x' = \left[\frac{1}{\sqrt{1 - u^2/c^2}}\right] x - \left[\frac{u}{\sqrt{1 - u^2/c^2}}\right] t \qquad (13.6)$$

so that

$$\frac{1}{\sqrt{1 - u^2/c^2}} \quad \text{and} \quad \frac{u}{\sqrt{1 - u^2/c^2}} \qquad (13.7)$$

are the analogs of $\cos\theta$ and $\sin\theta$. The same goes for the second equation involving t'. For a given u, these are just numbers depending on u, and (x', t') is a linear combination of (x, t). But I should warn you that the coefficients in square brackets are not cosine or sine of anything, because if they were, their squares should add up to 1, and they don't. It is, however, still a *linear homogeneous transformation*. Linear means the new coordinates (x', t') are related to the first powers of the old coordinates (x, t). They don't involve t^2 or x^3 and so forth. Homogeneous means that only the first power is present. This ensures that if $(x, t) = (0,0)$, then $(x', t') = (0,0)$ as well.

Suppose you wish to invert these equations and write (x, t) in terms of (x', t'). There are two options open for you. One is to say these are simultaneous equations. You have to find a way to solve for (x, t) in terms of (x', t') and all these funny coefficients involving u. You treat them all as constants and juggle them around, multiply by this, divide by that, and so on, to isolate x and t in terms of (x', t'). But you shouldn't do that because you know what the answer should be, namely, the old Lorentz transformation with the velocity reversed:

$$x = \frac{x' + ut'}{\sqrt{1 - u^2/c^2}} \qquad (13.8)$$

$$t = \frac{t' + \frac{u}{c^2}x'}{\sqrt{1 - u^2/c^2}}. \qquad (13.9)$$

(For rotations, you can express (x, y) in terms of (x', y') by reversing θ in Eqns. 14.4 and 14.5.)

A crucial step in getting some mileage out of the Lorentz transformation is to take *a pair of events*. Let event 1 have coordinates (x_1, t_1) according to me, (x_1', t_1') according to you, and similarly for event 2. If you write the Lorentz transformation for the two events and subtract the primed coordinates of 1 from those of 2 you will find

$$\Delta x' = \frac{\Delta x - u\Delta t}{\sqrt{1 - u^2/c^2}} \qquad (13.10)$$

$$\Delta t' = \frac{\Delta t - \frac{u}{c^2}\Delta x}{\sqrt{1 - u^2/c^2}} \qquad (13.11)$$

where $\Delta x = x_2 - x_1$ et cetera. Note that the *differences are not necessarily small*.

The differences in coordinates also obey the same Lorentz transformation as the coordinates, thanks to the linearity of the Lorentz transformation. If you want the differences that I get in terms of yours, you just have to reverse the sign of u:

$$\Delta x = \frac{\Delta x' + u\Delta t'}{\sqrt{1 - u^2/c^2}} \qquad (13.12)$$

$$\Delta t = \frac{\Delta t' + \frac{u}{c^2}\Delta x'}{\sqrt{1 - u^2/c^2}}. \qquad (13.13)$$

13.2 The velocity transformation law

The two events chosen above were unrelated. Now consider them to be related as follows:

- Event 1: I fire a gun.
- Event 2: The bullet hits the wall and gets embedded.

The bullet moves by an amount Δx in time Δt according to me, and by $\Delta x'$ in time $\Delta t'$ according to you.

Let us agree on the following notation for velocities once and for all:

$$v = \frac{\Delta x}{\Delta t} \quad \text{velocity of any object according to me} \quad (13.14)$$

$$w = \frac{\Delta x'}{\Delta t'} \quad \text{velocity of any object according to you} \quad (13.15)$$

$$u = \text{your velocity relative to me,} \quad (13.16)$$

where all Δ's should of course be infinitesimals approaching 0 as we are talking about instantaneous velocities. Suppose I say the bullet has velocity v. To find the velocity w according to you, we take the ratio of Eqns. 13.10 and 13.11:

$$\frac{\Delta x'}{\Delta t'} = \frac{\Delta x - u\Delta t}{\Delta t - \frac{u}{c^2}\Delta x} = \frac{\frac{\Delta x}{\Delta t} - u}{1 - \frac{u}{c^2}\frac{\Delta x}{\Delta t}} \rightarrow \frac{v - u}{1 - \frac{uv}{c^2}} = w \qquad (13.17)$$

where, in the end, we have let all the difference go to 0. For small velocities (dropping the terms that go as $1/c^2$) we get results agreeing with common sense. Equation 13.17 is the *relativistic velocity transformation law* .

Let us get used to going from your description to mine. Suppose you think a particle has velocity w. What will I think? Now we use Eqns. 13.12–13.13. Taking the ratios as before and recalling the definition of w we get

$$v = \frac{w + u}{1 + \frac{vw}{c^2}}, \qquad (13.18)$$

which amounts to changing the sign of u, and exchanging w and v in Eqn. 13.17. This velocity transformation law is what makes it possible to have an upper limit on velocity. If velocities added as in the old days, there can be no upper limit. If I tell you that according to relativity, nothing can go faster than the speed of light, you might try to beat the system as follows. You might ask if there can be a gun whose bullets go at three-fourths the velocity of light? I would say yes, and you would then ask, "How about a train that goes at $\frac{3c}{4}$?" I would say that seems to be allowed. Then you can say, "Let me get into this train at $\frac{3c}{4}$ and fire a bullet at $\frac{3c}{4}$; then, from the ground, it should appear to be going at 1.5c." Well, that's the naive pre-relativistic expectation. But if you put $w = \frac{3c}{4}$ and $u = \frac{3c}{4}$ into the correct formula Eqn. 13.18, you get the correct answer to be

$$v = \frac{\frac{3c}{4} + \frac{3c}{4}}{1 + \frac{9}{16}} = \frac{24}{25}c. \tag{13.19}$$

If I apply this to a light pulse seen by you ($w = c$), the speed I will find is

$$v = \frac{c + u}{1 + u/c} = c. \tag{13.20}$$

13.3   Relativity of simultaneity

We used to think simultaneity was absolute. If two events occurring in Los Angeles and New York are simultaneous for me, fixed on the planet, then for you, flying in a rocket, they should be simultaneous. How can they not be? If two things are happening right now in different places for me, the same *has* to be true for you. But it's not, as a consequence of the Lorentz transformation.

Before I explain that, I must point out that there is one case when simultaneity is absolute. If the two events occurred at the same time and the same place, $\Delta x = 0$, $\Delta t = 0$, then the transformation will tell you $\Delta x' = 0$, $\Delta t' = 0$. If I clapped my hands, my two hands were at the same place at the same time. If someone says my hands were not at the same place at the same time, she is saying I didn't clap. Relativity may change the coordinates of an event as we change observers, but it will not deny the very event.

Now for the general case of the relativity of simultaneity. You, S', are in the train, as shown in the upper half of Figure 13.2. To make two things happen simultaneously at the two ends of the train, you go to the middle of the train and send two flashes of light that go to the back (B) and front (F) of the train and set up two explosions. Because you're in the middle of the train, you know the explosions will be simultaneous for you. Now, I see you from the ground. What do I see? The front F of the train is moving away from the light pulse, while the back B is rushing to meet the light pulse. *Now, the velocity of light is the same for everybody and independent of the velocity of the source.* If the back end is rushing to meet the light pulse and the front end is running away from the light pulse, I know the explosion at the back of the train occurs

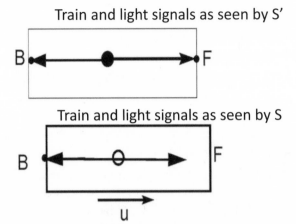

Figure 13.2   Top: The train as seen by you, observer S', at rest on the train. You emit a light pulse from the center of the train (big dot), which hits the front and back ends F and B at the same time, setting off two simultaneous explosions (small dots). Bottom: I, observer S, see you and the train moving to the right at speed u. I see the back end of the train rushing to meet the light pulse and the front end running away from it. The figure clearly shows the pulse reaching the back end B, while the pulse headed for F, the front, has not yet arrived.

first. Now, why do we bring in the two counter-propagating light pulses and not a pair of counter-propagating pigeons to trigger simultaneous events? Because the velocity of light is the same for everybody, independent of the motion of the source or the observer. That's a postulate. That's why many arguments in relativity involve doing things with light pulses or communicating with light pulses: we know the speed of light in any situation. (We know a lot less about pigeons.) Therefore, we know you couldn't have done any better in making them simultaneous, and I know for sure I will disagree with you. There's no answer to the question, who is right?

Here is an analogy. Look again at Figure 13.1. First, if two points have the same x and y coordinates in one set of axes, this will be true in any other set of axes because they are on top of each other. Next, if I pick two points that lie on a line parallel to the x-axis, then for me they have the same y. If you use rotated axes, you will say they have different y'-coordinates. There is nothing absolute about having the same y coordinate; it depends on the frame. In relativity, this happens to be true for time as well as space, and the two frames are related not by a relative rotation but by a relative velocity.

13.4 Time dilation

The next surprise has to do with clocks and time. You and I buy two identical clocks. Then you get into your train and start moving relative to me. I will find that your clock is running slow, and you will say the same about mine.

How do we turn this issue of the time period of the clock into a pair of events? I have my clock, and it goes tick, tick, tick. I pick two events. Event 1 is a tick of the clock, and event 2 is the next tick. Let me put the clock at the origin of my coordinates system. (It doesn't really matter where it is.) The spacetime coordinate of the first tick is $(x = 0, t = 0)$. What is the location of the clock during the second tick? Because I'm talking about a clock at rest in my frame, if the first event took place at $x = 0$, the second one also takes place at $x = 0$. The two successive ticks of the clock are separated in space by 0 and in time by τ_0, its time period. That means $\Delta x = 0$ and $\Delta t = \tau_0$. According to you the time between two successive ticks is

$$\Delta t' = \frac{\tau_0 - 0\frac{u^2}{c^2}}{\sqrt{1 - u^2/c^2}} = \frac{\tau_0}{\sqrt{1 - u^2/c^2}}. \tag{13.21}$$

Because the denominator is < 1, we find $\Delta t' > \tau_0$. For example, if $\frac{u}{c} = \frac{3}{5}$, you will say my clock has a time period $\frac{5}{4}\tau_0$. In fact, anyone moving relative to me will say my clock is slow.

Let us rederive this result by choosing the inverse Lorentz transformation, in which case my coordinates are written in terms of yours. Now we find

$$\Delta t = \tau_0 = \frac{\Delta t' + \frac{u}{c^2}\Delta x'}{\sqrt{1 - u^2/c^2}} \tag{13.22}$$

and bring in the other equation that states that the two events (successive ticks of my clock) took place at the same point for me:

$$\Delta x = 0 = \frac{\Delta x' + u\Delta t'}{\sqrt{1 - u^2/c^2}}, \tag{13.23}$$

which means $\Delta x' = -u\Delta t'$. (This just means you see me and my clock moving to the left at speed u.) Feeding this into Eqn. 13.22 we conclude that

$$\tau_0 = \frac{\Delta t'(1 - u^2/c^2)}{\sqrt{1 - u^2/c^2}} = \Delta t'\sqrt{1 - u^2/c^2}, \tag{13.24}$$

which agrees with Eqn. 13.21.

Now, here is a paradox. Just as I will say your clock is slow, you'll say my clock is slow. How can we both accuse each other of having slow clocks? (A related question for the psych majors: how can two people simultaneously look down on each other?) Here's the answer that's usually given. If I take a real clock like my watch and you ask why it looks slow to you when I'm moving relative to you, it's difficult to answer because it has electronics and stuff, and I am still working on setting the clock on my VCR. For all of us who are technically challenged, there is a clock that's particularly simple. It has two mirrors a distance L apart in the y-direction and a light pulse that bounces up and down between the mirrors as shown in part A of Figure 13.3. Every time the pulse completes a round trip, it sets off some detector at the lower end, and the clock goes "tick." The pulse travels a total distance $2L$ in the vertical direction between ticks, so that

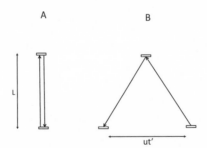

Figure 13.3 Part A shows my light clock in my frame. The light pulse goes straight up and down a total distance $2L$ between ticks. Part B shows the same clock as seen by you, to whom my clock seems to be moving to the left at speed u. You say the clock has slowed down because the zigzag path is longer than the straight up-and-down path, while the velocity of light is the same for both. A clock at rest in your frame will be moving to the right according to me, and its light pulse will be moving in a zigzag path and hence running slow according to me.

$\tau_0 = 2L/c$ is the time period of my clock. Now you are moving relative to me at velocity u, so according to you my clock is moving to the left at speed u, and it looks like part B in the figure. The light beam in my clock is going on a zigzag path along the hypotenuse. The time period t' according to you obeys the equation

$$ct' = 2\sqrt{L^2 + \left(\frac{ut'}{2}\right)^2},$$
(13.25)

which can be solved to give

$$t' = \frac{2L/c}{\sqrt{1 - u^2/c^2}} = \frac{\tau_0}{\sqrt{1 - u^2/c^2}}.$$
(13.26)

In this argument I am invoking the fact that the vertical distance between the mirrors in the y-direction is L for both observers. Here is one way to understand this. Suppose you and I paint an infinite line at height $y = y' = 1$ meter when we are at relative rest. As we now move relative to each other, our lines must match: there is no reason one should be below the other given the equivalence of inertial observers. The point is that the two lines, which extend forever, can be compared anywhere and at any time and do not run away from each other the way localized objects like clocks or rods do.

Why do we like this clock? Because we know everything about the operation of the clock. We know that the zigzag path is longer than the straight up-and-down path, because the transverse coordinate is known to be the same for both. Because the velocity of light is the same in all frames of reference, light going on a longer path is simply going to take longer. So, I know your clock would slow down, and you can say my clock has slowed down by the same factor because the zigzag paths have the same lengths.

What if you have some other clock with gears and wheels and teeth? How do I know that it too slows down when you go on your train? I don't know exactly why it appears slowed to me, but I know that it too should appear slowed just like the light clock. To see this, consider a light clock L and an electronic clock E of the same time period that you take on the train. You of course will find them to be in sync because you are inertial and think you are at rest. Let the clocks be at the same location $x' = 0$ with their ticks synchronized at $t' = 0$. The next ticks of the two clocks have the same spatial and time coordinates in your frame, and therefore in my frame. The clocks will be in sync according to me as well.

Therefore, all clocks must do the same thing as the light clock, regardless of their mechanisms, and that includes biological clocks. Now you yourself are a biological clock: I look at you, say over a period of 40 or 60 years, and I notice some changes. You become taller; your hair turns white; your teeth fall out. I don't care how your life systems work, but you are a clock and you've got to slow down in my view if you move relative to me. That's why physicists can make predictions about what happens to living systems even though that's not our main business, which happens to be staring at our shoes at cocktail parties.

13.4.1 Twin paradox

Say my twin goes on a space mission at some high speed for 20 years. Let's say he was 20 when he left, so that in the pre-relativity days, I would expect him to be 40 when he gets back. But he will come back younger, because as a clock, he has slowed down. What I think is 20 years could be 10 for him. So, he can come back 10 years younger than me. But what if he says the same thing and expects me to be younger?

Who will be younger? There can only be one answer to that question, and if I can get it one way, it has to be right.

Now, I have no reason to believe that I moved. I have been inertial the whole time, and I predict he will be younger. He cannot say the same, because he must have accelerated: he cannot go away and come back without first speeding up, then slowing down, reversing, speeding up toward home, and finally coming to rest on the earth. During all these periods of acceleration my twin has lost the right to argue he is equal to me.

Coming to the counter-moving clocks: suppose they both read zero when they cross. Now we wait for some time during which they are moving in opposite directions at uniform speed. A direct comparison will be impossible as long as they are both non-accelerated. To compare them, one or both of them have to change directions and of course velocities. If they did this symmetrically, they will agree at the second comparison, but if one was inertial the whole time and the other was not, the latter would be behind. If two of a set of triplets went away in opposite directions and returned on symmetric paths, they will both be younger than the one who stayed behind in the inertial frame. This is not science fiction at all. If you want to be alive for the year 3000, you can do it. You do the math; you find the appropriate speed (nearly that of light) so you will be, say, just 2 years older; you get into the rocket and off you go to a distant star (nearly 500 light years away); and you come back. Fortunately, you cannot foist pictures of your trip on any of your friends because they will be long dead.

Now, this experiment is done all the time with subatomic particles. They are accelerated in Fermilab, for example. They go around in a ring, and, just by virtue of their motion, they live a very long time. Particles that are supposed to have a short lifetime (which you calculate in their own rest frame) live much longer because they are moving.

Summary: Every clock runs the fastest in its own rest frame. (If it is a light clock, the light pulse follows the zigzag path in any other frame and appears to be slow.)

13.4.2 Length contraction

Length contraction refers to the fact that if you and I bought two identical meter sticks, and you set off in a plane or a rocket, I will claim your meter stick is actually shorter than a meter, and you will say the same about mine. In fact, way back, in the derivation of the Lorentz transformation, the fudge factor I calculated, taking your lengths into my lengths, is precisely connected with this factor. If you say the length of an object is something, I will tell you that the length is actually less because your meter sticks are short. And you will say the same about my length measurements.

Let me prove that to you. Let's take a rod that's moving at a speed u. You're carrying the rod, and I want to find its length. What should I do? I'm going to introduce a pair of events to find the length of the rod. Remember this rod is zooming past me. Event 1 is when the front end of the rod hits a certain marking on my graduated x-axis, and event 2 is when the back end of the rod crosses another point. The distance between those two is the length of the rod, provided one condition is met: I must locate the two ends simultaneously. Otherwise, I will screw up, right? If I find the front end now, go on a lunch break and come back, and then locate the other end, I will get a smaller result, even negative, because during this time period the rod has been sliding to the right.

Consequently, the two events have coordinates with $\Delta x = L$, $\Delta t = 0$ in my unprimed frame. In your primed frame the two measurements took place at the two ends of the static rod; hence they are separated by its rest length L_0. Thus

$$L_0 = \frac{L - u \cdot 0}{\sqrt{1 - u^2/c^2}}. \tag{13.27}$$

If you cross-multiply, you get the relation between the rest length L_0 and the length in the frame in which the rod is moving at speed u:

$$L = L_0\sqrt{1 - u^2/c^2}. \tag{13.28}$$

We could equally well consider the inverse transformation, and picture the rod to be at rest in the primed coordinates. In that case we find

$$L = \frac{L_0 + u\Delta t'}{\sqrt{1 - u^2/c^2}} \tag{13.29}$$

$$0 = \frac{\Delta t' + \frac{uL_0}{c^2}}{\sqrt{1 - u^2/c^2}} \tag{13.30}$$

where the second equation signifies that I measured the two ends simultaneously and tells us $\Delta t' = -uL_0/c^2$. Feeding this into the first equation, we find once again that $L = L_0\sqrt{1 - u^2/c^2}$.

So, a rod will appear longest in its rest frame and a clock will appear fastest in its rest frame.

In what sense is this length contraction real? In what sense is a meter stick contracted to half a meter the same as a half-meter stick that is not moving? The answer is that they will both fit in a static case half a meter long: the static half-meter stick for all times, and the moving one for just one instant.

I can say your meter stick is only half a meter long if you move fast enough. But you are inertial, and you can say my meter stick is half a meter long. How then do you explain my result that your stick is half as long as mine instead of the other way around? I will now discuss this paradox.

13.5 More paradoxes

13.5.1 Too big to fall

You and I bought identical rods one meter long, but you and your rod are now moving along the x-axis on top of my infinitely long table, at a velocity u such that $\gamma^{-1} = \sqrt{1 - u^2/c^2} = .5$, as shown in the top part of Figure 13.4. The table has a hole .5 m long. I will find that there will be an instant when the entire rod fits within the hole. But you say, "My rod is a meter long, I am at rest, you and your table are the ones moving to the left at speed u. The hole that you think is .5 meter long actually is .25 meters long. So, there will never be a time when the two ends of my meter stick

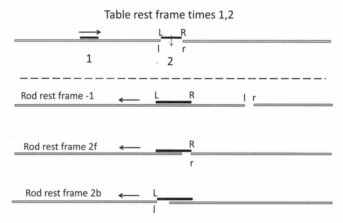

Figure 13.4   At the top are two snapshots (1,2) in my (table) rest frame: the rod has shrunk to .5 meters and is sliding over my infinite table, approaching a hole .5 meters long. At my time 1, the rod is to the left of the hole; at my time 2 it fits right over the hole with its right and left ends (R) and (L) aligned with the right and left ends (r and l) of the hole. In the lower part are three views from the rod (your) rest frame. At some earlier rod time, -1, the table is rushing to the left with a hole .25 meters long. The right end (r) of the hole passes the right end of the rod (R) at rod time $2f$. At a later time $2b$, the left end (l) of the hole passes the left end (L) of the rod. Events $2f$ and $2b$ took place simultaneously at my time 2. If we introduce a weak gravitational field near the hole (tiny vertical arrow), the rod will fall through according to me, and it must do so according to you as well. But how? The answer is provided in the main text.

lie entirely within that hole." When this experiment is performed, will the rod fit in the hole or not? That is the paradox. It is depicted and resolved in Figure 13.4.

Look at times labeled 1 and 2 in the top part of the figure. At my time 1, the rod is approaching the hole. At my time 2, its right and left ends (R and L) line up with the right and left (r and l) ends of the hole. So the rod does fit in the hole according to me. But you do not agree. Your description is shown in the lower parts of the figure. At your (and the rod's) time -1, the table is rushing to the left with a hole .25 meters long. Next, the right end (r) of the hole passes the right end of the rod (R) at your (rod) time $2f$. At a later time $2b$, the left end (l) of the hole passes the left end (L) of the rod. The rod is a full meter long, *and at no time does it lie entirely within the hole.* The events that took place at your times $2f$ and $2b$ took place simultaneously for me at my time 2.

This disagreement can be made more dramatic if we add a tiny gravitational field near the hole, pointing along the small arrow in the uppermost figure. Now the shrunken rod will fall through the hole in my table frame. Then it must also fall through according to you, because there cannot be two answers to whether or not it did. Here is the explanation of how a rod of length 1.0 m can fall through a hole of width 0.25 m, following W. Rindler, *American Journal of Physics* 29:365 (1961). It will help in this discussion to assume, without altering the paradox, that the hole is infinitesimally wider than the shrunken rod.

First, imagine that a trapdoor under the table supports the rod until the left end of the rod crosses the left end of the hole. This will support the rod until it is fully over the hole. Let this crossing of the two left ends define the origin in spacetime: $(x, t) = (x', t') = (0, 0)$.

I will refer to your inertial frame as the rod frame, though you will keep going at velocity u, even if the rod begins falling. You share only the rod's horizontal velocity after it begins to fall.

At $t = 0$, the trapdoor drops down rapidly, allowing the rod to fall unhindered under gravity, with a tiny acceleration a. In the table frame, the entire rod begins to accelerate at time $t = 0$. We may safely employ pre-relativistic, Newtonian theory to describe the downward acceleration under the weak gravitational force. At time $t > 0$, a point on the rod located at x has the vertical coordinate $z(x,t)$ (measured downward):

$$z(x,t) = \frac{1}{2}at^2 \cdot \Theta(t) \text{ where} \tag{13.31}$$

$$\Theta(t) = 0 \text{ if } t < 0 \tag{13.32}$$

$$= 1 \text{ if } t > 0. \tag{13.33}$$

The function $\Theta(t)$ is just a compact way of saying that the downward acceleration exists only for $t > 0$. That $z(x,t)$ is independent of x reflects the fact that all parts of the rod suffer the same acceleration under gravity and that the falling rod remains horizontal.

All we need to do now is to transcribe this result, spacetime point by spacetime point, to your frame, to find out what you will see. Using

$$t = \left(t' + \frac{ux'}{c^2}\right)\gamma \tag{13.34}$$

and the fact that the transverse coordinate of any event is the same for both frames, we find

$$z'(x',t') = z(x,t) = \frac{1}{2}a\left(t' + \frac{ux'}{c^2}\right)^2 \gamma^2 \cdot \Theta\left(\gamma\left(t' + \frac{ux'}{c^2}\right)\right) \tag{13.35}$$

where (x',t') and (x,t) correspond to the same spacetime point.

Though all points in the rod begin falling at the same time $t = 0$ in the table frame, *these events are not simultaneous in the rod frame*. While the left end begins falling at $t' = 0$, the rest of the rod begins falling at an *earlier* (negative) time, which I will call the *drop-time*, $t'_d(x')$ that depends on x'. Equation 13.34 tells us that $t = 0$ corresponds to

$$t'_d(x') = -\frac{ux'}{c^2}. \tag{13.36}$$

As expected, the left end $x' = 0$ starts dropping at $t' = 0$. As we move toward the right end, the drop-time turns more and more negative. The right end itself is seen to be the first one to start falling, at time

$$t'_d(x' = L_0) = -\frac{uL_0}{c^2},$$ (13.37)

where L_0 is the rest length of the rod.

To avoid negative times, let us define a shifted time in your frame:

$$t'' = t' + \frac{uL_0}{c^2}.$$ (13.38)

Now the right end starts falling at your time $t'' = 0$, the left end starts falling after $t'' = \frac{uL_0}{c^2}$, and intermediate points begin falling at intermediate times.

In your frame, the falling rod is not straight. In the time interval $0 < t'' < \frac{uL_0}{c^2}$, the rod is horizontal with $z' = 0$ from its left-most end up to some *drop-point* x'_d, and it droops downward beyond. We determine the drop-point x'_d as a function of t'' by turning to the same Eqn. 13.34, which previously gave us the drop-time $t'_d(x')$ as a function of x'. We find

$$x'_d(t') = -\frac{c^2 t'}{u} = -\frac{c^2}{u}\left[t'' - \frac{uL_0}{c^2}\right] = L_0 - \frac{c^2 t''}{u}.$$ (13.39)

At $t'' = 0$, the whole rod is straight, while at $t'' = uL_0/c^2$, even the left end has begun falling.

In your frame, the right edge of the hole moves leftward at speed u, and it is always at $z' = 0$. Since it was to the right of the right end of the rod to begin with, it will never overtake the drop-point, which moves leftward at a speed $\frac{c^2}{u}$. Since $\frac{c^2}{u} > c$, by the time the right edge of the hole crosses any point on the rod, that point will have already fallen below $z' = 0$. (That $\frac{c^2}{u} > c$ does not imply interactions or signals propagating at superluminal velocities. The points that make up the rod fall independently of each other and do not interact with each other in our treatment.)

If this discussion leads you to conclude that the notion of a rigid body does not exist in relativity, you are correct.

In summary, you and I can accuse each other of using shortened rods, and we can both be right. This is because the operational way to find the length of a moving body involves measuring the two ends simultaneously, and simultaneity is relative: you will say you measured the two ends of my rod at the same time and I will disagree, and vice versa.

13.5.2 *Muons in flight*

Muons are particles produced in the upper atmosphere and detected on earth. After an (average) lifetime of $2.2\mu s$ in their rest frame, they decay into an electron and a neutrino-antineutrino pair. Their lifetime multiplied by c equals roughly 660 meters, far short of the distance between their birth in the upper atmosphere and their detection on earth. How then do they make it to the ground? The answer depends on the point of view. Earthlings will say the muon lifetime of $2.2\mu s$ in its rest frame is prolonged by relativistic time dilation. In other words, the birth and death of the muon will coincide with successive ticks of a clock of period $2.2\mu s$ in the muon rest frame, and this clock will appear slowed to us on earth. The muon, on the other hand, will agree that it lives only $2.2\mu s$, but it will claim that the distance between the upper atmosphere and the ground is a lot less than we claim it is, due to length contraction. The dying muon is like a clock that has slowed down according to us, and the atmosphere is like a rod that has Lorentz contracted according to the muon.

CHAPTER 14

Special Relativity III:

Past, Present, and Future

In this chapter we continue to explore the consequences of the Lorentz transformation.

14.1 Past, present, and future in relativity

Let's take the equation for the time difference between two events numbered 1 and 2:

$$\Delta t' = \frac{\Delta t - \frac{u\Delta x}{c^2}}{\sqrt{1 - \frac{v^2}{c^2}}}.$$

(14.1)

First, something happens; then, something else happens, and $\Delta t = t_2 - t_1$ is a separation in time between them. Let's say $\Delta t > 0$ so that the second event occurred after the first event, according to me, using unprimed coordinates. How about according to you? Well, $\Delta t'$ doesn't have to have the same sign as Δt because you subtract from it this number, $\frac{u\Delta x}{c^2}$, which can be arbitrarily large and positive. Therefore, you can find that $\Delta t'$ could be negative. Surely you understand that's a big deal. I say this happened first and that happened later. Then, you say, no, it happened the opposite way. Now, this can lead to serious logical contradictions, especially if event 1 is the cause of event 2. Here is a standard example both in special and general relativity that people talk about: event 1 is the birth of some kid's grandmother, and event 2 is the kid's birth. The birth of the kid takes place long after the birth of the grandmother, according to me. What if, from some other point of view, the kid is born, but the grandmother is not yet born, and something is done to prevent her from being born. Where did the grandchild come from? Or consider this: event 1, I fired a bullet. Event 2, somebody is hit. And you go to the frame of reference in which the person has been hit, and I haven't fired the bullet. Now, you come and you finish me off. So now, we have a person wounded for no apparent reason because the cause (my firing) has been eliminated. That simply cannot happen. Einstein recognized that if A can be the cause of B, we better not find an observer for whom these events occur in reverse order, because if the cause occurs after the effect, then there is some time left for somebody to prevent the cause itself from happening, and we will have an effect with no perceivable cause. So, we want to make sure that the sign of $\Delta t'$ cannot be reversed whenever the first event is or *could have been* the cause of the second event. So, you ask Eqn. 14.1: if Δt is positive, when will $\Delta t'$ be negative? That is simple algebra; you want the second term in the numerator to beat the first term. That will happen in any frame with velocity u obeying

$$\frac{u\Delta x}{c^2} > \Delta t \tag{14.2}$$

$$\frac{u}{c} > \frac{c\Delta t}{\Delta x}. \tag{14.3}$$

Now compare the two distances:

- $c\Delta t$, the distance a light pulse can travel in the time Δt between the two events.
- Δx, the actual spatial separation between the events.

If $c\Delta t > \Delta x$, there is enough time for a light signal to go from event 1 to event 2. In this case the frame we are looking for is going faster than light: $\frac{u}{c} > 1$. Such a frame of course does not exist. So the order of events is the same for all possible observers. But this is the case when event 1 could have been the cause of event 2 because a light signal could have been used to cause event 2. By the same logic if $c\Delta t < \Delta x$, there isn't enough time for a light signal to go from event 1 to event 2. In this case the frame we are looking for moves at $\frac{u}{c} < 1$. Such a frame of course exists. So the order of events is different, with 2 occurring before 1. But this is the case when event 1 better not have been the cause of event 2. This is assured if the maximum speed with which one can influence other events is the speed of light. Because this pair of events is separated by a time that is too short even for a light signal to go from 1 to 2 in time, they could not have been causally connected.

In other words, we are going to say that if there isn't enough time for a light signal to go from event 1 to event 2, then event 1 could not have been the cause of event 2. Therefore, for the theory to make logical sense, no signal should travel faster than the speed of light.

When we heard rumors in 2011 about neutrinos traveling faster than light, most of us did not believe them. This is not because we believe in Einstein's infallibility (even he did not), but because we believe in the causal structure in our world. If it is absent, if events are not related by cause and effect, there is no point looking for laws correlating them.

Our current view of spacetime is shown in Figure 14.1. One axis is of course the x-axis. Along the other we measure ct rather than t so that both coordinates have the same dimension, a natural requirement if we want to treat space and time as symmetrically as possible. (Any velocity besides c would do the trick but would be unnatural, given the unique role played by c in the theory. It would also lead to inelegant formulas, polluted by the presence of this arbitrarily chosen velocity.)

Event 1 shows me at the origin of spatial coordinates when my clock says zero; I'm at $(0,0)$ in spacetime. In the Newtonian world, any point with $t > 0$, say event 2 or 4, is said to be in my absolute future, and any event with $t < 0$, say event 3 or 5, is called my absolute past, and all points on the x-axis with $t = 0$ are called "present." These labels are absolute because all observers will agree on the order (including simultaneity) of these events. Events in my future can be affected by me. For example, if I decide at event 1 that I want an explosion to occur at 4 or 2, I can make that happen. Events 3 and 5 are in my past—someone at 5, for example, can decide to harm me at 1 and do it. Thus, in the Newtonian world, we have three regions: future, past, and present, all labels being absolute, valid in any other frame.

After Einstein, we draw new lines at $x = \pm ct$, which describe light signals traveling past 1 in both directions. Thus, for example, the line $x = ct$ describes a light signal that originated to my left, reached me at $t = 0$, and kept going to the right for $t > 0$.

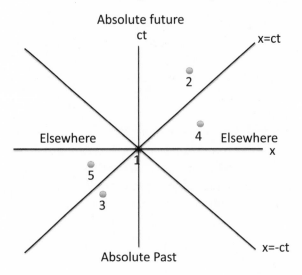

Relativistic spacetime

Figure 14.1   Relative to some event at the origin $(0,0)$, Newtonian spacetime is divided into the absolute future (points with $ct > 0$), absolute past ($ct < 0$), and absolute present $ct = 0$, where "absolute" means for all observers. Relativistic spacetime, depicted in the figure, is very different. I am at the origin $(0,0)$, named event 1. Event 2 lies to my absolute future, meaning that I can influence it and it will occur later according to all observers. Event 3 lies in my absolute past; it can affect me here and now at 1 and all observers will say it occurred earlier. Event 4 is in my future but not my absolute future, meaning I can find someone for whom it occurs before 1. However, no logical paradoxes will arise because I cannot affect 4 by using any signal that travels at or less than the speed of light. Likewise, event 5 is in my past but not absolute past; I can find someone who says it happened after 1.

Consider now event 2. Because $ct > x$, there is enough time for a light signal to go from 1 to 2. Not only is event 2 in the future of 1 according to me, it will be that way according to anybody else. In other words, Δt is positive for me, but, if you go back to Eq. 14.1, because $ct > x$, you will never find anybody who says 2 occurred earlier than 1. Another way to say it is that sitting at 1, I can make event 2, say an explosion, happen by sending a signal that travels slower than light. Therefore, 1 could have caused 2. And therefore, there is no messing around with the order of these events. Note that all we ask is that 1 *could have been the cause of* 2 and not that it actually was. If it could have been the cause, the theory automatically makes 1 occur before 2 for all observers. By this logic, all events lying above the 45° lines $x = \pm ct$ are called my *absolute future*. By "absolute," I mean lying in the future of 1 not only according to me but according to all observers. It may not be later by the same number of seconds, but it will be later. Events in the absolute future lie in the "forward light cone." I use the word *cone* because in higher spatial dimensions these points will lie inside a cone.

Similar ideas apply to points in the backward light cone, labeled *absolute past* in the figure. Thus event 3 could have been the cause of what's happening to me right now at 1, because from that event, a signal could have been sent to arrive where I am right now at a speed less than c.

Now consider events outside the light cone, called "elsewhere," such as 4. Suppose at event 1, I open an envelope and it says something terrible is going to happen at 4. Say the distance to 4 is two seconds times the velocity of light, and it is to occur one second later. There is nothing I can do, even though it has not happened yet (according to me). In the Newtonian days, there was something I could have done: tell someone else to really hurry up and get there and do something. But now I cannot do that because that would require that person to travel faster than light, and that's not allowed; it is outside the light cone. So, even though you know someone is planning to do something evil there, you cannot avert it. That's a very important thing for people who are going to law school. You know we got the DNA defense from biology, right? "My client's DNA doesn't match; you must acquit." Here's another defense from physics. If your client was accused of doing something at 4, and he was last seen at 1, you can argue that "My client was outside the light cone." The "outside the light cone defense" is absolutely watertight. If an event is outside your client's light cone, the client cannot be held responsible. Your client would have had to send a signal faster than light, and every jury knows that is impossible. As for an event like 4, the status is that I can actually find other observers moving at a speed less than light for whom 4 occurs before 1. So, the order of these events can be reversed. But it will not lead to logical contradictions because we know the two events could not have been causally connected. Similarly, 3 is to my absolute past, so it could be responsible for all the trouble I am having right now, but 5 cannot be blamed for what is happening to me at 1.

So, spacetime, which we used to divide into the upper half plane and lower half planes, the future and past, separated by a line called present or now, is now divided into three regions: the absolute future that you can affect, the absolute past that can affect you, and "elsewhere," which contains events that cannot affect you and cannot be affected by you (sitting at $(0, 0)$).

The consequences of the Lorentz transformation, whose equations are deceptively simple, a lot simpler than some equations for angular momentum, are really stupendous. If you invent a theory like this, you have to make sure that there are no contradictions. When you notice that the order of events can be reversed, you might panic. But the theory is so beautiful, it is so internally consistent; it says you can reverse the order of events only if they could not have been causally connected. Within the theory, "not being causally connected" means a signal would have to travel faster than light to go from the first event to the second.

14.2 Geometry of spacetime

Now we turn to something that is mathematically very pretty and physically rather profound.

The coordinates of a point are not sacred. They are dependent on who is looking at them and from what orientation. Recall the relation between new coordinates (x', y') and the old (x, y) when we rotate the axes:

$$x' = x\cos\theta + y\sin\theta \tag{14.4}$$

$$y' = -x\sin\theta + y\cos\theta. \tag{14.5}$$

But even in this world, something is sacred: $x'^2 + y'^2 = x^2 + y^2$. Namely, the distance from the origin (or more generally, the distance between two points) is unaffected by rotations. This distance is called an *invariant*. The same goes for the dot product between two vectors, because it involves their lengths and the angle between them, none of which is affected by the rotation of axes.

So, it is reasonable to ask, in the relativistic case, where people cannot agree on the time coordinate or space coordinate, whether the square of the time coordinate plus the square of the space coordinate will be the same for two people. We find out that that's not the case; $x^2 + t^2$ is not invariant. But even before you do that, you should shudder at the prospect of writing something like this. You cannot add t^2 and x^2, because they have different dimensions. We have to have both coordinates in spacetime with units of lengths or time. The standard trick is to introduce an object with two components with the same dimensions. I'm going to call the object X. The first component of X is going to be called x_0 which is just ct. The second is going to be called x_1, which is just our familiar x. In four dimensions you have $X = (x_0, x_1, x_2, x_3)$, which is sometimes written as

$$X = (x_0, \mathbf{r}) = (ct, x, y, z) \tag{14.6}$$

where $\mathbf{r}$ is the usual three-dimensional position vector. Why do I switch from ct and x to x_0 and x_1? If you're doing superstrings, you need ten coordinates: one will be x_0, the time, and the other nine will be spatial coordinates, $x_1 \ldots x_9$. Numerical indices rather than letters of the alphabet are preferred, because we may run out of letters but will never run out of numbers. It is also more natural to sum over a numerical index than letters.

What does the Lorentz transformation look like when I write it in terms of the components of X? I leave it to you to check that in terms of the *dimensionless velocity* β

$$x_0' = \frac{x_0 - \beta x_1}{\sqrt{1 - \beta^2}} \tag{14.7}$$

$$x_1' = \frac{x_1 - \beta x_0}{\sqrt{1 - \beta^2}} \tag{14.8}$$

$$\beta = \frac{u}{c}. \tag{14.9}$$

Now, you see the relationship is nice and symmetric. If you write it in terms of x and t, the coordinates transformation law is asymmetric because one has units of length and one has units of time. As for the other transverse coordinates in the y and z directions

$$x_2' = x_2 \quad x_3' = x_3. \tag{14.10}$$

In other words, the length perpendicular to the motion is something you can always agree on.

Let's return to the question of whether $(x_0')^2 + (x_1')^2 = x_0^2 + x_1^2$ as in rotations. It is not. However, the following is true:

$$(x_0')^2 - (x_1')^2 = x_0^2 - x_1^2 = s^2 \tag{14.11}$$

where s^2 is called the *spacetime interval*.

Let us verify this fundamental result starting with Eqns. 14.7 and 14.8:

$$(x_0')^2 - (x_1')^2 = \frac{(x_0 - \beta x_1)^2 - (x_1 - \beta x_0)^2}{1 - \beta^2}$$

$$= \frac{(x_0^2 - x_1^2)(1 - \beta^2)}{1 - \beta^2}$$

$$= x_0^2 - x_1^2. \tag{14.12}$$

In terms of more familiar quantities this becomes

$$(ct')^2 - (x')^2 = (ct)^2 - x^2. \tag{14.13}$$

If we consider the difference between two events (not necessarily infinitesimal), they also obey

$$(\Delta x_0')^2 - (\Delta x_1')^2 = \Delta x_0^2 - \Delta x_1^2 = \Delta s^2 \tag{14.14}$$

because coordinate differences transform just like the coordinates.

This result says that even though people cannot agree on the time or space coordinates of an event or time or space coordinate differences between two events, they can agree on this quadratic function of the coordinates or differences, which we called s^2. It is an example of an *invariant*, a name given to something unchanged under a given transformation (which here is the Lorentz transformation). Note that despite the notation, s^2 is not always positive. You can verify that s^2 is positive inside the light cones when it is described as *time-like*; s^2 is negative outside the light cone when it is referred to as *space-like*; and it vanishes on the light cone, when it is alluded to as *light-like*.

While in ordinary rotations you take the *sum of the squares*, here you must take the *difference of the squares*. And that's just the way it is. Even though time is like another coordinate that mixes with space, it's not quite the same. You can go forward or backward in the spatial direction, but not in the time direction. The three-dimensional space in which we live is Euclidean, and the invariant distance is given by the sum of the squares of all the coordinates. Spacetime is not Euclidean.

Including the transverse coordinates (perpendicular to the velocity u), s^2 becomes

$$s^2 = (x_0')^2 - (x_1')^2 - (x_2')^2 - (x_3')^2 = x_0^2 - x_1^2 - x_2^2 - x_3^2. \quad (14.15)$$

Most of the time, I'm not going to worry about transverse coordinates.

14.3 Rapidity

Just as trigonometric functions naturally enter the formulas relating old and new coordinates under rotations in the $x - y$ plane, *hyperbolic functions* are tailor-made for the Lorentz transformation. If you already know these functions you will appreciate their utility, and if not, you can use this opportunity to learn about them here, in which case I urge you to fill in the missing proofs of various identities. You can also skip this section without loss of continuity.

Let us begin with some properties of these functions and then turn to their application. The hyperbolic functions $\sinh \theta$ (pronounced "cinch" θ) and $\cosh \theta$ are the analogs of $\sin \theta$ and $\cos \theta$, and defined as follows in terms of $e^{\pm \theta}$:

$$\cosh \theta = \frac{e^\theta + e^{-\theta}}{2} \quad (14.16)$$

$$\sinh \theta = \frac{e^\theta - e^{-\theta}}{2}. \quad (14.17)$$

A key identity follows from this definition by straightforward computation:

$$\cosh^2 \theta - \sinh^2 \theta = 1. \quad (14.18)$$

Note the minus sign compared to $\cos^2 \theta + \sin^2 \theta = 1$. The addition formulas that also follow from the definitions are

$$\cosh(A + B) = \cosh A \cosh B + \sinh A \sinh B \quad (14.19)$$

$$\sinh(A + B) = \sinh A \cosh B + \cosh A \sinh B. \quad (14.20)$$

The hyperbolic analog of $\tan \theta$, pronounced "tanch" θ, is:

$$\tanh \theta = \frac{\sinh \theta}{\cosh \theta} = \frac{e^{\theta} - e^{-\theta}}{e^{\theta} + e^{-\theta}}. \tag{14.21}$$

It vanishes as $\theta \to 0$ and tends to ± 1 as $\theta \to \pm\infty$.
The addition formula

$$\tanh [A + B] = \frac{\sinh(A + B)}{\cosh(A + B)} = \frac{\tanh A + \tanh B}{1 + \tanh A \tanh B} \tag{14.22}$$

follows from the addition formulas Eqn. 14.19 and 14.20 for sinh and cosh.
Now for the application to relativity, starting with the velocity addition formula. We have seen that if an object has velocity w according to you and you are moving with a velocity u with respect to me, its velocity with respect to me is not $v = w + u$ as in non-relativistic physics but

$$\frac{v}{c} = \frac{\frac{w}{c} + \frac{u}{c}}{1 + \frac{u}{c}\frac{w}{c}} \tag{14.23}$$

where the denominator $1 + \frac{uw}{c^2}$ plays a crucial role in keeping velocities from exceeding c.
Suppose we express the dimensionless velocities $\frac{u}{c}, \frac{v}{c}$, and $\frac{w}{c}$ in terms of the *rapidities* θ_u, θ_v, and θ_w defined as follows:

$$\frac{u}{c} = \tanh \theta_u \tag{14.24}$$

$$\frac{v}{c} = \tanh \theta_v \tag{14.25}$$

$$\frac{w}{c} = \tanh \theta_w. \tag{14.26}$$

Observe that both sides lie in the range $[-1, +1]$. You can invert these formulas to find the θ corresponding to some dimensionless velocity. For example, in the case of $\beta = \frac{u}{c}$:

$$\theta_u = \ln \sqrt{\frac{1 + \beta}{1 - \beta}}. \tag{14.27}$$

The relativistic velocity addition formula now takes the form

$$\tanh \theta_v = \frac{\tanh \theta_w + \tanh \theta_u}{1 + \tanh \theta_w \tanh \theta_u} = \tanh \left[\theta_w + \theta_u \right] \qquad (14.28)$$

by virtue of Eqn. 14.22. This leads to the remarkably simple result

$$\theta_v = \theta_w + \theta_u. \qquad (14.29)$$

In other words, rapidities simply add when compounded, unlike velocities. Thus if you see an object moving at rapidity θ_w, and you are moving with respect to me with a rapidity θ_u, I will ascribe to the object a rapidity $\theta_v = \theta_w + \theta_u$. While this means rapidities can grow without limit when we jump from frame to frame, the property $|\tanh \theta| \leq 1$ ensures that the corresponding velocity never exceeds c.

Now consider the Lorentz transformation. Given that

$$\frac{u}{c} = \tanh \theta_u \qquad (14.30)$$

$$= \frac{\sinh \theta_u}{\cosh \theta_u} \qquad (14.31)$$

$$= \frac{\sqrt{\cosh^2 \theta_u - 1}}{\cosh \theta_u} \qquad (14.32)$$

we can readily invert the above to obtain the following expressions for $\cosh \theta_u$ and $\sinh \theta_u$:

$$\cosh \theta_u = \frac{1}{\sqrt{1 - \frac{u^2}{c^2}}} = \frac{1}{\sqrt{1 - \beta^2}} \qquad (14.33)$$

$$\sinh \theta_u = \frac{u/c}{\sqrt{1 - \frac{u^2}{c^2}}} = \frac{\beta}{\sqrt{1 - \beta^2}}, \qquad (14.34)$$

which satisfy $\cosh^2 \theta_u - \sinh^2 \theta_u = 1$.

We may rewrite the Lorentz transformation, Eqns. 14.7 and 14.8, in terms of θ_u as

$$x_0' = x_0 \cosh \theta_u - x_1 \sinh \theta_u \tag{14.35}$$

$$x_1' = -x_0 \sinh \theta_u + x_1 \cosh \theta_u, \tag{14.36}$$

which are analogous to the formulas for rotations. The invariance of the spacetime interval

$$(x_0')^2 - (x_1')^2 = x_0^2 - x_1^2 = s^2 \tag{14.37}$$

follows from $\cosh^2 \theta - \sinh^2 \theta = 1$, just as $x'^2 + y'^2 = x^2 + y^2$ follows from $\cos^2 \theta + \sin^2 \theta = 1$.

Suppose a Lorentz transformation parametrized by θ_1 relates X' to X and then a second one involving θ_2 relates X'' to X'. If we now directly express X'' in terms of X by eliminating X', we will find that it is given by a Lorentz transformation with rapidity $\theta_1 + \theta_2$. (This will require using the addition formulas Eqns. 14.19 and 14.20.) Again this just means that rapidities add under a sequence of two Lorentz transformations just like rotation angles add under a sequence of two rotations.

Even a cursory reading of this section should impress upon you how marvelously the inventions of mathematicians fulfill and often anticipate the needs of the physicist.

14.4 Four-vectors

Let me introduce some notation now. Our friend $X = (x_0, x_1, x_2, x_3) = (x_0, \mathbf{r})$ is the first example of a *four-vector*. Later we will see other four-vectors with four components each. If $A = (a_0, a_1, a_2, a_3)$ is a four-vector, its components will transform like those of X:

$$a_0' = \frac{a_0 - \beta a_1}{\sqrt{1 - \beta^2}} \tag{14.38}$$

$$a_1' = \frac{a_1 - \beta a_0}{\sqrt{1 - \beta^2}} \tag{14.39}$$

$$a_3' = a_3 \tag{14.40}$$

$$a_4' = a_4 \tag{14.41}$$

if the relative velocity is in the a_1 direction.

This transformation law is what defines a four-vector, more precisely a four-vector under Lorentz transformations, just as **r** transforms like a three-vector under ordinary rotations. Just as we took the three components of space and added one more of time to form the position four-vector X, every four-vector we encounter will be the union of a three-vector like **r** and a fourth component that, like time, is unaffected by ordinary rotations but gets mixed up with the others under Lorentz transformations.

Now we're going to define a *dot product of two four-vectors A and B*:

$$A \cdot B = a_0 b_0 - a_1 b_1. \tag{14.42}$$

I leave it to you to show that in a different frame

$$A' \cdot B' = a_0' b_0' - a_1' b_1' = a_0 b_0 - a_1 b_1 = A \cdot B \tag{14.43}$$

for the same reason that $X \cdot X$ was invariant under the Lorentz transformation.

14.5 Proper time

Now I am going to apply the notion of the spacetime interval to the study of a single particle. Previously, Δx and Δt were separations between two random, unrelated or arbitrary events. But now I want you to consider the following events. A particle moves in spacetime from point 1 to point 2, traveling a distance dx in time dt according to some generic observer. So, these are two events in the life of a particle, lying on its trajectory. Let's look at the infinitesimal spacetime interval $ds^2 = c^2 dt^2 - dx^2$ between the two events. We're going to rewrite this as follows:

$$ds^2 = c^2 dt^2 - dx^2 = (cdt)^2 \left(1 - \frac{1}{c^2} \left(\frac{dx}{dt} \right)^2 \right) \tag{14.44}$$

$$ds = cdt \sqrt{1 - \frac{v^2}{c^2}} \tag{14.45}$$

where $\frac{dx}{dt} = v$ is *the velocity of the particle*. Because ds is an invariant, the same for all observers, let us calculate it according to the particle itself. The particle does not think it is moving and so it sets $v = 0$. So the two events, particle sighted here and particle sighted there, have different x

coordinates for a generic observer, but they are the same point according to the particle. However, the particle does think some time $d\tau$ has elapsed. Therefore, in terms of the $d\tau$ of the particle

$$ds = c\,d\tau. \qquad (14.46)$$

The time τ, measured by a clock moving with the particle, is called *proper time*. In other words, if the particle had its own clock, $d\tau$ is the time it would say elapsed between these two successive points in its trajectory. It's not hard to understand why everybody agrees on proper time. You and I don't have to agree on how much time elapsed between when the particle was here and when it was there. But if we ask how much time elapsed *according to the particle*, we are all asking the same question, and we all get the same answer, $d\tau$.

Remember for future use that the proper time difference $d\tau$ between two events in the trajectory of a particle is related to the time difference dt according to an observer who ascribes to the particle a velocity v as follows:

$$d\tau = dt\sqrt{1 - \frac{v^2}{c^2}} \quad \text{which means} \qquad (14.47)$$

$$\frac{dt}{d\tau} = \frac{1}{\sqrt{1 - v^2/c^2}}. \qquad (14.48)$$

CHAPTER 15

Four-momentum

In Newtonian mechanics particles have coordinates, let us say (x, y), which could vary with time. From these we form a two-dimensional vector $\mathbf{r} = \mathbf{i}x + \mathbf{j}y$. In a rotated frame the components (x', y') are given by

$$x' = x\cos\theta + y\sin\theta \tag{15.1}$$

$$y' = -x\sin\theta + y\cos\theta. \tag{15.2}$$

An entity $\mathbf{V}$ is defined as a vector (in two dimensions) if it has two components (V_x, V_y) that, under rotation of the axes, go into V'_x and V'_y, related to (V_x, V_y) exactly as in Eqn. 15.1 and 15.2.

Now, I say to you, "Okay, that's one vector, the position vector $\mathbf{r}$. Can you point to another vector?" You might suggest the velocity $\mathbf{v} = \frac{d\mathbf{r}}{dt}$ as an answer. That is right, but why does taking the derivative of a vector produce another vector? Well, what's the derivative? You change the vector by some $\Delta\mathbf{r}$, and you divide it by the time difference Δt. Now, the change in the vector is obviously a vector, because the difference of two vectors is a vector. Dividing by time is like multiplying by the reciprocal of the time. That's like multiplying by a number or a scalar, because time does not respond to rotations. And I've told you multiplying a vector by a number also gives you a vector, maybe longer or shorter. Therefore, $\Delta \mathbf{r}/\Delta t$ is a vector because it's the difference of $\mathbf{r}$ later minus $\mathbf{r}$ now, and dividing by a tiny Δt is the same as multiplying by a large number—10,000 or 1,000,000, it doesn't matter. And the limit $\mathbf{v} = \frac{d\mathbf{r}}{dt}$ is also a vector. Therefore, when you take a derivative of a vector with respect to a parameter like time, *which does not respond to rotations*, you get a vector. Once you get this derivative, it becomes addictive. You can take second derivatives, and you get the acceleration $\mathbf{a}$. Then you can multiply that by mass or inertia, which is postulated to be a scalar or invariant under rotations. Thus the product $m\mathbf{a}$ is a vector, which Newton equates to the force $\mathbf{F}$, also a vector.

What I want to do is generate more four-vectors using this idea of starting with the position four-vector X. Now, take this X to be the coordinate in *spacetime* of an object that's moving. I want to take the derivative of that to get myself something I could call the velocity vector in relativity. But the derivative cannot be the time derivative. I can of course take the time derivative, but the time derivative of a vector in four dimensions is not a vector, because time is like any other component. It's like taking the y derivative of x for a moving particle to get the velocity. That doesn't give you a vector. You have to take a derivative with respect to something that does not transform under the Lorentz transformation, something that does not change from one observer to the other. Do you have any idea where I'm going with this? I mean, you can take the derivative with respect to τ, the time as measured by the particle. So, I'm going to form a new quantity V called *four-velocity*,

$$V = \frac{dX}{d\tau} = \left(\frac{dx_0}{d\tau}, \frac{dx_1}{d\tau} \right). \tag{15.3}$$

By construction, V will be a four-vector. By that I mean, its four components will transform when you go to a moving frame just like the four components of X. But we do not have a good intuition for the τ-derivative. While dx and dt are the separations measured by a generic observer, $d\tau$ is the time elapsed according to the particle. So we rewrite Eqn. 15.3 for V in terms of quantities that are directly measured by the generic observer. To this end, we replace τ-derivatives by t-derivatives by invoking Eqn. 14.48:

$$V = \frac{dX}{d\tau} = \frac{dX}{dt} \cdot \frac{dt}{d\tau} \tag{15.4}$$

$$= \frac{1}{\sqrt{1 - v^2/c^2}} \frac{dX}{dt} \tag{15.5}$$

$$= \frac{1}{\sqrt{1 - v^2/c^2}} \left(c\frac{dt}{dt}, \frac{dx}{dt} \right) \tag{15.6}$$

$$= \left(\frac{c}{\sqrt{1 - v^2/c^2}}, \frac{1}{\sqrt{1 - v^2/c^2}} \frac{dx}{dt} \right) \tag{15.7}$$

$$= \left(\frac{c}{\sqrt{1 - v^2/c^2}}, \frac{\mathbf{v}}{\sqrt{1 - v^2/c^2}} \right), \tag{15.8}$$

where the last equation applies when all four dimensions are exhibited.

The four-velocity has an unusual feature: its "length squared"

$$V \cdot V = V_0^2 - V_1^2 = c^2 \tag{15.9}$$

does not depend on how fast the particle is moving! You can verify this the hard way, by computing $V_0^2 - V_1^2$ starting with Eqn. 15.7, or the easy way by evaluating the invariant in a frame in which the particle is at rest and $V = (c, 0)$.

Now we are ready to define the four-momentum P as the mass m times the four-velocity V:

$$P = m\frac{dX}{d\tau} \tag{15.10}$$

$$= \left(\frac{mc}{\sqrt{1 - v^2/c^2}}, \frac{m\mathbf{v}}{\sqrt{1 - v^2/c^2}} \right) = (p_0, p_1) \tag{15.11}$$

in two dimensions

$$= \left(\frac{mc}{\sqrt{1 - v^2/c^2}}, \frac{m\mathbf{v}}{\sqrt{1 - v^2/c^2}} \right) = (p_0, p_1, p_2, p_3) \tag{15.12}$$

in four dimensions.

For this to be a four-vector, m should be the same in all frames, that is, invariant under Lorentz transformations.

So, we have manufactured a new beast with four components. What is it? Let me keep just the component of velocity in the x direction, but still call it a four-vector.

Consider first

$$p_1 = \frac{mv}{\sqrt{1-v^2/c^2}}. \tag{15.13}$$

For $\frac{v}{c} << 1$, we find (upon setting the denominator to unity)

$$p_1 = mv. \tag{15.14}$$

Thus we conclude that p_1 stands for the momentum of the particle in the relativistic theory. However, if the particle picks up speed, we need to take into account the denominator. As $v \to c$, p_1 grows without limit: that is, in this theory, there is a limit to the particle velocity but not its momentum.

Some people like to write

$$p_1 = \left(\frac{m}{\sqrt{1-v^2/c^2}} \right) v \equiv m(v)v \tag{15.15}$$

where $m(v) = m/\sqrt{1-v^2/c^2}$ is a new velocity dependent mass. They also refer to $m(0) = m$ as the rest mass m_0. Their point is that if you introduce a velocity dependent mass, then momentum can still be mass times velocity as in the old days. We will not do that: for us m is always the rest mass, and momentum is now a more complicated function of this mass and velocity.

Suppose $\frac{v}{c}$ is small but not utterly negligible. Then we can use a slightly better formula for momentum by using

$$(1+x)^n = 1 + nx + \ldots$$

for $x << 1$, to write

$$\frac{1}{\sqrt{1-v^2/c^2}} = \left[1 - \frac{v^2}{c^2} \right]^{-1/2} = 1 + \frac{v^2}{2c^2} + \ldots \tag{15.16}$$

and

$$p_1 = mv + m\frac{v^3}{2c^2} + \ldots \tag{15.17}$$

where the ellipses (dots) stand for even smaller corrections we are ignoring. We can calculate more such correction terms or simply use the exact expression with the $\sqrt{1 - \frac{v^2}{c^2}}$ in it.

It is also clear that if we bring in p_2 and p_3 we just get the vector

$$\mathbf{p} = \frac{m\mathbf{v}}{\sqrt{1 - v^2/c^2}}. \tag{15.18}$$

What does the 0-th component

$$p_0 = \frac{mc}{\sqrt{1 - v^2/c^2}} \tag{15.19}$$

stand for? If we set $v = 0$ we get the mass of the particle times c, which is a constant. Let us go to the next level of approximation and write as before,

$$p_0 = mc + \frac{1}{2c}mv^2 + \ldots \tag{15.20}$$

We see that if we multiply both sides by c finally something familiar emerges:

$$cp_0 = mc^2 + \frac{1}{2}mv^2 + \ldots \tag{15.21}$$

We see that the second term on the right is just the non-relativistic kinetic energy. So it must be that all higher powers of $\frac{v}{c}$ shown by the ellipsis stand for corrections to kinetic energy as we consider faster particles.

But it must then be that the first term mc^2 also stands for energy, but of a particle at rest. This is called its *rest energy*. Einstein did not tell us how to extract this energy (in contrast to kinetic energy of motion, which can be extracted, say, in hydroelectric power generators using turbines that slow down the water). Later on, when people discovered fusion or fission, they found that some amount of mass was missing at the end of the reaction and that this missing mass (upon multiplying by c^2) exactly equaled the additional kinetic energy of the final particles. For this reason P is called the *energy-momentum or momentum four-vector*.

To summarize, we have seen two four-vectors:

$$X = (x_0, x_1) = (ct, x) \quad \text{the position four-vector} \tag{15.22}$$

$$P = (p_0, p_1) = \left(\frac{E}{c}, p \right)$$

the energy-momentum or momentum four-vector

(15.23)

Here are some consequences of P being a four-vector.

- *The components of P transform as follows* when we go from one frame to another:

$$p_0' = \frac{p_0 - \beta p_1}{\sqrt{1 - \beta^2}} \tag{15.24}$$

$$p_1' = \frac{p_1 - \beta p_0}{\sqrt{1 - \beta^2}} \tag{15.25}$$

$$p_2' = p_2 \tag{15.26}$$

$$p_3' = p_3 \tag{15.27}$$

where $\beta = \frac{u}{c}$. We will ignore p_2 and p_3 from now on.

It will be instructive for you to verify that this transformation law is in accord with another way of finding p_0' and p_1'. To this end, consider a particle moving at speed v as seen by the unprimed observer. Write explicit formulas for p_0 and p_1 in terms of m and v and keep them handy. Now ask what velocity w this particle will have according to the observer moving at velocity u. Write down her expressions for p_0' and p_1' in terms of w, and now write w in terms of u and v. Check that the result agrees with what is above. To save paper you may set $c = 1$ in this exercise.

- $P_A \cdot P_B$ *is invariant, where P_A and P_B are any two four-momenta, which could refer to two different particles A and B or the same particle, in which case $A = B$.*

First consider just one particle and the value of $P \cdot P$. Because it can be found in any frame, find it in its own co-moving frame. In this case $p_1 = 0$ and $p_0 = mc$. Thus

$$P^2 \equiv P \cdot P = p_0^2 - p_1^2 = m^2 c^2. \tag{15.28}$$

You may verify that if you went to a generic frame and wrote down expressions for p_0 and p_1 in terms of v, you would get the same result. If there are two particles we may assert that

$$P_A \cdot P_B = p_{A0} p_{B0} - p_{A1} p_{B1} = \frac{E_A E_B}{c^2} - p_{A1} p_{B1} \tag{15.29}$$

will have the same value for all observers. In the rest frame of particle B, $P_A \cdot P_B = E_A m_B$ where by E_A, I mean the energy of A as measured in the rest frame of B.

A photon has no mass. This means

$$K \cdot K = 0 \tag{15.30}$$

where K is a common name for the four-momentum of any photon. The photon has no rest frame; in any frame it moves at c. However, its components (k_0, k_1) will undergo Lorentz transformation as we change frames of reference. The components of K again stand for E/c and momentum of the photon. Zero mass means that

$$0 = K \cdot K = k_0^2 - k_1^2 \longrightarrow k_1 = \pm k_0. \tag{15.31}$$

In the above equation, the spatial momentum k_1 can be positive or negative, but k_0, which corresponds to energy, is always positive.
The energy and momentum of photons are usually denoted by ω and k:

$$K = (k_0, k_1) = \left(\frac{\omega}{c}, k \right). \tag{15.32}$$

The zero-mass condition becomes

$$k = \pm \frac{\omega}{c}. \tag{15.33}$$

An equivalent way to show the zero-mass condition explicitly is the following:

$$K = (k, k) \quad \text{right moving photon} \tag{15.34}$$

$$K = (k, -k) \quad \text{left moving photon} \tag{15.35}$$

where it is understood that $k > 0$. Thus a single number, the photon momentum $\pm k$, fixes the K in one spatial dimension.

You could equally well write K in terms of ω instead of k:

$$K = \left(\frac{\omega}{c}, \frac{\omega}{c}\right) \quad \text{right moving photon} \tag{15.36}$$

$$K = \left(\frac{\omega}{c}, -\frac{\omega}{c}\right) \quad \text{left moving photon.} \tag{15.37}$$

- *Four-momentum, if conserved in one frame, is conserved in any frame related by a Lorentz transformation.*

  The virtue of momentum as we derived it is that if it is conserved in one frame, it will be conserved in any other. For example, if particles A and B turn into C, D, E, and in one frame we have

$$P_{initial} = P_A + P_B = P_C + P_D + P_E = P_{final} \tag{15.38}$$

then in any other frame we will have

$$P'_{initial} = P'_A + P'_B = P'_C + P'_D + P'_E = P'_{final} \tag{15.39}$$

because *if two vectors $P_{initial}$ and P_{final} are equal in one frame they are equal in any other.* This must be clear from analogy with usual vectors: if $\mathbf{A} + \mathbf{B} = \mathbf{C}$, that is, the three form a triangle in one frame, then they will also form a triangle in a rotated frame. Or one can say that because $\mathbf{A} + \mathbf{B} - \mathbf{C} = \mathbf{0}$, the null vector, the LHS will be the null (zero) vector in any frame because the rotated version of the null vector is the null vector. In the case of the Lorentz transformation, if $P_{final} - P_{initial} = 0$, the difference vector will vanish in any frame if it vanishes in one.

It also follows that if you did not like my definition of four-momentum and made up your own, yours may not have the property of being conserved in all frames if conserved in one. Conservation in all frames is what makes momentum an important quantity, and for that it has to be a four-vector.

15.1 Relativistic scattering

Let us consider some examples of relativistic kinematics.

15.1.1 Compton effect

Imagine a photon of momentum K moving along the x-axis, bouncing off a stationary electron of momentum P, as shown in Figure 15.1. This process is called the Compton effect. We consider the one-dimensional version of this process, in which the final electron and photon are forced to travel along the x-axis. What is the energy of the outgoing photon? I will show you how to do this problem using units in which $c = 1$. This makes the manipulations easier. I will also show you how we can eventually re-instate factors of c guided by dimensions. With these units, the initial and final photon momenta look as follows:

$$K = (\omega, \omega), \quad K' = (\omega', -\omega') \tag{15.40}$$

The electron's four-momenta are, before and after,

$$P = (m, 0) \quad P' = (E', p') \quad \text{with } E'^2 - p'^2 = m^2. \tag{15.41}$$

In preparation for what follows, let us compute some dot products of four-vectors using

$$A \cdot B = a_0 b_0 - a_1 b_1. \tag{15.42}$$

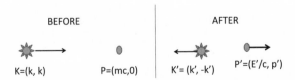

Figure 15.1   A photon bounces off an electron at rest. This is the one-dimensional version of *Compton scattering*.

The following will be needed:

$$K \cdot K' = (\omega\omega' - (\omega)(-\omega')) = 2\omega\omega' \tag{15.43}$$

$$P \cdot K = (m\omega - 0\,\omega) = m\omega \tag{15.44}$$

$$P \cdot K' = (m\omega' - 0\,\omega') = m\omega'. \tag{15.45}$$

Begin with the conservation law

$$K + P = K' + P'. \tag{15.46}$$

Because no one cares about the details of P', the scattered electron, we isolate and square it (take the dot product with itself) because we know that for any particle, no matter how it is moving, the square of the four-momentum will be $m^2c^2 = m^2$ (with $c = 1$). Here are the details.

$$P' \cdot P' \equiv (P')^2 = m^2 = (P + K - K')^2 \tag{15.47}$$

$$= P^2 + K^2 + K'^2 + (2P \cdot K - 2P \cdot K' - 2K \cdot K') \tag{15.48}$$

$$= m^2 + 0 + 0 + 2\left(m\omega - m\omega' - 2\omega\omega'\right) \tag{15.49}$$

$$0 = m(\omega - \omega') - 2\omega\omega' \tag{15.50}$$

$$\frac{1}{\omega'} = \frac{1}{\omega} + \frac{2}{m} \tag{15.51}$$

$$\frac{1}{\omega'} = \frac{1}{\omega} + \frac{2}{mc^2} \tag{15.52}$$

where, in the last equation, I restored the c^2 because ω is an energy and so is mc^2. In the three-dimensional case, the photon can emerge at an angle θ relative to the x-axis and we find

$$\frac{1}{\omega'} = \frac{1}{\omega} + \frac{1 - \cos\theta}{mc^2}. \tag{15.53}$$

(Equation 15.52 corresponds to $\theta = \pi$.) Compton did the scattering and confirmed this prediction, and this was very instrumental in convincing the community of the reality of the photon as a particle.

15.1.2 Pair production

What is the minimum energy E of the incident proton that strikes the proton at rest so that in the end we have a p, p, p , and $\bar{p}$, that is, three protons and an antiproton as in Figure 15.2? (Anti-particles have the same mass as particles.)

The energy-momentum of the incident proton is $P_1 = (E, p)$, that of the target is $P_2 = (m, 0)$. (Remember $c = 1$.) In the lab frame the total momentum is

$$P_{Tot}^{Lab} = P_1 + P_2 = (E + m, p). \tag{15.54}$$

In the minimal reaction the final four particles will have the minimum energy, but they can't all be at rest due to momentum conservation. So we go to the CM frame in which the two initial protons approach each other with opposing spatial momenta; that is, their four-momenta are (E_{cm}, p_{cm}) and $(E_{cm}, -p_{cm})$. Thus the initial total momentum is $P_{Tot}^{CM} = (2E_{cm}, 0)$. The final four-momentum of the four particles, now allowed to be created at rest, is

$$P_{Tot}^{CM} = (4m, 0). \tag{15.55}$$

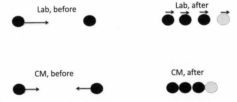

Figure 15.2   Top: The collision of a projectile proton and a static target proton, leading to three protons and an antiproton produced with the minimum energy. The final four particles cannot be at rest in the lab frame due to momentum conservation. Bottom: The minimal collision as seen in the CM frame, where the final four particles are allowed to be at rest.

Now recall that for any four-vector V, $V \cdot V$ is invariant, the same in all frames. Applying this to the total momentum we find, for the minimum energy process,

CM frame $P_{Tot} \cdot P_{Tot} = 16m^2$

Lab frame $P_{Tot} \cdot P_{Tot} = E^2 + m^2 + 2mE - p^2 = m^2 + m^2 + 2Em.$

Equating the results of $P_{Tot} \cdot P_{Tot}$ computed in two frames gives the desired result

$$E = 7m = 7mc^2. \tag{15.56}$$

If we thought only about energy, we would guess that the proton coming out of the accelerator should have an energy $E = 3mc^2$, which, along with the mc^2 of the target proton, would be able to produce the final four particles at rest. But momentum has to be conserved, of course, and the projectile has to have an energy $7mc^2$, not $3mc^2$, to pay for this extra mandatory kinetic energy of the final four particles. The Bevatron was constructed in the Lawrence Berkeley National Laboratory with the energy per proton slightly above $7mc^2$. Using it, Owen Chamberlain and Emilio Segrè successfully produced the antiproton in 1955.

At the Large Hadron Collider in CERN (European Organization for Nuclear Research), protons circulating one way collide with protons circulating the other way at the same energy, so that the lab frame is also the CM frame. Now all the beam energy goes to particle production. Each colliding proton would have to have just $E = 2mc^2$ to make the final three protons and an antiproton at rest. Of course the Large Hadron Collider is interested in producing the Higgs boson and bigger fish and not the antiproton.

15.1.3 Photon absorption

Finally, consider the case of photon absorption. If an atom absorbs a photon of energy ω, the atom will recoil, and its mass will increase due to the absorbed energy. We want to find the new mass. This confuses some students, who say, "But you said the mass of a particle is unaffected by its momentum: even if E and p change with velocity, $E^2 - c^2 p^2 = m^2 c^4$ always! So what does it mean to compute the new mass of the atom?" As long as a particle retains its identity, as the electron and photon do in Compton scattering, it will preserve its mass, but that is not the case here: a photon has disappeared and an "excited" atom has appeared. The atom in an excited internal state is to be viewed as a different particle, with its mass m' as a free parameter. We want to find the value of m'. (The electron does not seem to have internal states of different energy, which is why we did not have this issue in Compton scattering.)

Your guess may be $m' = m + \omega/c^2$. Let us work it out, following Figure 15.3, this time keeping the explicit factors of c and also writing down the conservation law for each component of the four-momentum.

Let

$$P = (mc, 0) \tag{15.57}$$

$$K = \left(\frac{\omega}{c}, \frac{\omega}{c}\right) = (k, k) \tag{15.58}$$

$$P' = \left(\frac{m'c}{\sqrt{1 - v^2/c^2}}, \frac{m'v}{\sqrt{1 - v^2/c^2}}\right) \tag{15.59}$$

be the four-momenta of the initial atom at rest, the incoming photon, and the final atom respectively. The conservation law for four-momentum or energy-momentum is

$$P + K = P'. \tag{15.60}$$

Figure 15.3 A photon is absorbed by an atom of mass m, which recoils and goes into an excited state. It is viewed as a different particle with a somewhat bigger mass m'.

In terms of components:

$$mc + \frac{\omega}{c} = \frac{m'c}{\sqrt{1 - v^2/c^2}} \qquad \text{conservation of energy of } E/c;$$

$$\text{(15.61)}$$

$$0 + k = \frac{m'v}{\sqrt{1 - v^2/c^2}} \qquad \text{momentum conservation.} \qquad \text{(15.62)}$$

Now you can juggle these equations and solve for m', but again there is a quicker way using four-vectors. Because we just want m', we need only calculate $P' \cdot P' = m'^2c^2$ as follows:

$$m'^2c^2 = P' \cdot P' \qquad \text{(15.63)}$$

$$= (P + K) \cdot (P + K) \qquad \text{(15.64)}$$

$$= P \cdot P + K \cdot K + 2P \cdot K \qquad \text{(15.65)}$$

$$= m^2c^2 + 0 + 2(mc\frac{\omega}{c} - 0 \cdot \frac{\omega}{c}) \qquad \text{(15.66)}$$

$$m' = \sqrt{m^2 + 2m\frac{\omega}{c^2}}. \qquad \text{(15.67)}$$

We can approximate as follows for small $\frac{\omega}{mc^2}$:

$$m' = m\sqrt{1 + 2\frac{\omega}{mc^2}} = m(1 + \frac{\omega}{mc^2} + \ldots) = m + \frac{\omega}{c^2} + \ldots \quad \text{(15.68)}$$

in accord with the naive expectation that ignores recoil. In other words, not all the photon energy can go into boosting the atom's rest mass, because it also needs to move to conserve the initial photon momentum. Thus the increased rest energy plus kinetic energy has to equal the photon energy.

Remember the following tricks when you deal with relativistic collisions: (1) In the four-vector equations square that four-momentum about which you know the least, because the answer for the square is always m^2c^2. (2) Sometimes the momentum you need to square may not be standing alone in one side of the equation. If this happens, isolate it (by moving other terms to the other side) and square it.

CHAPTER 16

Mathematical Methods

16.1 Taylor series of a function

I am going to introduce you to some mathematical tricks. As you've probably noticed by now, a lot of physics has to do with mathematics, and if you're not good in math, you're not going to be good in physics.

The first important trick is called the Taylor series. The philosophy of the Taylor series is the following: There is some function $f(x)$ depicted in Figure 16.1. But I'm going to imagine that you can only zero in on a tiny region near $x = 0$. And the question is, how will you write an approximation for this whole function, valid away from $x = 0$? Suppose I don't show you anything except what's happening at $x = 0$; I show you only $f(0)$. The value of the function is 92. What should you do away from $x = 0$? You have no additional information about this function; you don't know if it's going up or going down. The best approximation you can make is the flat line in Figure 16.1. There's no reason to tilt it one way or the other, given the information you have. If you do not pick the constant to be 92 you will even miss the answer at the one place ($x = 0$) where you were given the value.

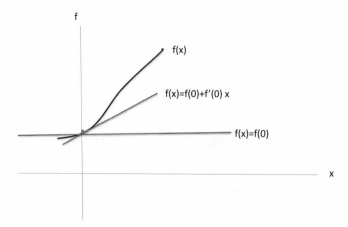

Figure 16.1   A sample function $f(x)$ and two approximations to it near $x=0$ based on the Taylor series. The flat line $f(x)=f(0)$ assumes $f(x)$ does not change as we move away from $x=0$. It is based on the first term in what is called the Taylor series. The second line $f(x)=f(0)+f'(0)x$ is the linear approximation based on two terms, and it matches the function and its derivative at $x=0$. It assumes the slope does not change.

So, the first approximation of the function you will say is

$$f(x)=f(0)+\dots \tag{16.1}$$

where the ellipsis means there may be corrections away from $x=0$ but you do not know them. The left-hand side stands for the actual function and the right to an approximation that matches the real thing at $x=0$ with possible unknown corrections as we move away. If it turns out that $f(x)$ is really constant, the horizontal line in Figure 16.1 will be the exact representation of the function.

To set $f(x)=f(0)$ is like saying "The temperature today ($x=0$) is 92. I don't know anything else, so my best guess away from $x=0$ would be 92, shown by the horizontal dotted line $f(x)=92$." But if you know that this is summer and that the temperature is going up from day to day at a known rate, and somebody tells you, "I know that the rate of change of temperature is $\frac{df}{dx}\big|_0$ today," you can use that information to predict what the temperature will be tomorrow or what it must have been yesterday. That is to say, for $x\neq 0$ you can improve our approximation to f using your knowledge of its derivative at $x = 0$:

$$f(x) = f(0) + \left.\frac{df}{dx}\right|_0 \cdot x + \ldots \qquad (16.2)$$

$$\equiv f(0) + f'(0) \cdot x + \qquad (16.3)$$

where I am using another popular notation

$$f'(x) = \frac{df}{dx}, f''(x) = \frac{d^2f}{dx^2}, \ldots \qquad (16.4)$$

where the number of primes denotes the number of derivatives. This notation is convenient if we do not plan taking too many derivatives. Since $f'(0)$ is the derivative at $x = 0$, when you multiply it by x you get the best guess for the change as we go away from $x = 0$. What we are approximating the function by is a straight line with the correct intercept and the correct slope, shown in Figure 16.1. If it turns out that the function really is a straight line, you are done. It's not even an approximation; it will track the function all the way to infinity. But it can happen, of course, that the function decides to curve upward, as I've shown in this example, and this linear approximation will not work if you go too far. For a while, you'll be tangent to the function, but then it will bend away from you. So, it's really good for a very small x and you can say, "Well, I want to do a little better when I go further out."

This approximation ignores the rate of change of the rate of change. The linear approximation assumes that the rate of change is fixed at the rate of change at the origin. And the rate of change of the rate of change is the second derivative $f''(0)$. Suppose you were given that as well. How do you use it? The answer is

$$f(x) = f(0) + f'(0) \cdot x + f''(0) \cdot \frac{x^2}{2!} + \ldots \qquad (16.5)$$

Why should you divide by 2!? *Because our goal is to generate an approximation that has whatever you know about the function built into it and the* $\frac{1}{2!}$ *ensures it has the right second derivative at $x = 0$.* Let's check that and the other features as well.

First of all, let's compare the two sides of Eqn. 16.5 at $x = 0$. The left-hand side is $f(0)$. On the right-hand side, when you put $x = 0$, you kill the

x and x^2 terms and are left with $f(0)$, and that matches the left-hand side. So, you certainly have the right value of the function here.

Then you say, "What if I take the derivative of the function and compare at $x = 0$?" Let's first take the derivative of both sides of Eqn. 16.5. We find

$$f'(x) = 0 + f'(0) + f''(0)x + \ldots \qquad (16.6)$$

The left-hand side is $f'(x)$. When we take the derivative of the right-hand side of Eqn. 16.5, the first term $f(0)$, being constant, gives a 0. The derivative of the x in the second term is 1, and finally the derivative of x^2 is $2x$, so I get $f''(0)$ times x. Now we evaluate the derivative at $x = 0$. That kills the term with a residual x, and the derivative of my approximate function matches the derivative of the actual function.

How about the second derivative of this function? Take two derivatives on the left-hand side of Eqn. 16.5 to get $f''(x)$ on the left. See what happens on the right-hand side. If you take two derivatives, the only survivor is the x^2 term, which leaves f'' with a coefficient 1. Now, if you set $x = 0$, the left becomes $f''(0)$ and so does the right. Thus with the $\frac{1}{2!}$, the function you have cooked up has the right value, the right slope, and the right rate of change of the slope at the origin.

It's very clear what you should do to go beyond. If you knew more derivatives, the approximation you would write would be

$$f(x) = f(0) + \left.\frac{df}{dx}\right|_0 \cdot x + \left.\frac{d^2f}{dx^2}\right|_0 \cdot \frac{x^2}{2!} + \left.\frac{d^3f}{dx^3}\right|_0 \cdot \frac{x^3}{3!} + \cdots \left.\frac{d^nf}{dx^n}\right|_0 \cdot \frac{x^n}{n!} + \ldots$$

$$(16.7)$$

And you go as far as you can. If you know 13 derivatives, put in the 13 before you surrender to the dot-dots. That's still an approximation, unless the function happened to be a polynomial of degree 13.

Now, sometimes you hit the jackpot, and you know all the derivatives. If someone tells you all the derivatives of the function, then why stop? Add them all up:

$$f(x) = \sum_{n=0}^{\infty} \left.\frac{d^nf}{dx^n}\right|_0 \cdot \frac{x^n}{n!} \qquad (16.8)$$

where the $n=0$ term is just $f(0)$.

And if you do the infinite summation at every value of x, and, if that summation is meaningful and gives you a finite number, that, in fact, is exactly the function you were given. That is the Taylor series: an infinite number of terms that, if they sum up to something finite, will actually be as good as the left-hand side.

Here's a famous example:

$$f(x) = \frac{1}{1-x}. \tag{16.9}$$

You and I know this function; we know how to put it in a calculator; we know how to plot it. You give me an x and I subtract it from 1 in the denominator, and that's the value of the function. But instead, suppose this function was revealed to us in stages. Say we were told just $f(0)$. What is $f(0)$ here? Set $x=0$ and get $f(0)=1$. Now, let's take the derivative of this function,

$$\frac{df}{dx} = \frac{1}{(1-x)^2}, \tag{16.10}$$

and evaluate it at $x=0$, to get $f'(0)=1$.

If you now take the second derivative of this function, which I don't feel like doing, and evaluate that at $x=0$, you will find it is equal to 2!. In fact, the n-th derivative at the origin is $n!$. That's very nice, because then the Taylor series becomes

$$f(x) = 1 + x + x^2 + x^3 + \ldots = \sum_{n=0}^{\infty} x^n. \tag{16.11}$$

The Taylor series is this infinite sum. In practice, you may be happy to just keep a couple of terms.

So, let's get a feeling for what those couple of terms can do for us. Let me take $x=0.1$; the real answer is $\frac{1}{.9} = 1.1111\ldots$, where the 1's go on forever. That's the target value. What do you get with the series? The series starts at 1, plus one-tenth, plus 1 over 100, plus 1 over 1000, and so on. And you can see, as I keep more and more terms, I keep just filling up these 1's. If you stop at 1 over 1000, you stop right there, at 1.111. I hope it is clear to you, perhaps from this simple example, that if you kept all the terms of the series, you really will get this infinite number of recurring 1's.

Consider the infinite Taylor series. Now, summing an infinite number of numbers is a delicate issue. I don't want to go there at all; I discuss that in my math book. But here are some caveats. Sometimes a sum makes no sense, and you have to quit. For example, put $x = 2$. The correct function is $\frac{1}{1-2}$, which is -1. Our approximation for $x = 2$ looks like $1 + 2 + 4 + 8\ldots$. First of all, this sum is going to grow to infinity, because the numbers are getting bigger and bigger. This sum seems to be all positive, while the correct answer is negative. Obviously, the series doesn't work. So, the next lesson of our Taylor series is this: you can write down the series, but it may not sum up to anything sensible beyond a certain distance from the starting point. So, if you're doing a Taylor series at $x = 0$, and you go to $x = 2$, you may find, as we did above, that it just doesn't work. So, you can ask, "How far can I go from the origin?" Well, in this simple example, we know that at $x = 1$ the function is going to infinity; that's why you couldn't go there or past that to the right of that point. The function is well defined on the other side for $x > 1$, but this series, this knowledge of the function and all its derivatives at the origin, is not enough to get you to the other side. So, this is a case where there are obvious problems at $x = 1$. It turns out that you cannot also reach or go to the left of $x = -1$, even though the function has no obvious problem there. Or consider a function like $1/(1 + x^2)$, with no evident troubles for any x. And yet, if you take the Taylor series for it, you will find that if you go beyond $|x| \geq 1$ the series blows up. (This problem arises because the function blows up at $x = \pm i$. Here you really need to look in the complex plane to see how far you can go before the series breaks down.)

I don't want to go into the mathematical theory of series. I just want to tell you that functions can be approximated by series. And, if you're lucky, you can get by with a few terms. If you are luckier and know all the derivatives, the whole sum may converge to give a finite answer within some interval, in which case, the series is as good as the function. One person can use $1/(1 - x)$, the other one can use the infinite series, and both are morally and mathematically equal in every sense, as long as they don't stray outside the region of validity of the infinite series.

Finally, note an obvious generalization: if f and its derivatives are given at $x = a$ and not at $x = 0$, by the same logic, we may write

$$f(x) = \sum_{n=0}^{\infty} \left. \frac{d^n f}{dx^n} \right|_a \cdot \frac{(x-a)^n}{n!} \tag{16.12}$$

where the $n = 0$ term is $f(a)$.

16.2   Examples and issues with the Taylor series

Now let's return to a popular example that I've been using a lot, $(1+x)^n$. That's a function for which we want the Taylor series. We see $f(0)$ is 1. What's the derivative of the function? It's $n(1-x)^{n-1}$, which reduces at $x = 0$ to just n. That's how we get the famous result

$$(1+x)^n = 1 + nx + \ldots \tag{16.13}$$

If x is small enough, you stop there, because the next term is going to involve an x^2 and then an x^3 will follow and so on. If x is tiny, we have no respect for big powers of x; we just cut them off. But if we want the next term, we'll have to take the second derivative

$$\frac{d^2(1+x)^n}{dx^2} = n(n-1)(1+x)^{n-2}, \tag{16.14}$$

which has a value $n(n-1)$ at $x = 0$ so that now

$$(1+x)^n = 1 + nx + \frac{n(n-1)}{2!}x^2 + \ldots \tag{16.15}$$

What happens if we consider the case when n is a positive integer like 2? After all, we know from the days in the nursery that

$$(1+x)^2 = 1 + 2x + x^2, \tag{16.16}$$

while the series above gives

$$(1+x)^2 = 1 + 2x + \frac{2(2-1)}{2!}x^2 + \ldots = 1 + 2x + x^2 + \ldots \quad (16.17)$$

What are the ellipses doing there when we have the whole answer already? Luckily, the next term in the series will have a factor $2(2-1)(2-2)$ that vanishes, as will all subsequent terms which will also contain the $(2-2)$ factor. By this argument, the expansion will terminate after the x^n term if n is an integer. For other powers, fractional or negative, this will not happen. (I suggest you play with $n = -1$.)

Now for an example from relativity. The energy of a particle is

$$E = \frac{mc^2}{\sqrt{1 - v^2/c^2}} = mc^2 \left[1 - \frac{v^2}{c^2}\right]^{-\frac{1}{2}} = mc^2 + \frac{1}{2}mv^2 + \ldots \quad (16.18)$$

We never knew the first term was there (it cancels out of all equations where we balance kinetic energy). We just kept the first nontrivial term $\frac{1}{2}mv^2$ and did mechanics this way for three hundred years. So, the approximations can really be useful. If you say, "Well, I want to be exact," you can go back and use the full $\sqrt{1 - v^2/c^2}$. Unfortunately, somebody or other will tell you, "That's not exact either. There is quantum mechanics, which tells you the whole thing is wrong, that particles do not move on trajectories." I have a lot of respect for approximations. If we could not describe the world approximately, we could not have accomplished what we have. No one knows the exact answer to a single question you can pose: if your question requires an answer to arbitrary precision, we just cannot provide it. Sometimes we do not even know if the question is meaningful in a more advanced theory, as was the case with trajectories when we moved from Newtonian to quantum mechanics. Newtonian mechanics works for small velocities. Relativistic mechanics works for any velocity, but not for really tiny objects for which you have to use quantum mechanics and then relativistic quantum mechanics. If we bring in gravity we need general relativity. So, as old theories always give way to new theories, approximations are very important as we inch forward.

16.3 Taylor series of some popular functions

Now I'm going to consider the following function: e^x. This is one you all know and love. Every child knows e^x is its own derivative. This means every derivative of e^x is e^x. Why do we like that here? Because all the derivatives at the origin are known and equal just 1. It follows that

$$e^x = \sum_{n=0}^{\infty} \frac{x^n}{n!}.$$ (16.19)

Now I need to know the value of e, because when I lock my suitcase and check it at the airport, I use either e or π, because they're the only two numbers I can derive, as compared to my anniversary. So, if I forget the value of e, I just say

$$e = e^1 = 1 + \frac{1}{1!} + \frac{1^2}{2!} + \frac{1^3}{3!} + \ldots$$ (16.20)

and keep finding terms till my suitcase opens. It is roughly 2.718, which is enough for most locks. Now, π is a good number too, but the rules for computing its digits are somewhat more difficult.

Here is the very nice property of the exponential series: It is good for any x, unlike the series for $1/(1-x)$ that crashed and burned at $|x| = 1$. This series is always good. You put $x = 37$ million, you have 1 plus 37 million, half of 37 million squared, a sixth of 37 million cubed, and so on; but don't worry, these factorials downstairs will eventually tame it down and make it converge, and the result will give you e to whatever number you chose. That's something that I'm not proving.

We all see that Eqn. 16.19 defines a function of x, but why call it e raised to a power x? Is the x in the series really a power? You know that when integer powers are involved, when you raise a number to a power and multiply it by the number to a different power, the product is a number to the sum of the two powers. Powers combine. Thus $2^5 2^6 = 2^{11}$. That's true for 2 or any number raised to an integer power; but how do we know this is true for e^x, especially when x is not an integer? Let us verify this property by multiplying the series for e^x and e^y to see if we get the series for e^{x+y}, working out the first few powers until we are convinced:

$$e^x e^y = \left(1 + x + \frac{x^2}{2} \ldots\right)\left(1 + y + \frac{y^2}{2} \ldots\right) \tag{16.21}$$

$$= 1 + \underbrace{x + y}_{\text{linear}} + \underbrace{xy + \frac{x^2}{2} + \frac{y^2}{2} + \frac{xy^2}{2} + \frac{yx^2}{2}}_{\text{quadratic}} + \ldots \tag{16.22}$$

$$= 1 + (x + y) + \frac{(x+y)^2}{2} + \ldots \tag{16.23}$$

$$= e^{x+y}. \tag{16.24}$$

The proof carries over even if x and y are replaced by complex numbers, a result we will be using shortly.

Now look at $\cos x$. What do I need to know to write the series for it at $x = 0$? We know the cosine of 0 is 1. If you take the derivative you get $- \sin x$, and its value at 0 is 0. You take one more derivative and get $(- \cos x)$, which is -1 at the origin, and so on. Every other derivative will vanish, and the surviving derivatives will alternate between ± 1 to give

$$\cos x = 1 - \frac{x^2}{2!} + \frac{x^4}{4!} - \frac{x^6}{6!} + \ldots \tag{16.25}$$

$$= \sum_{n=0}^{n} (-1)^n \frac{x^{2n}}{(2n)!}. \tag{16.26}$$

It is clear from the series that $\cos x$ is an even function: $\cos x = \cos(-x)$ because only even powers of x appear. Near $x = 0$, it starts at 1 and falls down quadratically. So a very useful approximation is

$$\cos = 1 - \frac{x^2}{2!} + \ldots \tag{16.27}$$

It is not clear from the series that $\cos x$ is bounded. If you cut it off after some number of terms, it's not going to work. In the beginning, $1 - \frac{x^2}{2!}$ looks very good as the cosine starts coming down from 1. But eventually this approximation will go bad on you and become too negative, but then the next term $\frac{x^4}{4!}$ will turn it around but soon drive the answer to values that are too positive, and so on. If, however, you add all the powers, remarkably, you will reproduce this nice function that will oscillate with period 2π. It is very hard to imagine that the series in Eqn. 16.26 is actually the cosine with all its properties, but it is.

By similar reasoning,

$$\sin x = \frac{x}{1!} - \frac{x^3}{3!} + \frac{x^5}{5!} + \dots \tag{16.28}$$

$$= \sum_{n=0}^{n} (-1)^n \frac{x^{2n+1}}{(2n+1)!}. \tag{16.29}$$

It is obvious that this is an odd function, $\sin(-x) = -\sin x$, but not obvious that it is bounded or periodic.

The series for $\sin x$ and $\cos x$ converge for all finite values of x, just like the one for e^x.

If you are stranded and bored at some airport, take cosine squared plus sine squared using the series and group terms with the same power of x. You'll first find a 1 from squaring the 1 in the $\cos x$, and the net coefficient of all non-zero powers of x will miraculously vanish.

16.4 Trigonometric and exponential functions

Let us introduce, without any preamble right now, the number $i = \sqrt{-1}$. The only property I need is that $i^2 = -1$, $i^3 = -i$, $i^4 = +1$, $i^5 = i$, and so on. Consider the following rather strange object, e^{ix}. Now e is some number, and, if you want to raise it to a power, say 2, that's fine. But now we want to raise e to a complex power, ix. What does that even mean? Multiply e by itself ix times? Well, that definition of powers is no good. But the series for e^x defines it for all x, and we boldly define e^x for even complex values of the exponent to be the same series with ix in place of x. So, the exponential function is simply defined by the power series and not the notion of raising e to some power. Thus

$$e^{\mathrm{dog}} = 1 + \mathrm{dog} + \frac{\mathrm{dog}^2}{2!} + \dots \tag{16.30}$$

This way we can raise e to various things: real numbers, complex numbers, matrices, whatever you want. If you have a pet, you can put that up in the exponent. Of course, you've got to be careful. You cannot raise e

to dog, because the units don't match: there's a dog here and dog^2 there, and so on. So you should divide it by some standard dog, like President Obama's dog. Take some standard and divide by it; then you have something dimensionless like this:

$$e^{\frac{dog}{Bo}} = 1 + \frac{dog}{Bo} + \frac{1}{2!}\left[\frac{dog}{Bo}\right]^2 + \ldots, \tag{16.31}$$

which converges for dogs of any size.

That's a fantastic leap of imagination. Let's consider the series for e^{ix}:

$$e^{ix} = \sum_{n=0}^{\infty} \frac{i^n x^n}{n!}. \tag{16.32}$$

Using what we know about the powers of i we get the following terms:

$$e^{ix} = 1 + ix + \frac{i^2 x^2}{2!} + \frac{i^3 x^3}{3!} + \frac{i^4 x^4}{4!} + \ldots \tag{16.33}$$

$$= \left[1 - \frac{x^2}{2!} + \frac{x^4}{4!} - \frac{x^6}{6!} + \cdots\right] + i\left[x - \frac{x^3}{3!} + \frac{x^5}{5!} + \cdots\right], \tag{16.34}$$

which leads to the following all-time hit due to Leonhard Euler:

$$e^{ix} = \cos x + i \sin x. \tag{16.35}$$

This is a super-duper formula worth memorizing. Life as we know it cannot go on without this formula. It says that the trigonometric and exponential functions are very intimately connected by the power series. A particularly beautiful case of this formula follows if you put $x = \pi$:

$$e^{i\pi} + 1 = 0. \tag{16.36}$$

Everybody agrees this has to be one of the most beautiful formulas we can imagine, involving all the key numbers in mathematics: π, defined from ancient times as a ratio of circle to diameter; i, the mother of all complex numbers; e, the base for the logarithm; and finally 0 and 1, from which we can build all numbers in binary.

Now let's do the following two other variations, and then we'll move on. If I change x to $-x$ in Eqn. 16.35, I get

$$e^{-ix} = \cos x - i \sin x, \tag{16.37}$$

using the even/odd nature of $\cos x$ and $\sin x$. I now combine Eqns. 16.35 and 16.37 to obtain

$$\cos x = \frac{e^{ix} + e^{-ix}}{2} \tag{16.38}$$

$$\sin x = \frac{e^{ix} - e^{-ix}}{2i}. \tag{16.39}$$

This means that if you have exponential functions, you can manufacture trigonometric functions out of them, provided you're not afraid to use complex exponents. And all the identities about sines and cosines will follow from this. For example, if you take cosine squared plus sine squared, you're supposed to get 1. Well, you can square the right-hand sides of Eqns. 16.38 and 16.39 and add them and you will get 1 provided you remember $e^{ix}e^{-ix} = e^{i(x-x)} = e^0 = 1$.

16.5 Properties of complex numbers

Now I'm going to do a little more with complex numbers. I introduced you to i by saying it's the square root of minus 1. Complex numbers entered our life even though we didn't go looking for them. You can write down equations with real numbers with no intention of invoking anything fancy, like this:

$$z^2 + 1 = 0, \tag{16.40}$$

and you find there is no solution to this equation. You can say, "I want $z^2 = -1$," and you can manufacture a number i with the property $i^2 = -1$; and then, of course, you can have $z = \pm i$ as your answer. So, complex numbers arose first in attempts to solve quadratic equations with real coefficients. Let me write you a slightly more interesting quadratic equation,

$$z^2 + z + 1 = 0. \tag{16.41}$$

Recall the answer

$$z = \frac{-1 \pm \sqrt{-3}}{2}. \tag{16.42}$$

But we don't know what to do with $\sqrt{-3}$, so we will write it as $\sqrt{-1} \cdot \sqrt{3}$ and end up with

$$z = -\frac{1}{2} \pm \frac{\sqrt{3}}{2} i. \tag{16.43}$$

These are formally solutions to Eqn. 16.40 in the following sense. Take one of the roots, say the one with the $+$ sign; put that into Eqn. 16.40, and it will work:

$$z^2 + z + 1 = \left(-\frac{1}{2} + \frac{\sqrt{3}}{2} i\right)^2 - \frac{1}{2} + \frac{\sqrt{3}}{2} i + 1 \tag{16.44}$$

$$= \frac{1}{4} - 2\frac{1}{2}\frac{\sqrt{3}}{2} i + \left(\frac{\sqrt{3}}{2} i\right)^2 - \frac{1}{2} + \frac{\sqrt{3}}{2} i + 1 \tag{16.45}$$

$$= \frac{1}{4} - \frac{\sqrt{3}}{2} i - \frac{3}{4} - \frac{1}{2} + \frac{\sqrt{3}}{2} i + 1 = 0. \tag{16.46}$$

All you have to know in these manipulations is that $i^2 = -1$. Using this one property, you can now solve any quadratic equation. People realized that if we enlarge numbers to include complex numbers, then we can solve any n-th order polynomial equation and obtain n roots. If it's quadratic, it'll have two roots; if it's cubic, it'll have three roots. The roots may be complex even if the coefficients in the equation are all real. Because complex numbers arose from equations with real coefficients, will the equations with complex coefficients perhaps lead to even crazier numbers? Luckily or unluckily, this is not so: a complex polynomial equation of degree n with complex coefficients will have n roots, generally complex.

Now, a very important point to notice is that this whole thing, $z = -\frac{1}{2} \pm \frac{\sqrt{3}}{2} i$, is a *single* complex number. Don't think of it as the sum of two numbers; it cannot be simplified any further, just as a two-dimensional vector $\mathbf{V} = 2\mathbf{i} + 3\mathbf{j}$ cannot be simplified further.

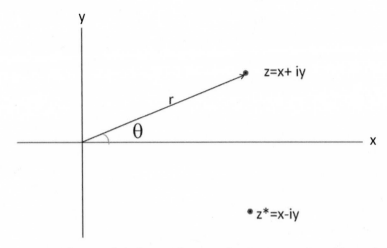

Figure 16.2 The complex plane with a typical point $z = x + iy$ and its conjugate $z^* = x - iy$. In polar form, z is assigned a length r and an angle θ. We will see that $z = re^{i\theta}$. Note that we measure y and not iy along the y-axis.

We are going to generalize this particular case and introduce now a complex number z as follows:

$$z = x + iy. \tag{16.47}$$

Just as x stood for a generic real number, z stands for a generic complex number. However, z has two parts: x, called the *real part*, and y, called the *imaginary part*. In the root of the quadratic equation $-\frac{1}{2} + \frac{\sqrt{3}}{2}i$, the real part is $-\frac{1}{2}$ and the imaginary part is $\frac{\sqrt{3}}{2}$. We're going to visualize the complex number $z = x + iy$ as a point in the complex plane as shown in Figure 16.2. Note that the point (x, y) stands for $z = x + iy$. We introduce a related complex number known as z^*, pronounced z-star, called the *complex-conjugate of z* and given by

$$z^* = x - iy. \tag{16.48}$$

It's obtained from z by changing the sign of i, or reflecting z on the real axis.

How does one add, multiply, and divide complex numbers? First, if $z_1 = x_1 + iy_1$ and $z_2 = x_2 + iy_2$, we will define their sum as follows:

$$z_1 + z_2 = (x_1 + x_2) + i(y_1 + y_2), \tag{16.49}$$

which is really just like adding vectors. Subtracting is similar.

If you want to get the real or imaginary part of z, you do the following:

$$Re[z] \equiv x = \frac{z + z^*}{2}$$

$$Im[z] \equiv y = \frac{z - z^*}{2i}. \tag{16.50}$$

Of course, given $z = x + iy$, you can spare all this and read off the real and imaginary parts by inspection. However, you will soon need the real and imaginary parts of more complicated expressions. In general, if you are given some expression f, its complex conjugate f^* *is obtained from f by changing every i to* $-i$ *and leaving all real numbers alone.* For example, if

$$f = (\cos A + i \sin B)^2 = \cos^2 A - \sin^2 B + 2i \cos A \sin B \tag{16.51}$$

then

$$f^* = (\cos A - i \sin B)^2 = \cos^2 A - \sin^2 B - 2i \cos A \sin B, \tag{16.52}$$

assuming A and B are real. The real and imaginary parts of f may be read off by inspection or generated by the more formal prescription

$$Re[f] = \frac{(f + f^*)}{2} \tag{16.53}$$

$$Im[f] = \frac{(f - f^*)}{2i}. \tag{16.54}$$

What is z_1 times z_2? Just open the brackets and remember $i^2 = -1$. Thus

$$z_1 \cdot z_2 = (x_1 + iy_1)(x_2 + iy_2) \tag{16.55}$$

$$= x_1x_2 - y_1y_2 + i(x_1y_2 + y_1x_2). \tag{16.56}$$

$$Re[z_1z_2] = x_1x_2 - y_1y_2 \tag{16.57}$$

$$Im[z_1z_2] = x_1y_2 + y_1x_2. \tag{16.58}$$

Something very nice happens when you multiply z by z^*:

$$zz^* = (x + iy)(x - iy) = x^2 + y^2 \equiv |z|^2. \tag{16.59}$$

We refer to

$$|z| = \sqrt{x^2 + y^2} \tag{16.60}$$

as the *modulus* of the complex number z, which is also the length r of z in Figure 16.2 by Pythagoras' theorem.

I'm going to use z^* to help *divide* z_1 by z_2. What is

$$\frac{z_1}{z_2} = \frac{x_1 + iy_1}{x_2 + iy_2}? \tag{16.61}$$

If I had only x_2 in the bottom, I could divide x_1 and iy_1 by it, because it is an ordinary real number, but now I have to divide by $x_2 + iy_2$. The trick is to multiply the top and bottom by the complex conjugate of the denominator and proceed as follows:

$$\frac{z_1}{z_2} = \frac{x_1 + iy_1}{x_2 + iy_2} = \frac{(x_1 + iy_1)(x_2 - iy_2)}{(x_2 + iy_2)(x_2 - iy_2)} \tag{16.62}$$

$$= \frac{(x_1 + iy_1)(x_2 - iy_2)}{|z_2|^2}. \tag{16.63}$$

The denominator is an ordinary real number, $|z_2|^2$, while in the numerator you can open out the brackets. Remember that if you have a complex denominator, and you don't like it, multiply top and bottom by the complex conjugate of the denominator, and the new denominator will become a purely real (positive) number. This whole thing, $|z_2|^2$, could be 36, for example. So, dividing by a complex number is not a problem.

16.6 Polar form of complex numbers

Now you're going to use Euler's result

$$e^{ix} = \cos x + i \sin x. \tag{16.64}$$

Let's take the complex number $z = x + iy$ in Figure 16.2. Let us introduce, just as we would for an ordinary vector, the angle θ and the length r, which is just $|z|$. Whenever you have a vector, you can talk about the Cartesian components x and y, or you can talk about the polar components, the length r and the angle θ it makes with the x-axis. Their interrelationship is as follows:

$$x = r\cos\theta \quad y = r\sin\theta \tag{16.65}$$

$$r = \sqrt{x^2 + y^2} \quad \theta = \tan^{-1}\frac{y}{x}. \tag{16.66}$$

Let us do the same starting with the *Cartesian form* of z and write

$$z = x + iy = r\cos\theta + ir\sin\theta = r(\cos\theta + i\sin\theta) = re^{i\theta}, \tag{16.67}$$

which is called the *polar form*. One refers to r as the *amplitude* of z and θ as the *phase* of z. The complex conjugate is

$$z^* = re^{-i\theta} \quad \text{and} \tag{16.68}$$

$$|z|^2 = z^*z = re^{i\theta} \cdot re^{-i\theta} = r^2. \tag{16.69}$$

Note that the modulus z is the same as the amplitude r:

$$|z| = r = \sqrt{x^2 + y^2}. \tag{16.70}$$

The Cartesian and polar forms both describe the same number. One displays transparently the real part and imaginary parts; the other, its amplitude and its angle relative to the real axis. It is easier to add complex numbers in the Cartesian form (just add the real and imaginary parts), while the polar form is best for multiplying and dividing, as I will show now.

Consider the product of two complex numbers, $z_1 = r_1 e^{i\theta_1}$ and $z_2 = r_2 e^{i\theta_2}$. It is

$$z_1 z_2 = r_1 e^{i\theta_1} r_2 e^{i\theta_2} = r_1 r_2 e^{i(\theta_1 + \theta_2)}. \tag{16.71}$$

If you want *to multiply two complex numbers, multiply the lengths to get the length of the product and add the phases to get the phase of the product, as shown in Figure 16.3.* It's a lot easier to multiply them in this form than in Cartesian form, as we did in Eqn. 16.56.

If the polar form is well suited for multiplying, it's even better suited for dividing:

$$\frac{z_1}{z_2} = \frac{r_1 e^{i\theta_1}}{r_2 e^{i\theta_2}} \tag{16.72}$$

$$= \frac{r_1}{r_2} e^{(i\theta_1 - \theta_2)}. \tag{16.73}$$

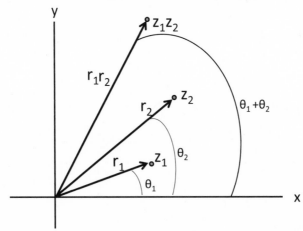

Multiplication in the complex plane

Figure 16.3   The rule for multiplying z_1 and z_2: add the phase angles θ_1 and θ_2 to get the phase of the product and multiply the amplitudes r_1 and r_2 to get the amplitude of the product.

To divide by a complex number, divide by its modulus and subtract its phase from the phase of the numerator.

Here is what I want you to carry in your head, because it's very, very important. Every complex number has a length and a direction. When you multiply by a second complex number, you're able to do two things at the same time. You're able to rescale it, and you're able to rotate it. You rescale by the length of the second factor, and you rotate by the phase angle carried by the second factor. The fact that two operations are done in one shot is one of the reasons complex numbers play an incredibly important role in physics, as well as in engineering and mathematical physics. I also find imaginary numbers useful when computing my tax deductions.

CHAPTER 17

Simple Harmonic Motion

We're now going to study what are called small oscillations, or simple harmonic motion. Take any mechanical system that is in a state of equilibrium. Equilibrium means the forces on the body add up to zero. It has no desire to move. If you give it a little kick, a push away from the equilibrium point, what will happen? There are two main possibilities. Imagine a marble on top of a hill. That is in *unstable equilibrium* because if you give the marble a nudge, it will roll downhill and never return to you. The other possibility involves *stable equilibrium*: if you push the system away from equilibrium, there are forces bringing it back. The standard example is a marble in a bowl: when it is shaken from its position at the bottom, it will rock back and forth until it settles again. A rod hanging vertically from the ceiling from a pivot, when pulled to the side and released, will swing back and forth. These are examples of simple harmonic motion, which results whenever any system is slightly disturbed from stable equilibrium.

The example that we're going to consider is a mass m, resting on a table, connected to a spring, which in turn is connected to the wall. The spring is not stretched or contracted; the mass is at rest, as shown in Figure 17.1. That's what I mean by equilibrium. Now let it be displaced by

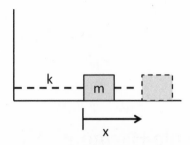

Figure 17.1   The mass m rests on a table and is connected to a spring of force constant k, which is anchored to the wall. The displacement from equilibrium is denoted by x. It is positive in the figure but it could also be negative if the mass were to be displaced the other way.

an amount x from this point of equilibrium. The spring force is $F = -kx$ and Newton's law says

$$m\frac{d^2x}{dt^2} = -kx. \tag{17.1}$$

If the mass strays to the right, x is positive and $-kx$ is to the left, so as to send it back toward its equilibrium position. If x is negative, the restoring force is positive, again pointing to the equilibrium position.

We want to understand the behavior of such a mass. How do we solve this problem? Our job is to find the function $x(t)$ that satisfies this equation, which we rewrite as follows:

$$\frac{d^2x}{dt^2} = -\omega^2 x \tag{17.2}$$

$$\omega = \sqrt{\frac{k}{m}}. \tag{17.3}$$

You can make it a word problem and say, "I'm looking for a function whose second derivative is minus itself, except for this number ω^2." Trigonometric functions have the property that if you take two derivatives, they return to minus themselves. So, you can guess that $x = \cos t$ but it won't work, as I showed you before. On the other hand, the guess

$$x(t) = A \cos \omega t \tag{17.4}$$

will obey this equation. While A is clearly the amplitude, ω is related to the frequency of oscillations as follows. If I start at $t = 0$ when $x = A$, how long do I have to wait until it comes back to A? I have to wait a time T, such that

$$\omega T = 2\pi \tag{17.5}$$

because that's when the cosine returns to 1. That means the time that I have to wait is

$$T = \frac{2\pi}{\omega}. \tag{17.6}$$

You can rewrite this as

$$\omega = \frac{2\pi}{T} = 2\pi f \tag{17.7}$$

where $f = \frac{1}{T}$ is what we would normally call frequency, which is how many oscillations it completes per second. It is the inverse of the time period. In physics talk, frequency usually means ω.

So, if you pull a mass and let it go, it oscillates with a frequency that is connected to the force constant and the mass. If the spring is very stiff and k is very large, the frequency is very high. If the mass is very big and the motion is very sluggish, f is diminished. So, all that stuff you expect intuitively is quantified by the solution to the equation, but there is more. For example, it is not intuitively obvious that if you make the mass four times as big, you will double the time period.

One remarkable part of the solution is that you can pick any A you like without changing ω or T. Think about what that means. The amplitude A is the amount by which you pulled the mass when you let it go. You find that whether you pull the spring by one inch or by ten inches, it takes the same time to finish a full back-and-forth motion. If you pull it by two inches, compared to one inch, it has a longer way to go. But if you pull it by two inches, the spring is going to be that much more tense, and it's going to exert a bigger force so that it will go faster for most of the time; that's very clear. But the fact that it goes faster in exactly the right way to complete the trip in exactly the same time is rather a miraculous property of Eqn. 17.2. If you tamper with it, if you add to the force even a tiny extra term, say proportional to x^3, then this feature is gone. It's like saying that planets move around the sun in *closed* elliptical orbits only under the $\frac{1}{r^2}$ force. It is not true if the force falls as $\frac{1}{r^{2.0000001}}$.

Now consider the following variant of this solution. You set your clock to 0 at the origin in the graph. Suppose I set my clock to 0, right there on the dotted vertical line in Figure 17.2 at $t = \frac{\pi}{2}$. When my clock says 0, x is not at the maximum; it vanishes. But it's the same physics, and it's the same equation. Where, then, is the solution that describes what I see? It is there and it comes from the fact that we had the latitude of adding a certain angle ϕ, called a *phase*, to the solution:

$$x(t) = A \cos[\omega t + \phi]. \tag{17.8}$$

Your choice is $\phi = 0$ and mine is $\phi = \frac{\pi}{2}$. You can verify that whatever we pick for ϕ, the above $x(t)$ will be a solution because two derivatives of the solution with the ϕ is also $-\omega^2$ times itself. And, whatever you pick for A, it will still work, because A cancels out of both sides in Eqn. 17.2. So, whenever you have an oscillator, say, a mass and spring system, and you want to know what x is going to be at all times, it is not enough to know that it obeys Eqn. 17.2; you need to know the amplitude and the phase. These are determined by knowing two things about the solution, which is usually the x and v at some time, usually $t = 0$. For this reason we refer to them as *initial value data*.

Let me give you an example. Suppose an oscillator has $x(0) = 5$ and velocity $v(0) = 0$, at $t = 0$. What does that mean? I pulled the mass by 5 and I let it go. I give you the values of the spring constant k and the mass m, and I say, "What's the future x?"

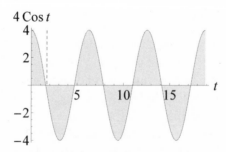

Figure 17.2   The function $A \cos \omega t$ for the case $A = 4$, $\omega = 1$. The dotted vertical line is another possible way to set the clock to zero, another choice of phase, namely $\phi = \frac{\pi}{2}$.

First observe that the velocity corresponding to our solution Eqn. 17.8 is

$$v(t) = -\omega A \sin(\omega t + \phi). \tag{17.9}$$

So we know two things at $t = 0$:

$$5 = A \cos[0 + \phi] \tag{17.10}$$

$$0 = -A\omega \sin[0 + \phi]. \tag{17.11}$$

The second equation gives us two choices: either $A = 0$, which is a trivial solution, or $\phi = 0$, which lets A survive this test. The first equation with $\phi = 0$ gives $A = 5$ leading to the solution $x(t) = 5 \cos \omega t$. This is a problem where we did not need a non-zero ϕ. But it could have been that when you joined the experiment, you were somewhere to the right of the origin, on the vertical dotted line in Figure 17.2, when you set your clocks to zero. Then you would have, as your initial conditions, $x = 0, v = -5\omega$, which means $\phi = \frac{\pi}{2}$. Of course $A = 5$ as before.

Let us agree that, if there's only one oscillator, it is perverse to set your clock to 0 at any time other than when the oscillator is at its maximum displacement, so that

$$x(t) = A \cos \omega t. \tag{17.12}$$

(If there are two oscillators oscillating out of step, it's impossible to make $\phi = 0$ for both of them: you can set your clock to 0 when one of them is at a maximum, but then the other may not be at its maximum.) Going forward, remember that the velocity and acceleration are, for all future times,

$$v(t) = -\omega A \sin \omega t \tag{17.13}$$

$$a(t) = -\omega^2 A \cos \omega t \quad \left(= -\frac{k}{m} x(t) \quad \text{in accordance with } F = ma\right). \tag{17.14}$$

So the velocity also oscillates sinusoidally but with an amplitude ωA. The acceleration also oscillates but with an amplitude $\omega^2 A$. These two results are true for any phase ϕ.

Let us explicitly verify the law of conservation of energy. Consider the total energy:

$$E(t) = \frac{1}{2}mv^2 + \frac{1}{2}kx^2 \qquad (17.15)$$

$$= \frac{1}{2}m\omega^2 A^2 \sin^2 \omega t + \frac{1}{2}kA^2 \cos^2 \omega t \qquad (17.16)$$

$$= \frac{1}{2}kA^2 \quad \text{because } \omega^2 = \frac{k}{m}. \qquad (17.17)$$

Thus, by magic, the time-dependent terms $\sin^2 \omega t$ and $\cos^2 \omega t$ have the same coefficient, and you find that $E(t)$ actually does not depend on time at all. Even though position and velocity are constantly changing, this combination will not depend on time. At the instant when the mass has reached one extremity and is about to swing back, it has no velocity; it only has an $x = A$, and the energy of the oscillator is all potential energy, $\frac{1}{2}kA^2$.

17.1   More examples of oscillations

If a body is in stable equilibrium, and you disturb it, it rocks back and forth, executing simple harmonic motion. The standard textbook example is the mass on a spring, which we just studied. But it is a very generic situation, as shown in Figure 17.3. Skipping the mass-and-spring example, let us go the top right, where we have a beam hanging from the ceiling by a cable that is fixed to its center of mass (CM). If you twist it by an angle θ, it will try to untwist itself. Now we don't have a restoring force but we have a restoring torque. What can be the expression for the restoring torque τ? When you don't do anything, the cable doesn't do anything, so τ vanishes when $\theta = 0$. If $\theta \neq 0$, it is some function of θ, and the leading term in the Taylor expansion would be proportional to θ:

$$\tau(\theta) = -\kappa\theta. \qquad (17.18)$$

The coefficient κ is the *torsion constant*, and the minus sign tells you it's a restoring torque. That means if you make θ positive, the torque will try to twist you the other way. The torsion constant, which is the restoring torque per unit angular displacement, is to rotations what the spring constant was to linear oscillations: the restoring force per unit displacement.

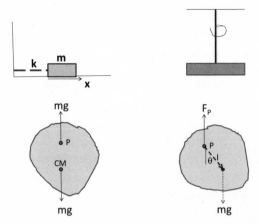

Figure 17.3   Some examples of simple harmonic motion. The left shows the
eternal favorite, the mass and spring; top right is a beam hanging from the ceiling
by a cable; the bottom left and right show a physical pendulum supported at the
pivot P, when it is in equilibrium, and when it is displaced by an angle θ. The
vector **g** represents the downward gravitational field of magnitude $9.8m/s^2$.

You have to find this κ, which is not given, the way k is. Once you do, you
can say

$$I\frac{d^2\theta}{dt^2} = -\kappa\theta \qquad\qquad (17.19)$$

where I is the moment of inertia of the beam about the point of suspen-
sion.

Mathematically, this equation is identical in form to

$$m\frac{d^2x}{dt^2} = -kx \qquad\qquad (17.20)$$

with the substitution $x \rightarrow \theta, m \rightarrow I, k \rightarrow \kappa$. So the answer follows:

$$\theta(t) = A\cos\omega t \qquad\qquad (17.21)$$

$$\omega = \sqrt{\frac{\kappa}{I}}. \qquad\qquad (17.22)$$

The mass-spring system executes linear oscillations, while the beam executes angular oscillations. Another example of the latter is the simple pendulum on the lower left of the figure. The pendulum has a bob of mass m hanging by a massless rod of length l. If you let it hang vertically, it will stay that way forever. No torque, no motion. Suppose you pull it by an angle θ and release it. To predict the future, you need to find I and κ. Now I is easy: for a single mass m at a distance l from the pivot point, $I = ml^2$. To find κ, you need to find the restoring torque per angular displacement. If you displace by θ, the torque about the pivot point is

$$\tau = -mgl\sin\theta \simeq -mgl\theta \qquad (17.23)$$

where I have approximated $\sin\theta$ by θ, which is the leading term in the Taylor expansion. With just this term, we can read off κ:

$$\kappa = -\frac{\tau}{\theta} = mgl. \qquad (17.24)$$

So

$$\omega = \sqrt{\frac{\kappa}{I}} = \sqrt{\frac{mgl}{ml^2}} = \sqrt{\frac{g}{l}}, \qquad (17.25)$$

from which follows the familiar formula

$$T = \frac{2\pi}{\omega} = 2\pi\sqrt{\frac{l}{g}}. \qquad (17.26)$$

Notice that if you displace the pendulum by large angles, when $\sin\theta$ cannot be approximated by θ, the frequency will no longer be independent of the amplitude.

Note that finding ω took some work. You had to disturb the system from equilibrium and find the restoring torque per unit angle $\kappa = -\frac{\tau}{\theta}$ and also compute I, whereas in the case of the mass-spring system, you were simply given m and k. In the case of the twisted cable, κ will be given to you, because computing it from first principles requires work beyond the scope of this course.

Let us move from a pendulum with all the mass concentrated in the bob to a *physical pendulum*, some irregularly shaped flat planar object, as shown in the middle of the second line of Figure 17.3. You drive a nail through it at some pivot point P and hang it on the wall. It will come to rest in a certain equilibrium configuration. Think about where the center of mass will be. It will lie somewhere on the vertical line going through P—otherwise the force of gravity, which is effectively acting at the CM, will produce a torque around P.

Let us look at the forces. This body, when hanging in its rest position, has two forces on it: the nail, which is pushing up, and the weight of the body $m\mathbf{g}$, which is pushing down, cancel each other. The nail will keep it from falling. The nail will not keep it from swinging, because the force of the nail, acting as it does at the pivot point, is unable to exert a torque, whereas the minute you rotate the body, $m\mathbf{g}$ is able to exert a torque, as is clear from the figure. That's why if you rotate it and let go, it will start swinging back and forth.

What will the torque be? It will be the same as before: $-mgl\sin\theta$, where l is now the distance between the pivot point and the center of mass. As far as the torque is concerned, it's as if all the mass were sitting at the CM. But the moment of inertia is not as if all the mass is sitting at the CM, in which case it would be ml^2. So don't make that mistake. All the mass is not sitting at the CM; it is all over the place. The moment of inertia is $I = I_{CM} + ml^2$ by the parallel axis theorem, where I_{CM} is hard to compute for an irregular object.

So, every problem that you will ever get will look like one of these two. Either something is moving linearly with a coordinate that you can call x, or something is rotating or twisting by an angle you can call θ. And if you want to find out the frequency of vibration, you have to disturb it from equilibrium—by pulling the mass, twisting the cable, or displacing the pendulum from its equilibrium position—in order to find the restoring force or torque per unit displacement.

17.2 Superposition of solutions

I will now go over more complicated oscillations using some of the formulas we learned in the last chapter. Here is the most important one:

$$e^{i\theta} = \cos\theta + i\sin\theta. \tag{17.27}$$

This is a formula worth memorizing. You should realize that given any expression involving complex numbers, you can get another equation by taking the complex conjugate of both sides, where every i is changed to minus i. That will give you

$$e^{-i\theta} = \cos\theta - i\sin\theta. \tag{17.28}$$

This is true because if two complex numbers $z_1 = x_1 + iy_1$ and $z_2 = x_2 + iy_2$ are equal, then the real and imaginary parts are separately equal, and so are their complex conjugates: $z_1^* = x_1 - iy_1 = x_2 - iy_2 = z_2^*$. The two previous equations can be inverted to give

$$\cos\theta = \frac{e^{i\theta} + e^{-i\theta}}{2} \tag{17.29}$$

$$\sin\theta = \frac{e^{i\theta} - e^{-i\theta}}{2i}. \tag{17.30}$$

You don't need trigonometric functions once you have the exponential function, provided you let the exponent be complex or imaginary. This is one example of unification. People always say Maxwell unified this, and Einstein tried to unify that. Unification means things that you thought were unrelated are, in fact, related, and they are different manifestations of the same thing. When we first discovered trigonometric functions, we were thinking right-angle triangles, opposite sides and adjacent sides, and so on. Then, we discovered the exponential function, which, by the way, was used by bankers who were trying to calculate compound interest continuously at every instant. The fact that those functions are related is a marvelous result, but it emerges only if you invoke complex numbers.

Finally, remember that there are two ways to write a complex number:

$$z = x + iy = re^{i\theta} \equiv |z|e^{i\theta}. \tag{17.31}$$

Now we use the new tools to attack the familiar equation

$$\ddot{x} \equiv \frac{d^2x}{dt^2} = -\omega_0^2 x \qquad \omega_0 = \sqrt{\frac{k}{m}} \tag{17.32}$$

where the second derivative of x is written as $\ddot{x}$, and ω_0, the natural frequency of vibrations of the oscillator, has been given a subscript to distinguish it from other ω's that will arise shortly. Earlier we solved this equation by turning it into a word problem: "What is the function, $x(t)$, with the property that two derivatives of the function look like the function itself, except for a proportionality constant?" We raked our brains and we remembered that sines and cosines had this property. One derivative is no good; it turns sine into cosine and vice versa. Two derivatives bring back the function you started with, which is why the answer could be sines or cosines. But now I'm going to solve the equation a different way. I know a function that is even better—it reproduces itself when it is differentiated *once*. If so, it's obvious that its 92-nd derivative will also look like the function. But recall why we rejected

$$x(t) = Ae^t. \tag{17.33}$$

I want to get something proportional to $-x(t)$ upon taking two derivatives, and this does not do it: I get $+x(t)$. It does not help to try something like Ae^{-t} because after two derivatives I again get $+x(t)$. So this function is no good. Also, it doesn't look like what I want. Even without doing much work, I know that if I pull this spring it's going to go back and forth, whereas these functions are exponentially growing or they're exponentially falling; they just don't do the trick. But now we have a way out: let the exponent be complex.

We are going to make a guess, called an *ansatz* in the business:

$$x(t) = Ae^{\alpha t} \tag{17.34}$$

where we will now allow α to be some general complex number.

The ansatz is a tentative guess with some parameters, A and α in this instance, the judicious choice of which may yield a solution. If you're lucky, it will work. If not, you move on and try another solution; it is just like speed dating.

So, let's take the ansatz in Eqn. 17.34, put it in Eqn. 17.32, and demand that it be satisfied:

$$\ddot{x} + \omega_0^2 x = 0 \tag{17.35}$$

$$\alpha^2 Ae^{\alpha t} + \omega_0^2 Ae^{\alpha t} = 0 \tag{17.36}$$

$$A(\alpha^2 + \omega_0^2)e^{\alpha t} = 0. \tag{17.37}$$

Our ansatz will work if we manage to get $(Ae^{\alpha t})(\alpha^2 + \omega_0^2)$ to vanish. How many ways are there to kill this beast? The choice $A = 0$ is called the trivial solution and corresponds to the oscillator sitting still forever. So A can be anything, except 0. Now $e^{\alpha t}$ never vanishes (even if α is complex), so it is not the cause of the zero. So it must be that

$$\alpha^2 + \omega_0^2 = 0 \quad \text{which means } \alpha = \pm i\omega_0. \tag{17.38}$$

(More generally, if in place of $e^{\alpha t}$, which *never vanishes*, we had any function that did not vanish identically, we can still cancel it by picking a time when it is non-zero, and deduce Eqn. 17.38.)

So now I have two solutions of the form $Ae^{\alpha t}$. For A you can pick any number you like, in fact, real or complex—it doesn't matter. The equation is satisfied. But α can be only one of two numbers: $\pm i\omega_0$.

How do we choose between the two solutions

$$x_+(t) = Ae^{i\omega_0 t} \quad \text{and} \tag{17.39}$$

$$x_-(t) = Ae^{-i\omega_0 t}? \tag{17.40}$$

It turns out that we can pick both, and I'll tell you what I mean by that. Let us begin with the fact that Eqn. 17.32 is a *homogeneous, linear* differential equation. I'll have to tell you what that means through an example:

$$17\frac{d^{96}x}{dx^{96}} + 16\frac{d^3x}{dx^3} + 2x = 0. \tag{17.41}$$

It is homogeneous because you only find a single power of x anywhere, which happens to be the first power here. It is a linear equation because you find either the function x or its derivatives, but never the squares of cubes or higher powers of x or the derivatives. Note that the 96-th derivative does not change this fact; it is still the 96-th derivative of x and not, say, x^3. By contrast,

$$\frac{d^2x}{dt^2} + 3x^2 = 0 \tag{17.42}$$

is a non-linear equation because of the x^2 term. A linear equation has a very important property that lies at the heart of so many things we do.

This is called the *principle of superposition*, and it states: *if $x_1(t)$ and $x_2(t)$ are two solutions of a homogeneous linear equation, then so is any linear combination of them with constant (t-independent) coefficients A and B:*

$$x(t) = Ax_1(t) + Bx_2(t).$$

Let us prove this for the oscillator case to understand where linearity comes in. Given

$$\ddot{x}_1 + \omega_0^2 x_1 = 0 \tag{17.43}$$

$$\ddot{x}_2 + \omega_0^2 x_2 = 0, \tag{17.44}$$

let us multiply the first by a constant A, the second by B, and add to get

$$A\ddot{x}_1 + A\omega_0^2 x_1 + B\ddot{x}_2 + B\omega_0^2 x_2 = 0 \tag{17.45}$$

$$\frac{d^2(Ax_1 + Bx_2)}{dt^2} + \omega_0^2(Ax_1 + Bx_2) = 0, \tag{17.46}$$

which clearly shows that $x(t) = Ax_1(t) + Bx_2(t)$ is also a solution. We used the fact that any derivative of a linear combination is the same linear combination of the derivatives and that the non-derivative term was linear in x. Try doing this for the non-linear case, say Eqn. 17.42, and you will find it does not work because $3Ax_1(t)^2 + 3Bx_2(t)^2 \neq 3(Ax_1(t) + Bx_2(t))^2$.

The bottom line is that if you give me two independent solutions to a homogeneous linear equation, I can manufacture an infinite number of solutions because I can pick A and B any way I like. The solutions x_1 and x_2 are like unit vectors $\mathbf{i}$ and $\mathbf{j}$, whose linear combinations with all possible coefficients yield an infinite number of vectors in two dimensions. A word of caution: $\mathbf{i}$ and $3\mathbf{i}$ are also two vectors, but by combining them you can only get solutions parallel to $\mathbf{i}$. These two vectors are said to be *linearly dependent*, which in this simple case means one is a multiple of the other. Likewise $e^{i\alpha t}$ and $5e^{i\alpha t}$ cannot be used to build anything other than multiples of $e^{i\alpha t}$. However, $e^{-i\alpha t}$ is an independent object because it is not a multiple of $e^{i\alpha t}$.

By the same analogy with **i** and **j**, if a linear combination of two linearly independent functions equals another linear combination, the coefficients have to match on both sides. Thus

$$Ae^{\alpha t} + Be^{5\alpha t} = Ce^{\alpha t} + De^{5\alpha t} \quad \text{implies} \tag{17.47}$$

$$A = C \qquad\qquad B = D. \tag{17.48}$$

17.3 Conditions on solutions to the harmonic oscillator

Let us then consider the general solution

$$x(t) = Ae^{i\omega_0 t} + Be^{-i\omega_0 t}. \tag{17.49}$$

How do we decide what A and B are? In general they are arbitrary. But on a given day, when you pull the mass by 9 cm and release it from rest, A and B have to be chosen so that at $t = 0$, $x(0) = 9$ and the velocity $v(0) = 0$. But I have a bigger problem. The answer is manifestly not real, and we know x is a real function. That is not a mathematical requirement of the equation, but a physical requirement. To say that x is real means the following. A complex number $x + iy$ has a complex conjugate $x - iy$, and the property of real numbers is that when you take the complex conjugate, nothing happens: it satisfies the condition $z = z^*$. There is no imaginary part whose sign you can flip. Real numbers are their own complex conjugates.

So, I'm going to demand that this solution, in addition to satisfying the basic equation, also is real. To do that, I'm going to demand $x(t)$ equals its complex conjugate $x^*(t)$:

$$x^*(t) = A^* e^{-i\omega_0 t} + B^* e^{+i\omega_0 t} = x(t) = Ae^{i\omega_0 t} + Be^{-i\omega_0 t}. \tag{17.50}$$

To find $x^*(t)$ given $x(t)$, I conjugated everything in sight. The complex conjugates of A and B became A^* and B^*. The complex conjugate of $e^{+i\omega_0 t}$ is $e^{-i\omega_0 t}$ and vice versa, because the i goes to minus i while t and ω_0 are real numbers and nothing happens to them.

So $x(t) = x^*(t)$ for all times t, if the coefficients of $e^{\pm i\omega_0 t}$ in Eqn. 17.50 match:

$$A = B^* \qquad B = A^*. \tag{17.51}$$

However, if $A = B^*$, then $B = A^*$ follows automatically because both are saying the same thing: A and B have equal real parts and opposite imaginary parts. This can also be seen another way. Conjugating both sides of $A = B^*$, we get $A^* = (B^*)^* = B$ because conjugating any complex number twice changes the sign of its imaginary part twice, which is equivalent to doing nothing: $(z^*)^* = z$.

The reality of x then leads to the solution

$$x(t) = A e^{i\omega_0 t} + A^* e^{-i\omega_0 t}. \tag{17.52}$$

In other words, B is not an independent number; it has to be the complex conjugate of A if x is to be real. I hope you can see at a glance that the solution above is real, because whatever the first animal is, the second is its complex conjugate and has the opposite imaginary part. When you add them, the answer will be real. But A is not necessarily real. In polar form it has a modulus $|A|$ and a phase ϕ, so that

$$x(t) = |A| e^{i\phi} e^{i\omega_0 t} + |A| e^{-i\phi} e^{-i\omega_0 t} = |A| e^{i(\phi + \omega_0 t)} + |A| e^{-i(\phi + \omega_0 t)}$$
$$= |A| \left[e^{i(\phi + \omega_0 t)} + e^{-i(\phi + \omega_0 t)} \right]. \tag{17.53}$$

Now, what is this function I have in brackets? You should be able to recognize this creature as a cosine. We have ended up with

$$x(t) = 2|A| \cos(\omega_0 t + \phi). \tag{17.54}$$

This describes an oscillator of amplitude $2|A|$ and phase ϕ. Notice how the amplitude and phase of the oscillator were encoded in a *single* complex number A.

Suppose you had chosen to use $\sin \omega_0 t$ and $\cos \omega_0 t$ as the two basic solutions instead of $e^{\pm i\omega_0 t}$. The general solution would have been

$$x(t) = A \cos \omega_0 t + B \sin \omega_0 t \tag{17.55}$$

where A and B are arbitrary. However, demanding that x be real will force them both to be real. No matter how you slice it, a physical oscillator will have in its solution just two free parameters: they could be two real numbers A and B as above or one complex number $A = |A| e^{i\phi}$ as before.

Now, this is a long and difficult way to get back the old answer. Your reaction may be, "We don't need these complex numbers. We have enough problems in life; we're doing well with sines and cosines, thank you." But now I'm going to give you a problem where you cannot talk your way out by just turning it into a word problem.

17.4 Exponential functions as generic solutions

Here is the problem: a mass m, connected to a spring of force constant k, is moving on a surface with friction. The minute there is friction, you have an extra force. We know that if you're moving to the right, the force of friction is to the left, and, if you are moving to the left, the force is to the right, that is, the frictional force is velocity dependent. The equation that crudely models this velocity dependence is

$$m\ddot{x} = -kx - \gamma m\dot{x} \tag{17.56}$$

where I include a factor m in the frictional coefficient to simplify subsequent algebra. Dividing by m, our equation becomes

$$\ddot{x} + \gamma\dot{x} + \omega_0^2 x = 0. \tag{17.57}$$

Can you solve this as a word problem? It's going to be difficult, because you want a function that, when you take two derivatives, add some amount of its own derivative, and then some of itself, gives zero. It is not clear a trigonometric function can do that. However, an exponential has to work because it reproduces itself no matter how many derivatives you take. Thus we make the ansatz

$$x(t) = Ae^{\alpha t}. \tag{17.58}$$

Note that I do not explicitly use a complex exponential. If α is meant to be complex, it will come out that way; we are not forcing it to be real in making this ansatz. When we feed it into Eqn. 17.57 we find, because every derivative brings a factor of α,

$$A(\alpha^2 + \gamma\alpha + \omega_0^2)e^{\alpha t} = 0. \tag{17.59}$$

Once again, A cannot be the cause of the zero, because if A vanishes you've killed the whole solution and $e^{\alpha t}$ is not going to vanish, so the only way is for the stuff in brackets to vanish:

$$(\alpha^2 + \gamma\alpha + \omega_0^2) = 0 \qquad (17.60)$$

That means the α that you put into this guess must be one of the roots

$$\alpha_\pm = \frac{-\gamma \pm \sqrt{\gamma^2 - 4\omega_0^2}}{2} = -\frac{\gamma}{2} \pm \sqrt{\frac{\gamma^2}{4} - \omega_0^2}. \qquad (17.61)$$

The general solution is

$$x(t) = Ae^{\alpha_+ t} + Be^{\alpha_- t} \qquad (17.62)$$

$$= A\exp\left[\left(-\frac{\gamma}{2} + \sqrt{\frac{\gamma^2}{4} - \omega_0^2}\right)t\right]$$

$$+ B\exp\left[\left(-\frac{\gamma}{2} - \sqrt{\frac{\gamma^2}{4} - \omega_0^2}\right)t\right]. \qquad (17.63)$$

The motion described by the solution depends on the value of $\frac{\gamma}{2\omega_0}$.

17.5 Damped oscillations: a classification

Let us classify the different kinds of behavior that emerge as we vary $\frac{\gamma}{2\omega_0}$.

17.5.1 Over-damped oscillations

We first consider the *over-damped case*

$$\frac{\gamma}{2} > \omega_0. \qquad (17.64)$$

In this case both roots $\alpha_\pm$ are real and both are negative: α_- is negative being a sum of two negative numbers, while α_+ is negative because the positive square root is smaller than $\gamma/2$. This means that $x(t \to \infty) \to 0$, which is in accord with our expectation that friction will eventually bring the oscillations to an end.

How about A and B? First of all, they are both real as can be seen by equating $x(t)$ to its conjugate. Because the exponentials are real they do not respond to conjugation and we require $A = A^*$ and $B = B^*$.

To find A and B, we need two pieces of data, which I will take to be initial position, $x(0)$, and the initial velocity, $v(0)$. If we put $t = 0$ in Eqn. 17.62 we find

$$x(0) = A + B. \tag{17.65}$$

Next I take the derivative of Eqn. 17.63 *and then set $t = 0$* to find

$$v(0) = A\alpha_+ + B\alpha_-. \tag{17.66}$$

Solving these simultaneous equations will yield A and B. To test yourself, try showing that if the oscillator is displaced to some $x(0) > 0$ and released from rest, that is, $v(0) = 0$, then $x(t)$ never becomes 0 and hence cannot become negative. This means the mass will simply relax to its equilibrium position without any oscillations.

17.5.2   Under-damped oscillations

In turning on friction we got carried away: from being 0 in the very first example, γ jumped to a value greater than $2\omega_0$. Consider now the intermediate case when $0 < \gamma < 2\omega_0$. What do the solutions look like now? We should be able to guess that, at least for very tiny values of γ, the oscillator will oscillate as before, but with a slowly diminishing amplitude. Let us verify and quantify this expectation.

The roots now become

$$\alpha_\pm = -\frac{\gamma}{2} \pm \sqrt{\frac{\gamma^2}{4} - \omega_0^2} \tag{17.67}$$

$$= -\frac{\gamma}{2} \pm i\sqrt{\omega_0^2 - \frac{\gamma^2}{4}} \tag{17.68}$$

$$\equiv -\frac{\gamma}{2} \pm i\omega'. \tag{17.69}$$

We have introduced yet another frequency

$$\omega' = \sqrt{\omega_0^2 - \frac{\gamma^2}{4}} < \omega_0, \qquad (17.70)$$

which describes the oscillatory part of the motion. Note that the roots are complex conjugates

$$\alpha_+ = \alpha_-^* \qquad (17.71)$$

and the general solution becomes

$$x(t) = A e^{\alpha_+ t} + B e^{\alpha_- t} \qquad (17.72)$$

$$= e^{-\frac{1}{2}\gamma t} \left[A e^{i\omega' t} + B e^{-i\omega' t} \right]. \qquad (17.73)$$

I leave it to you to verify that once again $x = x^*$ implies $A^* = B$ because the A and B terms get exchanged under complex conjugation. Repeating the analysis for the case $\gamma = 0$, this solution may be rewritten as

$$x(t) = C e^{-\frac{1}{2}\gamma t} \cos \left[\omega' t + \phi \right] \quad \text{where} \qquad (17.74)$$

$$C = 2|A| \quad \text{and} \quad A = |A| e^{i\phi}. \qquad (17.75)$$

Figure 17.4 shows what the damped oscillation looks like for $A = 2$, $\gamma = 1$, and $\omega' = 2\pi$. This is typically what you will see if you excite any

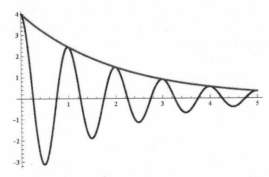

Figure 17.4   Damped oscillations with $x(t) = 4e^{-.5t} \cos(2\pi t)$, i.e, $A = 2$, $\gamma = 1$, and $\omega' = 2\pi$. The falling exponential shows the decay of the amplitude.

system with some modest amount of frictional loss. If γ is very small, you may not realize the oscillations are being damped.

17.5.3 Critically damped oscillations

Having considered the cases $\gamma > 2\omega_0$ (over-damped) and $\gamma < 2\omega_0$ (under-damped), we turn to the *critically damped* case $\gamma = 2\omega_0$. In this case $\alpha_+ = \alpha_- = -\frac{\gamma}{2}$. Where is the second solution to accompany $Ae^{-\frac{\gamma t}{2}}$? We know in every problem there must be two solutions, because we should be able to pick the initial position and velocity at will. That's an area of mathematics I don't want to enter now, but you can verify that the second solution is $Bte^{-\frac{\gamma t}{2}}$, which is not a pure exponential. You will find the derivation of this solution in my math book. The general solution for the critically damped case is thus

$$x(t) = e^{-\frac{\gamma t}{2}}[A + Bt]. \tag{17.76}$$

Try to show in this case that $A = x(0)$ and $B = v(0) + \frac{\gamma}{2}x(0)$.

17.6 Driven oscillator

Next we turn to a more challenging problem. I have, as before, the mass, the spring, and friction. But now I'm going to apply an extra force, $F_0 \cos \omega t$. This is called a *driven oscillator*. Imagine that I am actively shaking the mass with my hand, exerting the force $F_0 \cos \omega t$. Now there are three ω's: $\omega_0 = \sqrt{\frac{k}{m}}$, the natural frequency of the undamped free oscillator; $\omega' = \sqrt{\omega_0^2 - \frac{\gamma^2}{4}}$, which entered the under-damped oscillator; and finally ω, the frequency of the driving force, which is completely up to me to choose. The equation to solve is

$$m\ddot{x} + \gamma m\dot{x} + kx = F_0 \cos \omega t, \tag{17.77}$$

which we rewrite as

$$\ddot{x} + \gamma \dot{x} + \omega_0^2 x = \frac{F_0}{m} \cos \omega t. \tag{17.78}$$

This problem is difficult because you cannot guess the answer to it by turning it into a word problem: neither $x(t) \propto \cos(\omega t)$ nor $x(t) \propto \sin(\omega t)$ is a good ansatz because you cannot have all four terms be functions of the same kind as the ansatz. In fact, only an exponential can lead to all four terms being the same functional form (exponential) because taking any number of derivatives will leave it alone. *But our driving force is a cosine and not an exponential.*

Here is a clever trick to beat this problem. Recall that with no driving force if

$$\ddot{x}_1 + \gamma \dot{x}_1 + \omega_0^2 x_1 = 0 \quad \text{and} \tag{17.79}$$

$$\ddot{x}_2 + \gamma \dot{x}_2 + \omega_0^2 x_2 = 0, \tag{17.80}$$

then multiplying the first by a constant A and the second by a constant B, and adding we found that $Ax_1 + Bx_2$ was also a solution:

$$\frac{d^2[Ax_1 + Bx_2]}{dt^2} + \gamma \frac{d[Ax_1 + Bx_2]}{dt} + \omega_0^2 [Ax_1 + Bx_2] = 0. \tag{17.81}$$

I have used the fact that the derivatives of a linear combination $Ax_1 + Bx_2$ are the same linear combination of the derivatives.

Suppose now that there is a driving force behind x_1 and x_2:

$$\ddot{x}_1 + \gamma \dot{x}_1 + \omega_0^2 x_1 = \frac{F_1(t)}{m} \tag{17.82}$$

$$\ddot{x}_2 + \gamma \dot{x}_2 + \omega_0^2 x_2 = \frac{F_2(t)}{m}. \tag{17.83}$$

It follows by the same manipulations that

$$\frac{d^2[Ax_1 + Bx_2]}{dt^2} + \gamma \frac{d[Ax_1 + Bx_2]}{dt} + \omega_0^2 [Ax_1 + Bx_2]$$

$$= A\frac{F_1(t)}{m} + B\frac{F_2(t)}{m}. \tag{17.84}$$

In other words, *in a linear equation, the response to a linear combination of forces is the corresponding linear combination of responses.*

Now for the trick. Let $x(t)$ be the solution to

$$\ddot{x} + \gamma \dot{x} + \omega_0^2 x = \frac{F_0}{m} \cos \omega t \qquad (17.85)$$

and $y(t)$ the solution to

$$\ddot{y} + \gamma \dot{y} + \omega_0^2 y = \frac{F_0}{m} \sin \omega t. \qquad (17.86)$$

(We could call these two solutions x_1 and x_2, but there is a reason for this nomenclature.) Multiply the second equation by i and add it to the first to obtain

$$\frac{d^2[x+iy]}{dt^2} + \gamma \frac{d[x+iy]}{dt} + \omega_0^2 [x+iy] = \frac{F_0}{m}(\cos \omega t + i \sin \omega t)$$

$$= \frac{F_0}{m} e^{i\omega t} \qquad (17.87)$$

$$\ddot{z} + \gamma \dot{z} + \omega_0^2 z = \frac{F_0}{m} e^{i\omega t} \quad \text{where} \qquad (17.88)$$

$$z(t) = x(t) + iy(t). \qquad (17.89)$$

This is a special case of Eqn. 17.84 with $A = 1$ and $B = i$.

So, in Eqn. 17.88 I have manufactured a problem in which the thing that's vibrating is not a real number, but $z = x + iy$. The force driving it is also not a real number; it is $\frac{F_0}{m} e^{i\omega t}$. *The point is that if I can solve the problem somehow, I can get $x(t)$ as the real part of the answer.* (The imaginary part of it, $y(t)$, will be the solution to the fictitious Eqn. 17.86 I concocted.)

And I can solve Eqn. 17.88 for $z(t)$ very easily because I can now make the ansatz

$$z(t) = z_0 e^{i\omega t}. \qquad (17.90)$$

Because every derivative pulls out an $i\omega$ we have

$$[-\omega^2 + i\omega\gamma + \omega_0^2] z_0 e^{i\omega t} = \frac{F_0}{m} e^{i\omega t}. \qquad (17.91)$$

We may safely cancel $e^{i\omega t}$ because it is not identically zero to obtain the equation for z_0:

$$z_0 = \frac{F_0/m}{[-\omega^2 + i\omega\gamma + \omega_0^2]} \tag{17.92}$$

$$= \frac{F_0/m}{Z(\omega)} \quad \text{where we have defined} \tag{17.93}$$

$$Z(\omega) = [-\omega^2 + i\omega\gamma + \omega_0^2]. \tag{17.94}$$

The magic of the exponential is that the differential equation 17.88 has reduced to an algebraic equation for the (complex) amplitude z_0

$$Z(\omega)z_0 = \frac{F_0}{m}, \tag{17.95}$$

which is solved by dividing both sides by $Z(\omega)$:

$$z_0 = \frac{F_0/m}{Z(\omega)}. \tag{17.96}$$

It follows that

$$z(t) = z_0 e^{i\omega t} = \frac{[F_0/m]\, e^{i\omega t}}{Z(\omega)}. \tag{17.97}$$

All we need to do now is take the real part to get $x(t)$. If you thought that this means replacing $e^{i\omega t}$ by $\cos\omega t$ you are wrong, because

$$Z(\omega) = [-\omega^2 + i\omega\gamma + \omega_0^2] \tag{17.98}$$

is itself a complex number whose real and imaginary parts can mix with the real and imaginary part of $e^{i\omega t}$. So here is the correct way to do this. Take $Z(\omega)$ in Cartesian form

$$Z(\omega) = [\omega_0^2 - \omega^2] + i\omega\gamma \tag{17.99}$$

and write it in polar form

$$Z(\omega) = |Z|e^{i\phi} \quad \text{where} \tag{17.100}$$

$$|Z| = \sqrt{\left[\omega_0^2 - \omega^2\right]^2 + \omega^2\gamma^2} \tag{17.101}$$

$$\phi = \tan^{-1}\left[\frac{\omega\gamma}{\omega_0^2 - \omega^2}\right]. \tag{17.102}$$

Figure 17.5 shows Z in the complex plane.

Return to Eqn. 17.97 with this result to obtain

$$z(t) = \frac{[F_0/m]e^{i\omega t}}{Z(\omega)} = \frac{[F_0/m]\,e^{i\omega t}}{|Z|e^{i\phi}} \tag{17.103}$$

$$= \frac{F_0}{m|Z|}e^{i(\omega t-\phi)}. \tag{17.104}$$

Now we can take the real part easily because $\frac{F_0}{m|Z|}$ is real. Here is the final answer:

$$x(t) = \frac{F_0}{m|Z|}\cos(\omega t - \phi) \equiv x_0\cos(\omega t - \phi). \tag{17.105}$$

Notice that the cause, $\frac{F_0}{m}\cos(\omega t)$, produces an effect that is reduced in magnitude by $|Z|$ and shifted in phase into $\cos(\omega t - \phi)$. While there is a way to obtain both these transformations with real numbers, it is so much

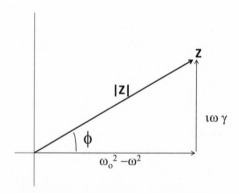

Figure 17.5   The complex number $Z(\omega)$ in its Cartesian and polar forms.

easier with complex numbers: dividing the force by a complex number $Z = |Z|e^{i\phi}$ achieves both these effects in one shot. Bear in mind that the phase ϕ cannot be eliminated by choice of the origin in time because it is the phase *relative* to that of the applied force $F_0 \cos \omega t$.

Let us pause to analyze Eqn. 17.105. Keeping $\frac{F_0}{m}$ fixed, let us vary ω, the frequency of the driving force, to see what happens to x_0, the amplitude of vibrations. When $\omega = 0$, that is, when the force does not vary with time, we find

$$|Z(0)| = \sqrt{\left[\omega_0^2 - \omega^2\right]^2 + \omega^2\gamma^2}\,\Big|_{\omega=0} = \omega_0^2 \qquad (17.106)$$

so that

$$x_0 = \frac{F_0}{m\omega_0^2} = \frac{F}{k}, \qquad (17.107)$$

which makes sense: a constant force F will produce a displacement $\frac{F}{k}$.

When $\omega \to \infty$, we find $x_0 \to 0$. Somewhere in between these extremes, the response peaks. It is clear that if γ is very small, we get the biggest response when $\omega = \omega_0$: this is when $|Z|$ is the smallest. This is called *resonance*. It tells us that the response of the system to a driving force is greatest when the driving frequency equals the natural frequency. Imagine you are pushing a kid on a swing, by periodically supplying the force. If you are not paying attention and pushing at your own frequency, sometimes you will slow the kid and sometimes you will speed up the kid. It is best to push exactly when the kid is moving away from you. Note that in a real swing $\gamma > 0$, and there is no danger of the kid flying off to infinity.

Radios exploit the phenomenon of resonance. Right now this room is filled with electromagnetic signals from many stations, and yet you are able to listen to the one you want. The trick is that you can adjust the natural frequency of the electrical circuits picking up the signal by turning the dial to match that of the station of interest. For this plan to succeed, you need the graph in Figure 17.6 to be extremely sharp. Imagine that there are just two stations at two frequencies. Even if you tune the radio to resonate with one, you will be getting a tiny response from the tail of the other one. The goal is to keep this interference to a minimum.

Where are the free parameters in this problem? Everything seems determined in Eqn. 17.105. What if this solution does not agree with some

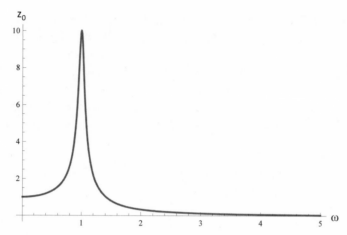

Figure 17.6   The amplitude $z_0(\omega)$ for a system with $\omega_0 = 1$ and $\gamma = 0.1$ driven by an external force with $\frac{F_0}{m} = 1$.

initial condition, such as a specified $x(0)$ or $v(0)$? The answer is that we can add to this solution (called the *particular solution*) any solution to the equation with $F(t) = 0$, referred to as the *complementary solution* and given in Eqn. 17.73. Thus the most general solution to the driven oscillator is

$$x(t) = \frac{F_0}{m|Z|} \cos(\omega t - \phi) + e^{-\frac{1}{2}\gamma t}\left[Ae^{i\omega' t} + Be^{-i\omega' t}\right]. \qquad (17.108)$$

Even after adding this term $x(t)$ satisfies Eqn. 17.78 because the added terms disappear when we compute the left-hand side. Another way to see this is to invoke superposition: consider the right-hand side of Eqn. 17.78 to be $\frac{F_0}{m}\cos(\omega t) + 0$ and add the response due to 0, which is the complementary function. The numbers A and B can once again be chosen to match the initial conditions, say the initial position and velocity. One may forget about the complementary function at large times, because it dies out exponentially.

Finally, consider the force pushing the kid as described earlier. It is periodic but not simply the function $\cos \omega t$. (For example, the force on the kid acts only for a small part of each period, while the cosine is non-zero except twice in a period.) Amazingly, we can use the technique described above to find the response to any periodic force, not necessarily a simple cosine or oscillatory exponential function. This is thanks to the mathematician Joseph Fourier, who showed that any function $F(t)$ with period T

may be written as a sum of oscillating exponentials, with suitable periods, multiplied by suitable coefficients F_n:

$$F(t) = \sum_{n=-\infty}^{\infty} F_n e^{2\pi int/T} \equiv \sum_{n=-\infty}^{\infty} F_n e^{i\omega_n t} \text{ where} \qquad (17.109)$$

$$\omega_n = \frac{2\pi n}{T}. \qquad (17.110)$$

In the right-hand sides of Eqns. 17.109 and 17.110, we have a sum of forces with frequencies $\omega_n = \frac{2\pi n}{T}$. I state without proof that the coefficients are determined by the given $F(t)$ as follows:

$$F_n = \frac{1}{T} \int_0^T F(t) e^{-i\omega_n t} dt. \qquad (17.111)$$

We are done because we know the response $z_0(n)$ due to each oscillating term $F_n e^{i\omega_n t}$ in the sum, and by previous linearity arguments the total response is the corresponding sum over responses:

$$z(t) = \sum_{n=-\infty}^{\infty} z_0(n) e^{i\omega_n t} \text{ where} \qquad (17.112)$$

$$z_0(n) = \frac{\frac{F_n}{m}}{Z(\omega_n)}. \qquad (17.113)$$

If the driving force $F(t)$ is real, the $z(t)$ above will automatically turn out to be real. If you want more practice using complex numbers, you are invited to read the following proof.

First note that

$$-\omega_n = \frac{2\pi(-n)}{T} = \omega_{(-n)} \qquad (17.114)$$

$$F_n^* = \frac{1}{T} \int_0^T F(t) e^{+i\omega_n t} dt \qquad (17.115)$$

$$= \frac{1}{T} \int_0^T F(t) e^{-i(-\omega_n)t} dt = \frac{1}{T} \int_0^T F(t) e^{-i\omega_{(-n)}t} dt \qquad (17.116)$$

$$= F_{-n}. \qquad (17.117)$$

Now pair the contributions from the terms in Eqn. 17.112 with any n and $-n$:

$$z_0(n)e^{i\omega_n t} + z_0(-n)e^{i\omega(-n)t} = \frac{\frac{F_n}{m}}{Z(\omega_n)}e^{i\omega_n t} + \frac{\frac{F_{-n}}{m}}{Z(\omega_{-n})}e^{i\omega(-n)t}$$

$$(17.118)$$

$$= \frac{\frac{F_n}{m}}{Z(\omega_n)}e^{i\omega_n t} + \left[\frac{\frac{F_n}{m}}{Z(\omega_n)}e^{i\omega_n t}\right]^*,$$

$$(17.119)$$

which is manifestly real, being a sum of something plus its conjugate. We have also used

$$Z(\omega_{(-n)}) = Z(-\omega_n) = Z^*(\omega_n),$$

$$(17.120)$$

because for any ω we have

$$Z(\omega) = \left[-\omega^2 + i\omega\gamma + \omega_0^2\right]$$

$$(17.121)$$

$$Z^*(\omega) = \left[-\omega^2 - i\omega\gamma + \omega_0^2\right] = Z(-\omega).$$

$$(17.122)$$

CHAPTER 18

Waves I

We are moving to another topic: waves. Everyone has a good intuitive feeling for waves. Suppose you drop some object in a lake and you see ripples traveling outward from the center. If you keep your eye level with the water you will find these ups and downs going outward. This is why one says a wave is some displacement of a medium. That's not a perfect definition because electromagnetic waves travel in a vacuum. For this course you should imagine waves as what happens when you excite a medium. Once you have the example of water waves, you can latch onto that example every time I say "wave."

Here's the way to think about the wave. You understand the harmonic oscillator pretty well, right? A mass and spring system in equilibrium will sit there forever. If you give it a kick, it will start vibrating around its equilibrium position. There you have to keep track of only one variable, $x(t)$, which is the location of the mass at time t. The wave is an oscillation of an entire medium, and that means that at every point in space (where there is some medium) there is something that's ready to oscillate. You have then a system with an infinite number of degrees of freedom because the height of the water at each point is an independent variable. If you don't do anything to the water, it will stay at the height, but, if you fiddle with it or drop something, it will start vibrating. I'm going to use the symbol $\psi(x, t)$ to denote the change in the height of the water from its undisturbed value at the point x at time t, or more generally, the displacement of the medium from its equilibrium configuration. Unlike the case of the single oscillator, x is not the dynamical variable here. Instead it labels the points in the medium, and $\psi(x, t)$ is the dynamical variable: it is the thing jumping up and down. For simplicity I am considering waves in one dimension; in general I will have to invoke $\psi(x, y, z, t)$.

The wave travels with some velocity v. If you drop a rock in a lake, you see when the waves get to the shore, you measure the distance to the shore divided by the time taken, and that's the velocity. If someone lights a firecracker, you can find the velocity of the sound by determining how far away you were from the firecracker and how long it took for the sound to reach you. That's the velocity of sound in air. Unlike the velocity of light, which doesn't depend on anything, all other wave velocities depend on various conditions. The velocity of sound in air depends on the temperature, for example. Sound can also travel in a solid. You can take an iron rod, which you know is made up of a lot of atoms, and hit it with a hammer. You basically compress these atoms and they in turn compress the atoms next to them, and the shock wave travels through this rod; that's also called sound. That travels at a speed much faster than the speed of sound in air. Thunder and lightning are the most famous examples of different velocities for different phenomena: the light gets to you first, and then the sound.

Waves can be *longitudinal* or *transverse*, as shown in Figure 18.1.

When I talk to you, this is what happens. The air in this room is at some constant pressure; the pressure is pushing your eardrum from the outside and also transmitted through your Eustachian canal and pushing the drum from the inside. The two pressures balance and you don't feel anything. As I talk, my vocal chords move back and forth and they increase and decrease the air pressure. These pressure waves propagate through the air to your ear. They hit your eardrum, which feels the changing pressure outside, relative to the fixed pressure inside. So the eardrum goes back and forth in response. Behind the eardrum are the three little bones that transmit the motion of the eardrum to a fluid inside. There are little hairs shaking in this fluid, sending signals to your brain, which in turn tries to figure out what I am saying. That's a long and impressive chain of transducers of sound energy to your brain. I draw the line at the eardrum; I don't want to go on the other side. But what I do know is that a sound wave is a *longitudinal wave*: the back-and-forth motion of air is in the same direction as the sound signal—both are along the line joining

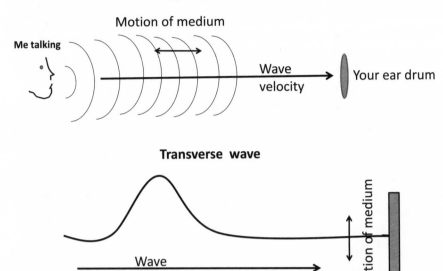

Figure 18.1   Top: A sound wave is longitudinal: the sound goes from left to right and the air also vibrates back and forth. Bottom: The wave on a string is transverse: the wave goes from left to right, but the string moves up and down.

you to me. In a longitudinal wave like sound, the motion of the medium is in the same direction as the motion of the wave.

This is not always the case: In a *transverse wave* the medium moves perpendicular to the wave velocity. Take a string, clamp it down at the right end, pull it tight to give it some tension, and then give it a little flick at the left end. The blip you create will travel along the string, but the displacement of the string (the blip) is perpendicular to the velocity, as shown in Figure 18.1. This is a transverse wave.

Do not confuse the motion of the medium with the motion of the signal. The string, at any given point, is either at rest or going up and down, but the signal is going from left to right. Even in the case of sound, when I talk to you, the air molecules may wiggle back and forth in the direction of the velocity, but their average position does not change; the molecule hitting your ear is not the one that was next to my mouth. So, the medium doesn't actually propagate along with the wave. The game called Telephone illustrates this: kids line up and each kid tells the next kid some secret; the secret is eventually transmitted all the way to the end. What has traveled is the secret. The kid at the head of the line has not physically moved to the other end. Each kid hears something from the neighbor on one side and passes it on to the neighbor on the other side. Then each goes back to doing nothing. That's what happens in the string as the blip travels. A section of the string that is not doing anything initially jumps up and down for a while when the blip passes through. Then it settles down and the section next to it starts moving.

18.1 The wave equation

There are many, many waves: water waves, electromagnetic waves, elastic waves, and so on. I want to consider one concrete example of a wave so you can develop a feeling for how to handle them. I'm going to discuss waves on a string.

Imagine a string that's been clamped at two ends ($x = 0$ and $x = L$ in Figure 18.2). The thin horizontal line is the x-axis, and that is the string's position in static equilibrium. Each point on the string is labeled by the value of x that it will have when the string is in the equilibrium position. The displacement of the string at the point labeled x at time t is denoted $\psi(x, t)$, and this is our new dynamical variable. It is the one for which we would like to write the equations of motion.

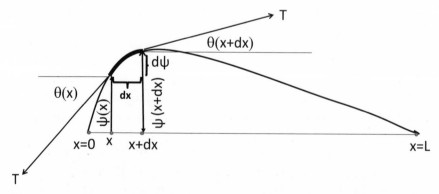

Figure 18.2   The string under tension T has mass μ per unit length and is fixed at $x = 0$ and $x = L$. The highlighted segment has a width dx, with the same tension T pulling the two ends but at slightly different angles. The displacement ψ and angles θ are exaggerated for clarity. The derivation is valid only when all these are very small.

The string is under some tension T. That means you hang some weights at the ends or you tighten it with some screws, as in a violin. The tension means that the string is dying to break apart if you cut it. Without the tension, none of what follows would work, as you will see. Another essential parameter here is μ, the mass per unit length. To find it you put the string on a weighing scale, you find the mass, and you divide by the length. For example, if you have a ten-meter string and it weighs one-hundredth of a kilogram, then the mass per unit length is $\mu = 10^{-3}$ kilograms per meter.

Now, I pull or pluck this string in some way, given by the solid curve $\psi(x, 0)$ in the figure, and I want to know what the whole string will do. Compare this to the mass and spring system. There you pull the mass out to some new location $x(0)$, you let it go, and you want to know $x(t)$. There was just one degree of freedom, the location of the mass, $x(t)$. The answer was $x(t) = x(0) \cos \omega t$. Here, at *every point x between 0 and L*, I have some segment of the string. The displacement of each segment from equilibrium is a degree of freedom, $\psi(x)$. I displace all those infinite degrees of freedom to $\psi(x, 0)$ at time t, and I let them go. I want to know $\psi(x, t)$. For this we need to find the equation satisfied by $\psi(x, t)$.

What authority will decide the behavior of this string? Newton's law is the answer. There are no new laws that I'm going to invoke. I'm not going to say, "We studied masses and springs before; today it's time to study strings, and here is the new law of motion." There's only one law of motion. That's $F = ma$. My whole purpose is to show you that this law really does control everything; that's why it's a super law.

The string is a long, extended, and complicated object. I isolate a tiny segment of length dx highlighted in the figure. I am going to calculate the total force on it and equate it to its mass times acceleration. Gravity is not necessary for vibrations, and we will neglect its effect.

The figure shows the two forces on the little segment. Both equal the tension T, which doesn't change from point to point in magnitude. But the angle at which the tension acts is not necessarily the same. It is tangent to the string, and the direction of the tangent (measured from the horizontal) is changing from $\theta(x)$ to $\theta(x + dx)$. The string is curving in general; therefore, the tangents to the string at the two ends of the tiny bit are not quite the same, and there is generally a net force on the bit.

So, I'm going to find out the vertical component of the two forces and take the difference. The upward force at $x + dx$ will be $T \sin(\theta(x + dx))$, and the downward force on the left side will be $-T \sin(\theta(x))$, yielding a

total of $T\left[\sin(\theta(x+dx)) - \sin(\theta(x))\right]$. That's going to be mass times acceleration. The mass of this little segment is the mass per unit length μ times
the length of the segment, which is dx. Now, what is the acceleration in
the language of calculus? No, it is not $\frac{d^2x}{dt^2}$ but $\frac{\partial^2 \psi(x,t)}{\partial t^2}$ because $\psi(x,t)$ is the
vertical coordinate of the string bit. What's jumping up and down is ψ,
so the acceleration is its second derivative, and I use the partial derivative
because $\psi(x,t)$ can vary with x and t. So $F = ma$ becomes

$$T\left[\sin(\theta(x+dx)) - \sin(\theta(x))\right] = \mu dx \frac{\partial^2 \psi(x,t)}{\partial t^2}. \qquad (18.1)$$

Now, come to the left-hand side and assume the angles involved are
very small, that is, that the string does not deviate too much from being
horizontal. If you remember the series

$$\sin\theta = \theta - \frac{\theta^3}{3!} + \dots \qquad (18.2)$$

$$\cos\theta = 1 - \frac{\theta^2}{2!} + \dots \qquad (18.3)$$

$$\tan\theta = \frac{\theta - \frac{\theta^3}{3!} + \dots}{1 - \frac{\theta^2}{2!} + \dots} = \left(\theta - \frac{\theta^3}{3!} + \dots\right)\left(1 + \frac{\theta^2}{2!} + \dots\right) = \theta + \dots$$

$$(18.4)$$

and keep only terms up to order θ, you may then approximate as follows:

$$\sin\theta \simeq \theta \simeq \tan\theta = \frac{\partial\psi}{\partial x}. \qquad (18.5)$$

Equation 18.1 becomes

$$T\left[\frac{\partial\psi}{\partial x}\bigg|_{x+dx} - \frac{\partial\psi}{\partial x}\bigg|_{x}\right] = \mu dx \frac{\partial^2 \psi(x,t)}{\partial t^2}. \qquad (18.6)$$

Dividing both sides by T and dx and letting $dx \to 0$, we finally obtain the
wave equation

$$\frac{\partial^2 \psi(x,t)}{\partial x^2} = \frac{\mu}{T}\frac{\partial^2 \psi(x,t)}{\partial t^2}. \qquad (18.7)$$

This is a *partial differential equation*. It is usually rewritten as

$$\frac{\partial^2 \psi(x,t)}{\partial x^2} = \frac{1}{v^2}\frac{\partial^2 \psi(x,t)}{\partial t^2} \tag{18.8}$$

$$v = \sqrt{\frac{T}{\mu}}. \tag{18.9}$$

You should verify that v has dimensions of velocity. It will turn out to be the velocity of waves on the string.

In summary, when you pull a string up, it comes down because the tensions at the two ends of the string bit have vertical components that don't quite cancel. So, the net force depends on the rate of change of $\sin\theta \simeq \tan\theta = \frac{\partial \psi(x,t)}{\partial x}$, that is, the rate of change of the rate of change, and that's why you get $\frac{\partial^2 \psi(x,t)}{\partial x^2}$ on the left-hand side. The second time-derivative on the right is just the acceleration of the string bit.

Two questions are usually asked at this point. From the figure we see that dx is actually the horizontal projection of the string bit. Shouldn't we use its full length in finding the mass of the bit? Shouldn't we worry that because different parts of the string have stretched by different amounts the mass density μ is really $\mu(x)$? We do not worry, because of the small angle approximation. By Pythagoras' theorem the length of the string bit is (see the figure)

$$dl = \sqrt{(dx)^2 + (d\psi)^2} = dx\sqrt{1 + \left[\frac{\partial \psi}{\partial x}\right]^2} \tag{18.10}$$

$$\simeq dx\left(1 + \frac{1}{2}\left[\frac{\partial \psi}{\partial x}\right]^2\right) \tag{18.11}$$

$$= dx(1 + \frac{1}{2}\tan^2\theta) \simeq dx \tag{18.12}$$

because $\tan^2\theta$ is at least quadratic in θ. Thus in this approximation every part of the string is stretched in the horizontal direction by the same amount as before it was plucked. That is why the tension has a constant magnitude T along its length.

The second question is why we do not worry about the difference in the *horizontal* components of T at the two ends. The answer is that these components are proportional to $\cos\theta = 1 + \frac{\theta^2}{2} + .. \simeq 1$ if we stop at first order in θ.

Equation 18.8 is ubiquitous and describes any elastic medium disturbed a little bit from equilibrium. The small disturbances ψ obey such an equation and v is the velocity of their propagation.

18.2   Solutions of the wave equation

What are the consequences of this wave equation? Why is v the velocity of the wave? To find out, let us go from the finite string to an infinite medium, like an infinite lake.

To proceed we need to guess a solution to this equation. In the case of the oscillator we said, "I have an equation: the second derivative of x is equal to some number times x. The answer is $\cos\omega t$ or $\sin\omega t$." But now we have the second derivatives with respect to space and second derivatives with respect to time. I know that this ψ is going to oscillate up and down in space and time. To break down the problem, imagine a water wave that is traveling; the ripples are traveling. If you took a snapshot of the wave at a given time, it would go up and down in space. Or, if you stood at one point in the water and let the wave go past you, the water would go up and down in time. So the wave oscillates in space and time, and we have to guess the answer. I could tell you how to deduce the answer, but I do not have room here. I'm just going to write down a solution and verify that it obeys the wave equation, and then I wil analyze it. Now there are many, many solutions to the equation. The solution I'm going to write down is

$$\psi(x,t) = A\cos(kx - \omega t). \tag{18.13}$$

Beware! This k is not a force constant. It is a new symbol called the *wave number*. We know that ω must have units of inverse time and k must have dimensions of inverse length in order that the argument of the cosine is dimensionless.

Let's verify that the Eqn. 18.13 is a solution to the wave equation. Consider $\frac{\partial^2 \psi}{\partial x^2}$. When you take a partial derivative with respect to x it says forget about time; treat x as the only variable. So

$$\frac{\partial^2 \psi}{\partial x^2} = -Ak^2 \cos(kx - \omega t). \tag{18.14}$$

Likewise

$$\frac{\partial^2 \psi}{\partial t^2} = -A\omega^2 \cos(kx - \omega t).$$ (18.15)

The wave equation

$$\frac{\partial^2 \psi}{\partial x^2} = \frac{1}{v^2}\frac{\partial^2 \psi}{\partial t^2} \quad \text{then demands that}$$ (18.16)

$$-Ak^2 \cos(kx - \omega t) = -A\frac{\omega^2}{v^2}\cos(kx - \omega t).$$ (18.17)

We cancel A and the $\cos(kx - \omega t)$ because neither is identically zero. This means that a solution of the form Eqn. 18.13 exists for any amplitude A. (Of course, A cannot be anything. The calculation assumed $\sin\theta \simeq \tan\theta$ and $\cos\theta \simeq 1$. So, once you make an approximation and you get an answer, you should not blindly apply the answer to circumstances where the equation itself is not valid. Even though the answer says you can have any A, in practice you shouldn't use it for an A for which the small angle approximation fails.)

Canceling $A\cos(kx - \omega t)$ from both sides of Eqn. 18.17, we find a solution of the assumed form exists provided the parameters ω and k satisfy

$$k = \pm\frac{\omega}{v}.$$ (18.18)

We will always choose ω to be positive, and the two signs of k will be seen to correspond to the two possible directions of propagation.

Now let us write the solution that respects the condition Eqn. 18.18, considering first the case $k = \frac{\omega}{v}$:

$$\psi(x,t) = A\cos(kx - kvt) = A\cos\left[k(x - vt)\right].$$ (18.19)

This equation says that $\psi(x,t)$, which could have depended on x and t separately, depends on them only through this combination $x - vt$. In other words, x and t do not separately determine $\psi(x,t)$; only this combination $x - vt$ does. If the combination has a certain value, ψ has a certain value. If I change x and t keeping $x - vt$ constant, *the function doesn't change*. I will now argue that this really means it's a wave that's traveling at a velocity v.

Let's take the bell-shaped function

$$f(x) = Ae^{-k^2 x^2}.$$

(handwritten: $f(x) = e^{x^2}$ peak $x=0$)

(18.20)

It peaks at $x = 0$. Take the same function with x replaced by $x - vt$. Where does this function

$$f(x, t) = Ae^{-k^2(x - vt)^2}$$

(handwritten: $f(x) = e^{(x-vt)^2}$ peak $x=vt$; vt)

(18.21)

have its maximum? It has a maximum when $x = vt$. So, if you followed this $f(x, t) = f(x - vt)$ as a function of time, the peak that was at $x = 0$ at $t = 0$ moves to $x = vt$ at time t: the shape moves undistorted at a velocity v.

Returning to our wave, it just looks like $A \cos kx$ at $t = 0$, and we all know it has a maximum or crest at $x = 0$. If you wait a bit, where is this maximum? If you increase time a little bit by dt, you have to increase x by dx such that the two changes kdx and ωdt cancel, keeping the argument of the cosine at zero. But that means that

$$\frac{dx}{dt} = \frac{\omega}{k} = v.$$

(18.22)

That means the crest is moving at the velocity v.

It follows that $\psi\left[k(x - vt)\right]$ describes a wave traveling to the right at speed v. If you want a wave traveling to the left, you should pick $k = -\frac{\omega}{v}$, and the solution will be of the form $\psi\left[k(x + vt)\right]$.

Calculus buffs can verify that *any* function $f(x - vt)$ (not just the cosine) satisfies the wave equation. To see this, use the chain rule: if $w = x - vt$ then $f = f(w)$ and

$$\frac{\partial f}{\partial x} = \frac{df(w)}{dw} \cdot \frac{\partial w}{\partial x} = \frac{df(w)}{dw} \cdot 1$$

(18.23)

$$\frac{\partial^2 f}{\partial x^2} = \frac{d^2 f(w)}{dw^2} \cdot 1^2$$

(18.24)

$$\frac{\partial f}{\partial t} = \frac{df(w)}{dw} \cdot \frac{\partial w}{\partial t} = \frac{df(w)}{dw} \cdot (-v) \tag{18.25}$$

$$\frac{\partial^2 f}{\partial t^2} = \frac{d^2 f(w)}{dw^2} \cdot (-v)^2 \quad \text{so that finally} \tag{18.26}$$

$$\frac{1}{v^2}\frac{\partial^2 f}{\partial t^2} = \frac{d^2 f(w)}{dw^2} = \frac{\partial^2 f}{\partial x^2}, \tag{18.27}$$

from which it follows that not only $f(x - vt)$ but also $f(x + vt)$ satisfies the wave equation. Even

$$\psi(x,t) = A e^{-k^2(x-vt)^2} \tag{18.28}$$

obeys the wave equation, even though it does not readily come to your mind when you think of a wave, the way the cosine does.

Note that every wave in a given medium has a definite velocity, but only plane waves have a definite wave number and frequency. In particular the wave in Eqn. 18.28 has a velocity v but does not repeat in time or space. Once the peak zips past you, it is all over.

18.3 Frequency and period

We now ask, "What is k and what is ω in $A\cos(kx - \omega t)$?" We are all good at visualizing a function of one variable, but this is a function of two variables. Let us first understand this function at a fixed time, say $t = 0$. At $t = 0$, it looks like $A \cos kx$, which starts out at A at $x = 0$ and finishes a full cycle when we move a distance x, such that $kx = 2\pi$. This distance is by definition the *wavelength* λ and obeys $k\lambda = 2\pi$. Thus k, the *wave number*, is related to the wavelength λ that we understand more intuitively by the formula

$$\lambda = \frac{2\pi}{k} \quad \text{or} \quad k = \frac{2\pi}{\lambda}. \tag{18.29}$$

Figure 18.3 shows the plot for $A = 2, k = 1$ ($\lambda = 2\pi \simeq 6.28$).

Now, you can ask if I would have obtained a different relation between k and λ had I picked some other time than $t = 0$. Think about it: ωt_0 is some angle ϕ_0 inside the cosine; it's just going to shift the whole pattern by some amount. Changing the phase won't change the fact that the peak-to-peak distance is λ.

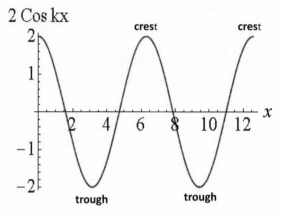

Figure 18.3   The wave $A\cos(kx - \omega t)$ at $t = 0$ as a function of x for the choice $A = 2, k = 1$. For $k = 1$, λ should be $2\pi \simeq 6.28$ according to Eqn. 18.29, and indeed it is. The maxima are called *crests* and the minima are called *troughs*.

Next we ask, "What does it look like to a person sitting at a certain location, as a function of time?" For example, the waves could be manufactured in the ocean and sent toward the shore. You stand at one place near the shore; the wave is going past you. You will bob up and down, and that's what I want to describe mathematically. So, I pick some location $x = 0$ for convenience, and then $\psi(0, t) = A\cos \omega t$. I didn't miss the minus sign; for a cosine, it doesn't matter if you change the angle to minus the angle. So, consider the same graph, this time as a function of time, and ask what happens. You start with a maximum at $t = 0$, and as t increases, the function oscillates and reaches its next maximum, a *crest* at a time T called the *time period* such that

$$\omega T = 2\pi \quad \text{or} \quad \omega = \frac{2\pi}{T} = 2\pi f \tag{18.30}$$

where f is the frequency. This relation between ω and T should be familiar from the single oscillator. The minima are called *troughs*.

So, if you like, you can write the plane wave as

$$\psi(x, t) = A\cos\left(\frac{2\pi x}{\lambda} - \frac{2\pi t}{T}\right). \tag{18.31}$$

It's strictly equivalent to $A\cos(kx - \omega t)$, and you may write it either way. Equation 18.31 makes it more obvious that when x changes by λ or t changes by T, nothing happens to the cosine because you are changing the argument by 2π.

Let us now rewrite the relation $\omega = kv$ in terms of λ and $f = \frac{1}{T}$:

$$v = \frac{\omega}{k} = \frac{2\pi f}{\frac{2\pi}{\lambda}} = \lambda f, \tag{18.32}$$

which can be understood as follows. Suppose I am at the origin and hold an infinitely long string. When I start wiggling my end with frequency f, the pulses will travel to the right. Let's wait one second. In one second, I would have manufactured f of these full cycles, and each has length λ. Thus the wavefront would have advanced to $x = \lambda f$ after one second, which is the velocity by definition. I am making λ-sized objects, f per second, and pushing them out, so the front of the wave advances a distance λf in one second.

Figure 18.4 shows the wave varying in both x and t for the choice $A = k = 1$ and $\omega = 2$ for x and t in the interval $[0, 4\pi]$. If you mentally slice it at fixed x or t, shown by the dotted lines, you can count two periods in x and four in t.

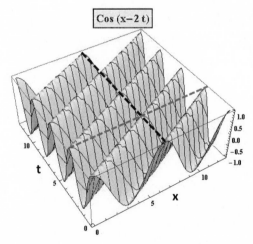

Figure 18.4   The wave $A\cos(kx - \omega t)$ with $A = 1, k = 1, \omega = 2$ or $\lambda = 2\pi, T = \pi$ in the range $[0, 4\pi]$ for both x and t. If you slice it at fixed t, you will see two full periods in x (see dotted line at $t = 2\pi$), and if you slice it at fixed x, you will see four periods in t (see dotted line at $x = 2\pi$). The lines track the maxima or crests, but one could pick any point within the cycle.

CHAPTER 19

Waves II

19.1 Wave energy and power transmitted

It is obvious that a vibrating string has more energy than a string that is not vibrating. I want to calculate the energy in a string vibrating with displacement

$$\psi(x,t) = A\cos(kx - \omega t). \tag{19.1}$$

Now, if it's an infinitely long string, the energy in it is infinite, so you define the energy per unit length. Take a portion of the string, associated with a segment of width dx, and ask, "How much energy does it have?"

The energy has a kinetic part and a potential part. The kinetic part is simple:

$$dK = \frac{1}{2}\mu dx \left[\frac{\partial\psi}{\partial t}\right]^2. \tag{19.2}$$

This is just the mass of the segment times the square of the velocity. For the solution in Eqn. 19.1 this becomes

$$dK = \frac{1}{2}\mu dx A^2 \omega^2 \sin^2(kx - \omega t). \tag{19.3}$$

The potential energy dU has a more subtle origin. The string is under tension T. When the string is displaced from equilibrium, a segment associated with the interval dx has a length (see Figure 18.2)

$$dl = \sqrt{dx^2 + (d\psi)^2} = dx\sqrt{1 + \left[\frac{\partial \psi}{\partial x}\right]^2} \simeq dx\left(1 + \frac{1}{2}\left[\frac{\partial \psi}{\partial x}\right]^2\right).$$

(19.4)

Thus it has expanded by an amount

$$\delta l = dl - dx = \frac{1}{2}\left[\frac{\partial \psi}{\partial x}\right]^2 dx$$

(19.5)

against a tension T. So the work pumped in and stored as potential energy is

$$dU = T\delta l = \frac{1}{2}T\left[\frac{\partial \psi}{\partial x}\right]^2 dx.$$

(19.6)

For the solution in Eqn. 19.1 this becomes

$$dU = \frac{1}{2}Tk^2 A^2 \sin^2(kx - \omega t) \cdot dx = \frac{1}{2}\mu A^2 \omega^2 \sin^2(kx - \omega t) \cdot dx = dK$$

(19.7)

upon using

$$Tk^2 = T\frac{\omega^2}{v^2} = T\omega^2 \frac{\mu}{T} = \mu\omega^2.$$

(19.8)

Notice that the kinetic and potential energies are equal at every value of x and t. In particular, when the string bit is maximally displaced, it has no velocity and hence no dK. It also has no dU because the slope is zero and so is the stretch δl. When the displacement is zero, it has the greatest velocity and slope and hence the greatest dK and dU. This is unlike the harmonic oscillator, where due to the conservation of energy, K and U add up to a constant, and when one is large the other is small. This is not a problem for the string bit because it is not an isolated system, and energy is pumped into (or out of) it by its neighbors.

The total energy is

$$dE = dK + dU = 2dK = \mu dx A^2 \omega^2 \sin^2(kx - \omega t). \tag{19.9}$$

The energy per unit length or *energy density* u is then

$$u(x, t) = \mu A^2 \omega^2 \sin^2(kx - \omega t). \tag{19.10}$$

Of interest is the *average energy density* $\bar{u}$, where the average is over a full cycle in space (a wavelength) or a full cycle in time (a full period T). The average of $\sin^2$ is $\frac{1}{2}$ in either case, and we have the final result

$$\bar{u} = \frac{1}{2}\mu A^2 \omega^2. \tag{19.11}$$

What is the average power sent into this wave by the person exciting the medium? Imagine that the string is tied to the point at infinity and that I have been wiggling it from $x = 0$. Let us say at this time the waves have gone out to some point $x = L$. The string is inert beyond that point. After one more second, an extra segment of length v has begun vibrating. What is the average energy contained in the extra segment? It is v times the average energy per unit length, $\frac{1}{2}\mu A^2 \omega^2$. So the average energy I pump in per second, or the average power, is

$$P = \frac{1}{2}\mu A^2 \omega^2 v = \bar{u}v. \tag{19.12}$$

A special property of life in one dimension is that this wave goes undiminished in amplitude. You can go 10 miles or 100 miles from me, and the amplitude is still the A that I produced at the source, because all the energy goes along this line. There is no escape or spreading out. What happens in three dimensions is more typical. If you have a very tall tower on top of which you put a speaker, the energy radiates out in concentric spheres as time goes by. The power at the source is now spread over bigger and bigger spheres as you go farther out. In three dimensions we have the notion of *intensity*, which is the *power per unit area*. If there is a speaker sending out sound waves, and I take a one-meter by one-meter window and hold it in front of me and ask how much power crosses it on average, that is the intensity I in W/m^2. Of course, the window doesn't have to be one square meter big. You can take a tiny square, provided you divide the power by its area to find I. You don't have that notion in one dimension because you cannot hold a window in one dimension perpendicular to the velocity.

If you are r meters from the source of sound in three dimensions, the power P is spread over a sphere of area $4\pi r^2$ and the intensity is

$$I = \frac{P}{4\pi r^2} \ W/m^2. \tag{19.13}$$

This is true also for electromagnetic waves.

The intensity of sound, β, is measured in *decibels* (dB) defined as follows:

$$\beta = 10 \, \text{Log}_{10} \frac{I}{I_0} \ \text{dB} \tag{19.14}$$

where $I_0 = 10^{-12} W/m^2$ is the *reference intensity*, the minimum an average human ear can detect. Thus the minimum intensity we can hear is zero dB according to this formula.

A whisper is 15 dB, while a rock concert or jet engine is around 120 dB. In a rock concert

$$10 \, \text{Log}_{10} \frac{I}{I_0} = 120 \tag{19.15}$$

$$\frac{I}{I_0} = 10^{12} \tag{19.16}$$

$$I = I_0 \cdot 10^{12} = 1 W/m^2. \tag{19.17}$$

Due to the logarithmic scale, an increase in intensity by a factor of $100,000$ causes an increase of $50dB$. Our sense of loudness seems to grow as the log of the intensity rather than the intensity itself. We need the logarithmic scale because the ratio of the largest intensity we can tolerate to the smallest we can hear is one trillion!

The difference in decibels between two different intensities I_1 and I_2 is equal to $10 \, \text{Log}_{10} \frac{I_1}{I_2}$.

19.2 Doppler effect

Now we will consider a totally different property of wave propagation that I'm going to illustrate with sound: the *Doppler effect*. This term refers to the well-known phenomenon that a source of definite frequency, like a siren on a fire truck, will be heard as a higher frequency when the fire truck is coming toward the observer and a lower one when it's going away from the observer. We want to know why, and by how much, the frequency changes. The answer is based on $v = \lambda f$.

Let us take a source S in the left half of Figure 19.1 that is sitting still, emitting waves that spread out spherically. Three equally spaced crests are shown at a particular time. You are the observer O, standing off to the right listening to the sound. The waves go by you and you observe a certain wavelength. That's the distance from one crest to the next; call it λ_0. (We need a subscript because soon another λ will enter.) As usual,

$$\lambda_0 f_0 = v \tag{19.18}$$

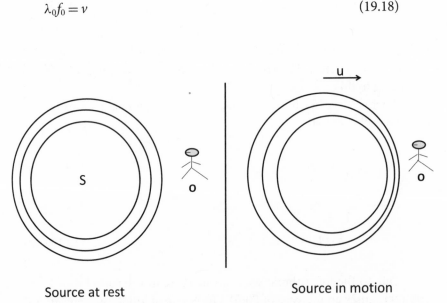

Source at rest   Source in motion

Figure 19.1   At the left we have a source at rest and observer O at rest. The three crests emitted are equally spaced in the air and also as perceived by O. At the right we have the source moving at velocity u. The wavefronts now get crunched in the forward direction because the source travels a bit to the right between emitting one crest and the next. In the backward direction the opposite happens.

where f_0 is the frequency. Now suppose the source is moving to the right at a certain speed u as shown in the right half. It's not hard to visualize that the waves in the air will get squashed in the forward direction to the right: having emitted one crest, the source moves to the right a little bit before emitting the next crest. The new λ will be λ_0 minus the distance the source travels in a time $T = \frac{1}{f_0}$. Thus

$$\lambda = \lambda_0 - uT = \lambda_0 - \frac{u}{f_0}. \tag{19.19}$$

What is the new frequency f that you will hear? It is determined by the usual relation

$$\lambda f = v. \tag{19.20}$$

You must understand that, even though the source is moving to the right, the velocity of sound is not altered by that process. A moving truck does not emit sound at an increased speed in the forward direction. The speed of sound with respect to a medium is controlled by the properties of the medium, not the velocity of the source. If the medium is air, the sound can travel only at a certain speed. On the other hand, if there is a machine gun on a moving truck that sprays bullets in all directions at the same speed *in its frame*, then according to a person on the ground, the ones in the forward direction would be moving faster than the ones in the backward direction. In any case, the new frequency is

$$f = \frac{v}{\lambda} = \frac{v}{\lambda_0 - \frac{u}{f_0}} = \frac{v}{\frac{v}{f_0} - \frac{u}{f_0}} = \frac{f_0}{1 - \frac{u}{v}}. \tag{19.21}$$

The observed frequency f is the normal frequency f_0 divided by $1 - \frac{u}{v} < 1$. So the observed frequency will go up. If you want to see what happens to the left of the source, you simply reverse the velocity of the source (moving to the right) relative to the sound waves (now moving to the left to meet the observer on the left) in the same formula. So the answer covering both cases is

$$f = \frac{f_0}{1 \mp \frac{u}{v}}. \text{ [source coming toward} \tag{19.22}$$
$$\text{(away from) static observer].}$$

Your common sense should tell you which sign applies when.

What happens with light and at relativistic speeds u? The velocity of light will be c in vacuum, but the frequency will change for many reasons. First of all, the ambulance clock slows down according to you by the standard relativistic time dilatation, even if it's not coming toward you or going away from you, but going around you, so that $f < f_0$. This is called the *transverse Doppler effect*. There is, in addition, the more familiar Doppler effect due to source motion toward or away from you, while will reduce to Eqn. 19.22 when $\frac{u}{c} << 1$.

Now for sonic booms, which can be traced back to

$$f = \frac{f_0}{1 \mp \frac{u}{v}}. \tag{19.23}$$

Relativity was chock-full of formulas where the denominator could vanish but never did because c was the limiting speed. But the speed of sound v is not an upper limit at all, and we have planes that travel at $u > v$. As $u \to v$, the crests are getting crunched more and more in the forward direction until $u = v$, and all the crests pile right on top of each other; that's when a sonic boom occurs. If you could go faster than that, you would leave behind a trail of waves, with nothing in front of you. So, if there were a jet plane coming toward you faster than the speed of sound, you would not have time to get out of the way—the jet would hit you before the sound waves hit you.

Now, here's one more variant of the Doppler effect: the observer is moving toward the source, which is at rest. The siren (at rest in the air) is emitting nice spherical waves, but you are rushing to meet them with velocity $-u$. What's the frequency at which you will see crests? In the frame of reference of the air, the sound is moving to the right at speed v, and you are moving to the left at speed u. So, you are zipping past the waves at a speed $u + v$, and the crests are spaced at a distance λ_0 apart. (Nothing happens to the spacing between them; they are just as they are in the left half of Figure 19.1.) The frequency according to you will be the distance you move relative to the medium in one second divided by the length of each cycle:

$$f = \frac{u+v}{\lambda_o} = \frac{v}{\lambda_0}\left(1 + \frac{u}{v}\right) = f_0\left(1 + \frac{u}{v}\right). \tag{19.24}$$

If you are running away from the source, you should change the sign. The general formula is then

$$f = \frac{v}{\lambda_0}\left(1 \pm \frac{u}{v}\right) = f_0\left(1 \pm \frac{u}{v}\right) \tag{19.25}$$

[observer moving toward (away from) static source].

So, when you move toward the source, the correction factor appears in the numerator; when the source moves toward you, the correction factor appears in the denominator. You have to use common sense in choosing the $\pm$ sign. For example, it's obvious that if you're rushing to meet the waves you should get a higher frequency. You can go on and ask what happens if you are moving and the ambulance is also moving. The fun never ends.

19.3 Superposition of waves

The rest of this chapter focuses on one essential property of waves called *interference*. Look at the wave equation. It's a linear equation. Don't be fooled by the 2's in $\frac{\partial^2 \psi}{\partial x^2}$ or $\frac{\partial^2 \psi}{\partial t^2}$. That doesn't make it quadratic; it's still the first power of ψ on both sides. Then verify the following in your mind: If ψ_1 and ψ_2 are solutions, so is $\psi_1 + \psi_2$. What that really means is the following. Suppose you emit some sound and you're sending some wave ψ_1 that travels through space. Then, I turn you off and I turn on another speaker; let's say the speaker emits another sound described by ψ_2. If you and the speaker emit sound at the same time, then the air disturbance will be simply the sum $\psi_1 + \psi_2$. So, if one cause produces one wave and a second cause produces a second wave, then, when both are turned on, the wave they produce will simply be the sum. This is the *principle of superposition*, and it follows from the linearity of the wave equation. From here until the end of the chapter we will analyze a variety of situations where we add two waves.

The simplest problem is the following. Pick a certain spot, say $x = 0$, and listen to two plane waves that are now functions of just t: $\psi_1 = A \cos \omega_1 t$ and $\psi_2 = A \cos \omega_2 t$. I've chosen the two waves to have the same amplitude, but not necessarily the same frequency. So, the signal you will hear will be

$$\psi = \psi_1 + \psi_2 = A \cos \omega_1 t + A \cos \omega_2 t. \tag{19.26}$$

Now, you can use the trigonometric identities and figure out what happens, but I want you to think a little bit first. Think about these two waves. If they have the same frequency, it's trivial, right? If $\omega_1 = \omega_2 = \omega$, then $\psi = 2A \cos \omega t$. That just means they reinforce each other. At every instant you get double the ψ you got before. Say the frequencies are not equal. Initially they are in step. This cosine is 1, and that cosine is 1, and they add up to 2A. As time goes by, $\omega_1 t$ and $\omega_2 t$ start differing, and the cosines are not in step anymore. After a while the two cosines are off by half a cycle, or the angles inside them differ by an odd multiple of π. Then they will completely cancel. If you wait longer they will again be in step and so on. One way to think about it is to imagine two runners going around on a circular track. They start out together, but they have slightly different speeds. As they go around, one starts lagging behind the other, and they will be on different parts of the circle. If you wait long enough, they will again line up at the start line but with one difference: the fast runner will have done an integer number of laps more than the other.

Consider two frequencies $\omega_1 = 2$ and $\omega_2 = 3$, which are small enough for us to follow the oscillations, as depicted in Figure 19.2. At $t = 0$ they are in step (runners start off together), at $t = \pi$ their phases differ by π (runners are on opposite sides of the track) and they cancel, and finally at $t = 2\pi$ they are in step (the slow one has done two laps and the fast one three). This repeats every 2π seconds.

Cos 2t + Cos 3t

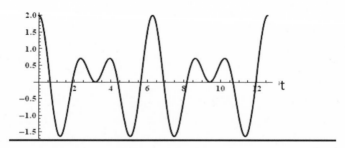

Figure 19.2   Superposition of two waves of unit amplitude $A = 1$ and angular frequencies $\omega_1 = 2$, and $\omega_2 = 3$. At $t = 0$ they are in step, at $t = \pi$ their phases differ by π and they cancel, at $t = 2\pi$ they are in step, and this repeats every 2π seconds.

Next consider the case where ω_1 and ω_2 are large but differ by a tiny amount. Now we are better off invoking

$$\cos A + \cos B = 2 \cos\left[\frac{1}{2}(A+B)\right] \cos\left[\frac{1}{2}(A-B)\right]$$

in Eqn. 19.26 to find

$$\psi(t) = 2A \cos\left[\frac{1}{2}(\omega_1 - \omega_2)t\right] \cos\left[\frac{1}{2}(\omega_1 + \omega_2)t\right]. \qquad (19.27)$$

Suppose for example $\omega_1 - \omega_2 = 2$ while $\omega_1 + \omega_2 = 2 \cdot 10^6$. Then Eqn. 19.28 reads

$$\psi(t) = [2A \cos t] \cos\left[10^6 t\right]. \qquad (19.28)$$

In the time it takes the first cosine to finish one cycle, the second has completed a million. Conversely, in the time it takes the second cosine to finish 100 cycles, the first has hardly changed. Thus we can treat $[2A \cos t]$ as the slowly varying amplitude for the second rapidly oscillating cosine. The rise and fall of this amplitude is called the *beat* and can be picked up by the ear. Figure 19.3 shows what happens when $\omega_1 = 41$ rads/s and $\omega_2 = 39$ rads/s. Notice that the envelope of the rapid oscillations is itself a cosine and that the time T between two maxima is π and the beat frequency is $\omega_b = \frac{2\pi}{T} = 2 = \omega_1 - \omega_2$. You might have thought it should be $\omega_b = \frac{1}{2}(\omega_1 - \omega_2)$, based on the argument inside the first cosine in Eqn. 19.28. It is true that only after a time 2π the amplitude returns to $2A$, and that after a time $T = \pi$ it returns to $-2A$. However, $-2A$ is still an amplitude maximum, just as loud as $2A$. Alternatively, think of the time period of the beats as the interval between two successive zeros of the amplitude. These occur *twice* in every full cycle of $\cos\left[\frac{1}{2}(\omega_1 - \omega_2)t\right]$.

Piano tuners use beats as follows. They hit the tuning fork, and it vibrates at some prescribed frequency, say $f = 440$ Hz. Imagine that your piano is slightly off; maybe it's at 438 Hz. The beat frequency will be $2Hz$. The tuner will keep fiddling with the piano until the beat (not the total sound) disappears.

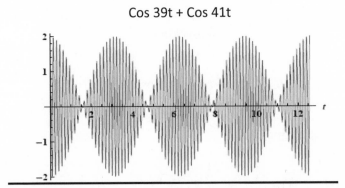

Cos 39t + Cos 41t

Figure 19.3   Beats due to two waves of unit amplitude $A = 1$ and large but nearly equal angular frequencies $\omega_1 = 41, \omega_2 = 39$. The (angular) beat frequency is supposed to be $\omega_b = 41 - 39$ rads/s, which means a time $T = \frac{2\pi}{\omega_b} = \pi$ between beats, in accord with the figure.

19.4   Interference: the double-slit experiment

Now we turn to a more complicated problem where I'm going to add two waves. In the case of beats I sat at one location and added two waves coming to me as a function of time. Now I am going to see how two waves interfere at different points in space in what is called the *double-slit experiment*, depicted in Figure 19.4. It shows a huge tank of water as seen from the top. There is a vibrating source E at the left end that emits water waves. These wavefronts will be concentric circles near the source, but if you go far to the right of the source, you can treat the wavefronts as just parallel lines. The figure shows three crests and two troughs, the wavelength λ and amplitude A. Next these waves hit an impenetrable barrier with two slits in it, labeled S_1 and S_2. These two slits will themselves start generating their own waves that radiate out from them to the right. The figure shows a crest and trough from each using solid and dotted lines.

Suppose you have a detector of water waves D (say a floating piece of cork skewered on a vertical rod), which you can place anywhere along the right side of the rectangular experimental region to measure the amplitude of water waves. What amplitude will it measure at different points?

First take a point M exactly midway between these two slits, lying on the perpendicular bisector of the line joining the slits. What are you getting here? You are getting a signal from each slit. They will arrive in step or in phase, because they were both generated by the same plane wave from the

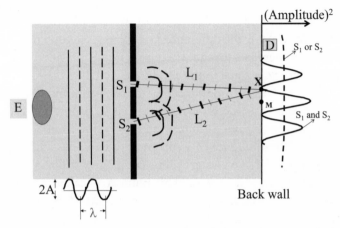

Figure 19.4   The double-slit experiment, from left to right. We are looking down at a tank of water. Waves are emitted from E, far enough away that the crests and toughs are parallel lines. Three crests (solid lines) and two troughs (dotted lines) are shown. They hit the two slits S_1 and S_2, and the semicircular waves emerge from the slits. A few maxima (and minima) are shown by solid (and dotted) lines. The thin and thick notches on the lines marked L_1 and L_2 correspond to maxima and minima along these two paths. The interference pattern at the far right is measured by a detector D, which can slide along the back wall. The point M is a maximum and X is a minimum. The maxima and minima measured by the detector refer to the variation of the *amplitude* squared due to interference of two waves, not to be confused with the oscillations of ψ itself, as shown to the left of the slits along with its λ. The amplitude squared with just one slit open is shown by a dark dotted line and is assumed to be essentially constant.

left that hit the two slits in phase and have traveled the same distance to reach M. So, when a crest from S_1 reaches M, a crest from S_2 will reach M, and likewise for the troughs. Assuming each slit generates a wave of amplitude A, their sum at M will have an amplitude $2A$. (We ignore the slight decrease in amplitude as the waves spread out from the slits.) So the water will be very choppy at M. If you shut off one slit, the amplitude will go back to just that from one hole (shown by the nearly constant dotted line). It makes sense that two open holes give you $2A$.

But consider the point X above M, where the figure shows a snapshot of the two waves arriving from the two slits, with crests and troughs indicated by thick and thin markers. Let us denote the distance from the two slits to X by L_1 and L_2. Whereas at M we had $L_1 = L_2$ by symmetry, that is not so at X. At X, the difference $L_1 - L_2 = \frac{\lambda}{2}$. As you know, between every two crests there is a trough where ψ is minus the amplitude, unlike at the crest where ψ is equal to the amplitude. So when $L_1 - L_2 = \frac{\lambda}{2}$, when a crest from S_2 arrives, a trough from S_1 arrives, canceling it. Later, when a trough from S_2 arrives, a crest from S_1 cancels it. *In fact, at every instant, whatever signal comes from S_2, minus that will come from S_1.* This is the case because a path difference of $\frac{\lambda}{2}$ corresponds to a phase difference of π in the function $\cos \frac{2\pi x}{\lambda}$. The water will be completely still at X at all times. The amplitude there will be zero. This point X is called an *interference minimum*, while M is an *interference maximum*. At M the interference is *constructive*, and at X it is *destructive*.

That point X is very interesting. Look what the theory is telling you. It says that if you sit there, the water, which was bobbing up and down with just one slit open, won't move at all when there are two slits open. It's the property of waves: two waves can lead to nothing, but one wave cannot— it has nothing to neutralize it. This is something you have to get straight. For example, if a hole in your mosquito net is letting mosquitoes in, it will not help to make another hole. Interference does not exist for particles like mosquitoes, because mosquitoes are never negative in number.

Now, if I go further up the right side of the rectangle, I reach a point where the difference $L_1 - L_2$ is the full wavelength, and then I'm back to being in step. Of course, a full wavelength means that the 13-th crest from S_1 and the 14-th from S_2 arrive in step, but we don't care, as long as a crest comes with a crest; we have constructive interference. As we move up and down the back wall, the amplitude of the sum of the two waves oscillates, as indicated by the solid wiggly line, which has a maximum at M, a minimum at X, and so on. Remember that the maxima and minima measured by the detector refer to the variation of the *amplitude* (squared) due to interference of two waves, not to be confused with the oscillations of ψ itself, as shown to the left of the slits along with its λ. (The dotted line is the amplitude squared due to just one slit, and it is roughly constant. At a point where this amplitude is, say, 5 cm, the water goes up and down by 5 cm.)

For those who want to go a little deeper, the *semi-circular waves* emanating from the slits have the form

$$\psi_1(\mathbf{r}, t) = A \cos(kr_1 - \omega t) \quad \text{and} \quad \psi_2(\mathbf{r}, t) = A \cos(kr_2 - \omega t)$$

$$(19.29)$$

where r_1 and r_2 are the distances from the slits to the point $\mathbf{r}$ we are interested in. (Actually, in two dimensions, A itself falls off with distance, but that is a much smaller effect, which we ignore.) At any given time, the lines of constant phase for ψ_1 and ψ_2 are semi-circles centered at S_1 and S_2 respectively, because the phase depends only on r_1 and r_2. Pick a point for the detector. It has some value of $r_1 = L_1$ and $r_2 = L_2$. At any given time t we can ignore the time-dependent phase ωt common to both waves. Their phase difference is then simply $k(L_1 - L_2) = \frac{2\pi}{\lambda}(L_1 - L_2)$. It follows that a path-length difference $m\lambda$, where $m = 0, \pm1, \pm2...$, causes phase difference of $\pm 2\pi m$ and hence constructive interference, while a difference of a half-integer wavelength causes a phase difference equal to an odd multiple of π and hence destructive interference.

The exact condition for destructive and constructive interference is

$$L_2 - L_1 = m\lambda \quad m = 0, \pm1, \pm2 \text{ for} \tag{19.30}$$

$$\text{constructive interference}$$

$$L_2 - L_1 = \left(m + \frac{1}{2}\right)\lambda \quad m = 0, \pm1, \pm2 \text{ for} \tag{19.31}$$

$$\text{destructive interference.}$$

To find the point of constructive or destructive interference is a matter of simple geometry. For example, to locate the *first* point of destructive interference, you measure the distances from the slits, using Pythagoras' theorem, take the difference, and set it equal to $\frac{\lambda}{2}$. However, we often employ the following approximation for computing $L_2 - L_1$, when the back wall and detector are at a distance far greater than d and λ. Then the lines going from the slits to the detector are nearly parallel, as shown in Figure 19.5. Say they leave at an angle θ from the forward direction to reach the detector. Then you can see that the extra distance from the lower slit is $d \sin\theta$ where d is the distance between the slits. (The equality of the two θ's in the figure should be familiar from the dark days of the inclined plane.) Thus we may set

$$L_2 - L_1 = d \sin\theta \tag{19.32}$$

in Eqns. 19.30 and 19.31. Given the values of d and λ, one can say at what angle θ one will observe the first minimum and so forth. Conversely,

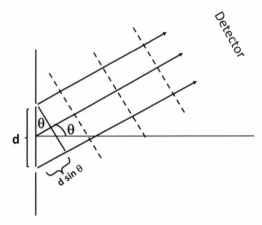

Figure 19.5   Computing the path difference in the double-slit experiment when the detector is very far from the slits. The dotted lines are the troughs.

given d and the angle θ at a maximum or minimum, one can deduce the wavelength λ.

In the early nineteenth century it was not clear whether light was a wave or a stream of particles. Thomas Young performed a double-slit experiment with light that showed an interference pattern and thereby nailed it: it is impossible to reduce the number of particles coming to a point like X (to zero) by opening a second slit. He also managed to deduce the wavelength from knowing d and θ, *without knowing exactly what medium was supporting light waves and what it was that was oscillating!*

19.5   Standing waves and musical instruments

I take a string, attach one end to a wall, and start shaking the free end. Let's say I shake it just once and send a pulse moving to the right as in part A of Figure 19.6. When this blip gets to the wall it finds the string cannot vibrate there. Thus the wall will exert a suitable force to ensure that the signal the wall sends out, along with the signal I send in, add up to zero there. The result of that is that the signal generated by the wall becomes the reflected wave, reversed in direction and sign as shown in part B.

Here is another way to understand the reflected wave. Suppose there is no wall, and I live to the right of where it used to be. I see you send a blip to the right as in Part A of the figure, and I manufacture an inverted mirror image of that traveling toward the reflection point R, as shown by

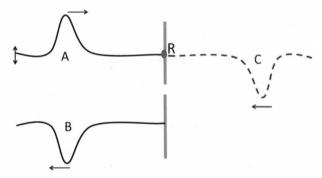

Figure 19.6   Part A: The blip I generate (up-and-down arrow at left) goes to the wall at the right (solid vertical line). Part B: The blip gets reflected from the wall at point R, reversed in direction and in sign. Part C: There is no wall, but a blip comes from the right and cancels the original blip at the wall for all times. The former keeps moving to the left and becomes the reflected blip of part B.

the dotted curve in part C. I arrange it so that when the two blips cross, they exactly cancel at the putative wall location for all times. Then, they go through each other (by the superposition principle), and my pulse will cross over to the left as your reflected wave. Because there is no movement of the string at R under this combined influence, it means that even if I clamp the string there with a wall that doesn't allow the string to move, it doesn't change the outcome. That is how we convince ourselves that the incoming pulse gets reflected as described.

Now I want to study, not a single incoming one-shot pulse, but an incoming periodic wave $\psi_i = A\cos(kx - \omega t)$. A continuous signal goes in, and how does it come out? It comes out as

$$\psi_r(x,t) = -A\cos(-kx - \omega t) = -A\cos(kx + \omega t) \qquad (19.33)$$

where I've chosen the wall to be at $x = 0$ and reversed the amplitude and the direction of motion of the incoming wave to get the reflected wave. I keep sending stuff to the wall, and the wall keeps reflecting that back to me. Both waves coexist in the region to the left of the wall to yield

$$\psi(x,t) = \psi_i + \psi_r = 2A\sin\big[kx\big]\sin\big[\omega t\big] \quad \text{using} \qquad (19.34)$$

$$\cos x - \cos y = 2\sin\left[\frac{x+y}{2}\right]\sin\left[\frac{y-x}{2}\right]. \qquad (19.35)$$

So the string oscillates with a frequency ω everywhere, but the amplitude at a point x is $2A \sin kx$. This is called a *standing wave*. Whenever $kx = 0, -\pi, -2\pi\ldots$ there is no vibration. (Remember x is negative to the left of the wall.) Such points are called *nodes*. Exactly in between two nodes at $kx = -\frac{\pi}{2}, -\frac{3\pi}{2}, \ldots$, et cetera, are *anti-nodes* where the amplitude has the maximum of $2A$. Figure 19.7 shows a standing wave on a string for the case $A = .5$ and $k = 1$ or $\lambda = 2\pi$. The vibrating string is a blur, shown by the shaded regions. Note that the distance between two nodes is $\frac{\lambda}{2}$.

If you grab a node with your fingers, it won't matter because the string was not planning to vibrate there anyway. You can also grab two nodes and it still will not matter. Using this fact, we can solve the following problem of a string that is clamped at both ends, $x = 0$ and $x = L$. What are the possible frequencies of vibration? To solve this, we first ask, "What are the possible standing waves?" The answer: the string length L has to be equal to an integer multiple n of $\frac{1}{2}\lambda$. But the string length has already been given to us as L, so it is the possible values of λ_n or k_n, labeled by n, that have to adjust themselves to obey

$$L = n\frac{\lambda_n}{2} \quad n = 1, 2, 3, \quad \text{which means } k_n = \frac{2\pi}{\lambda_n} = \frac{n\pi}{L} \quad (19.36)$$

$$\omega_n = k_n v = \frac{n\pi v}{L} \quad (19.37)$$

$$f_n = \frac{nv}{2L} \equiv nf_1 \quad (19.38)$$

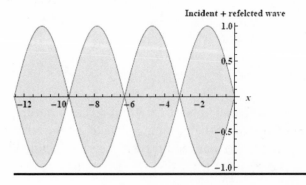

Figure 19.7   Nodes and anti-nodes on a string with $A = 0.5$ and $k = 1$ or $\lambda = 2\pi$. The incoming and reflected waves form a *standing wave*, with nodes at $0, -\pi, -2\pi, \ldots$ and anti-nodes at $-\frac{\pi}{2}, -\frac{3\pi}{2}, \ldots$ The vibrating string is a blur, shown by the shaded region.

where I have used $\omega = vk$ and where f_1 is called the *fundamental frequency*. The displacement of the string in a mode of vibration labeled by n is

$$\psi_n(x,t) = A \sin\left[\frac{n\pi x}{L}\right] \sin\left[\frac{n\pi vt}{L}\right], \qquad (19.39)$$

which I got from Eqn. 19.34 using $\omega_n = k_n v$. The first three modes are shown in Figure 19.8.

A string clamped at two ends can only vibrate in either the fundamental frequency $\omega_1 = \frac{\pi v}{L}$ or an integer multiple of it, where the integer gives the number of anti-nodes over the length of the string.

Now, this reasoning can be bodily lifted for the following problem. Take a tube of length L, blocked by walls at the two ends. The longitudinal sound wave can go back and forth and form standing waves. The equation obeyed is still the wave equation, and the two ends are again nodes because the walls will not permit any longitudinal vibration there. We can use the same picture as in Figure 19.8 and the frequencies ω_n will be given by the same formula. The only subtlety is that now the graphs in Figure 19.8 do not represent transverse motion of the molecules, but instead they measure the longitudinal displacement from equilibrium. If the height of the graph is *2mm* somewhere, it means the molecules there are moving *back and forth* (and not up and down) by *2mm*. Out of habit, we plot everything on the y-axis, but that doesn't mean the motion is perpendicular to the tube.

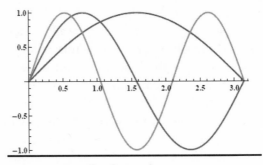

Figure 19.8   The plots of $A \sin\left[\frac{n\pi x}{L}\right]$ for $n = 1, 2, 3$ and $L = \pi$. The mode with n half-wavelengths has a frequency $\omega_n = \frac{n\pi v}{L}$.

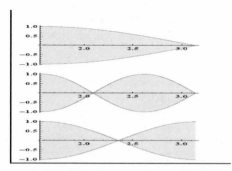

Figure 19.9   From the top: First two modes of vibration of a tube of length π closed at the right end and open at the left end followed by the first mode of a tube of length π open at both ends. You can read off the λ's by inspection and deduce the frequencies from $v = \lambda f$.

Here is a challenge. I take a tube, closed at one end and open at the other, like a Pepsi bottle, and blow into it. I hear a whistle at a certain frequency. What determines that? Because I'm making the noise at the open end, it is an anti-node. That is where the amplitude will be the biggest, and of course there is a node at the closed end where longitudinal motion is impossible. So, what pattern can I draw? I want maximum vibration at the open end and nothing at the other end. The lowest one I can draw looks like the top part of Figure 19.9. Because we see a quarter of a cycle on the interval of length L, we deduce $\lambda = 4L$ and the frequency $f = \frac{v}{4L}$. The middle picture corresponds to $f = \frac{3v}{4L}$. When one end is open and one end is closed, the frequencies are *odd multiples* of the fundamental frequency $f_1 = \frac{v}{4L}$.

Finally, consider a tube open at both ends, with anti-nodes at the ends. The vibration is depicted in the lowest part of Figure 19.9. Clearly $\lambda = 2L$ and $f = \frac{v}{2L}$. And again, you will find here all frequencies are integer multiples of the fundamental. So, the story is this: as long as the two ends are both open or both closed (both nodes or anti-nodes), all frequencies are integer multiples of the fundamental one. If one end is open and the other is closed, you'll get odd multiples of the fundamental.

A challenge: How does a trumpeter vary the pitch by varying the length of the vibrating air column or a violinist by moving her fingers?

CHAPTER 20

Fluids

20.1 Introduction to fluid dynamics and statics

This is a relatively simple topic. If you took any kind of high school physics, you would have seen fluids. Whenever I say fluid, you are free to imagine water or oil.

20.1.1 Density and pressure

Let us begin with a basic property of the fluid, the density, denoted by ρ. The density of water is $\rho_w = 1,000 \ kg/m^3$. The more subtle concept is the one of *pressure*. If you dive down to the bottom of a swimming pool, the pressure goes up. What is the formal definition of pressure? That's what I want to explain.

If we pick a point in the fluid and say the pressure there is such and such, we mean the following. Say you get into that fluid and you want to carve out a little space for yourself, maybe a glass cube, and you want to live inside that cube. The water is trying to push you in from all sides and compress this cube. You therefore have to push out on all the walls. If the force you exert on a wall is some F and the area of that wall is A, that ratio is the pressure. The pressure will not depend on which wall you choose, provided the cube you are in is infinitesimal. The pressure is an intensive measure of how hard the water is trying to push in. Even if you don't insert the cube, that pressure is still there, but one way to measure the pressure

335

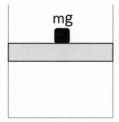

Figure 20.1   A gas in a cylinder with a massless piston of area A on top. The mass m exerts a force $F = mg$ and a pressure $P = \frac{mg}{A}$ Pascals.

is to try to go in there and push the fluid out and ask how hard it pushes back. The unit of pressure is N/m^2 and is called a *Pascal*.

Here's another example of pressure. You have a gas inside a cylinder as shown in Figure 20.1, with a massless piston at the top. The pressure of the gas and the pressure of the outside world are the same. But if you want to increase the pressure in the gas, you can put some extra weights on the piston. That mg will push down, and mg divided by the area of the piston A will be extra pressure $P = \frac{mg}{A}$ Pascals that you apply. That's also the pressure of the gas if there is no atmosphere outside. If there is atmosphere outside, the total pressure is the atmospheric pressure plus $\frac{mg}{A}$. The atmospheric pressure is everywhere. So, when you push down on the piston to compress the gas further, you are adding to the atmospheric pressure this extra force divided by area. The total is called the *absolute pressure*, and the extra bit due to the mg is called the *gauge pressure*. For example, when your car has a flat, the bright side of it is that the pressure inside the tire is not zero; it equals the atmospheric pressure. But that does not help you because the same pressure is also outside. When you stick a gauge in the valve stem and you measure something, say 32 psi, that's the gauge pressure.

20.1.2   Pressure as a function of depth

Pressure is a condition in a fluid. One important property of pressure in a fluid is that all points at a given depth have the same pressure. We understand that as follows. I imagine a little horizontal cylindrical section of the same fluid, shown by dotted lines in the lower part of the container in Figure 20.2. Remember this is not a real cylinder. This is the same fluid,

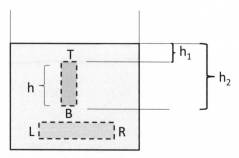

Figure 20.2   The horizontal dotted cylinder marks a part of the fluid in equilibrium under forces from the left L and right R. The vertical dotted cylinder marks a part of the fluid in equilibrium under forces from the top T and bottom B and its weight Mg.

and I have mentally isolated a part of the fluid that looks like a cylinder; this mental image is represented in dotted lines.

Can the pressure on the two sides—L and R—be different? No, because if the pressure on the left were bigger than the pressure on the right, the pressure times area on the left would exceed the pressure times area on the right, and the fluid would move to the right. But it's not doing anything. It's in equilibrium, and the only way that can happen is if it is being pushed equally from both sides. So, the pressure cannot change at a given depth.

Now let's take a cylinder that is vertical, with the top and bottom labeled T and B. It has a base area A and height h. The argument I just gave to exlain why the pressures at L and R in the horizontal cylinder had to be equal will now tell you that the pressures at the top T and bottom B cannot be equal. If they were equal, acting on equal areas, they would produce a net upward force of zero, and there would be nothing to keep the cylinder of water from falling down. So there must be a net upward force to equal the weight of that cylinder of the fluid. We are now going to calculate the pressure difference using this notion.

Let's call the pressures P_1 and P_2 at depths h_1 and h_2 and equate the net upward force to the weight of the marked body of fluid, which is g times its mass $M = \rho Ah$:

$$A(P_2 - P_1) = Mg = \rho Ahg = \rho Ag(h_2 - h_1), \qquad (20.1)$$

which may be rewritten as

$$P_2 = P_1 + \rho g(h_2 - h_1). \tag{20.2}$$

The area A cancels out, as it should, because the pressure difference cannot depend on what body of water I choose to focus on.

How about forces on the sides of the cylinder? They cancel at every height, because at every height the push from the left and right is equal.

It's standard to take 1 to be the point right at the surface (so that $h_1 = 0$), to take point 2 to be any point inside, and to call the *depth* of that point simply h. Thus h *increases as we move down* the fluid. The pressure at any point in the fluid is the pressure at the top, which is usually atmospheric pressure P_A, plus ρgh:

$$P = P_A + \rho gh. \tag{20.3}$$

The pressure at the surface of the lake is due to the atmosphere. If you dive to some depth h, the pressure goes up by ρgh. If you go to the bottom of the ocean, there will be an incredible amount of pressure outside, while inside there is just the atmospheric pressure in your lungs. That's why the human body cannot survive at the bottom of the ocean. That's why a submarine must be engineered to withstand the pressure difference. But fish don't have that problem, because fish are breathing the water. The water is going into their system and pushing out with the same pressure as the water outside.

Now, how about atmospheric pressure? The atmospheric pressure arises because we are ourselves living in the bottom of a pool, but a pool filled with air. The density of the air above our heads decreases with altitude and essentially vanishes beyond a height $h \simeq 10^5$ m. The atmospheric pressure at the bottom of this pool, due to the entire air column above, is $P_A = 10^5$ Pascals. So, we are living at the bottom of a pool where the pressure is 10^5 Pascals, relative to interstellar space. That pressure doesn't kill us because it acts both from the outside pushing in and from inside, pushing out. But you must have seen the dramatic experiment where a can is heated until some of the air expands and escapes, and then the can is sealed and cooled: the resultant drop in pressure is big enough to cause the can to implode.

Now, ask yourself the following question: If our atmosphere, instead of being a column of air, was a column of water, how high would it be to exert the same pressure? The answer is

$$10^5 Pa = \rho_w g h. \tag{20.4}$$

The density of water is $\rho_w = 10^3 \ kg/m^3$, and let's pretend $g = 10ms^{-2}$. Cancel all the powers of 10, to get $h \simeq 10m$, or roughly 32 feet.

Now we are going to put the formula $P = P_0 + \rho g h$ to work. The first thing you can do is to build yourself a *barometer*. The barometer tells you what the pressure is right now. When I said that the atmospheric pressure is 10^5 Pa, I was using average pressure. Atmospheric pressure doesn't really stay locked at that value. Each day there are fluctuations. That's why the weather report tells you the pressure is going up or the pressure is going down. Here is one way to measure atmospheric pressure. You take a container, fill it with some liquid, take a test tube, evacuate it completely, and stick it head first into the fluid as shown in Figure 20.3. There is a complete vacuum at the top of the tube. The atmosphere is pushing down on the surface (point X), so the fluid will rise up to some height h. What's going to be the height? It will be such that the pressure at the point Y, level with the fluid outside, equals the pressure at the surface of the fluid outside the tube (point X). We are simply equating the pressure at two points X and Y, which are at the same height.

Now X is at the atmospheric pressure P_A, which we are trying to measure. At Y the pressure is the zero pressure at the top of the tube, plus the $\rho g h$ of the fluid. Thus

$$P_A = 0 + \rho g h. \tag{20.5}$$

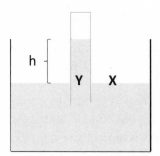

Figure 20.3   A barometer. The pressure at X, the atmospheric pressure P_A, equals the pressure at Y, which equals 0 from the evacuated region, plus $\rho g h$ of the vertical column.

So, if you build this gadget, this barometer out of water, the water column will rise to the height of roughly 32 feet. But nobody wants a gadget 32 feet high, so what you use instead is mercury, because it's very dense. You want to get the same atmospheric pressure, but you want to have a bigger ρ and a smaller h, which is close to 760 millimeters. That's why the weather guy says that the pressure today is so many millimeters, that the mercury is dropping, and so forth. Now, I'm not sure why they bother to give the numbers, because for most of us, those numbers don't mean anything. Here's a number: 746 millimeters. Does it speak to you? Not to me. It's not like saying that the temperature today it is 67 degrees Fahrenheit, for which I have a real feeling. So the weather report on millimeters of mercury goes through me like a beam of neutrinos. You can use any fluid you like, but you have to remember your choice. Thus, if the reading is 760 millimeters, be sure it is mercury, because if it's water the person is talking about, you are in serious trouble.

Imagine you are trying to drink from a straw. Now, remember the fluid is now water. It's not mercury now, because you're doing a different experiment. You know that when you drink from a straw you create a partial vacuum in your mouth, like the top of the evacuated test tube in the figure, except the pressure is not zero in your mouth—it's just lower than the atmosphere. So, you have to reduce the pressure more and more in your mouth until this fluid can climb up to your mouth. If you just want the water to make it to your mouth, the reduced pressure you need in your mouth depends on the height of the straw. If you want to drink water from a well that is more than 32 feet deep, you are out of luck. Even if your head is a complete vacuum, you cannot get the water to climb more than 32 feet.

Next, I give you a fluid that doesn't mix with water and I tell you to find its density. There are many ways to do this. One is to find the mass and volume of that fluid and divide. But here is another option. You take what's called a U-tube, shown in Figure 20.4. No, that's not where they post all the embarrassing videos, or the lectures on which this book is based; it is a physics contribution to pop culture that somehow did not do as well. Let's say you fill it up with one fluid in one leg and the other fluid in the other: oil on the left and water on the right. The figure is supposed to tell you right away that oil is less dense than water. We can quantify that by equating the pressure at two points at the same depth, say along the dark dotted line. The pressures must be equal at all points on that line: if

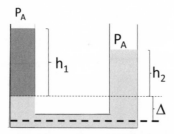

Figure 20.4   A U-tube, used to find the density of a fluid relative to another with which it does not mix, such as oil (left) and water (right). Equating the pressures at two points that lie at the same height in the liquid (on the dark dotted line), one deduces $\rho_1 g h_1 + \rho_2 g \Delta = \rho_2 g h_2 + \rho_2 g \Delta$. One can use the light dotted line as well for comparison by subtracting $\rho_2 g \Delta$ from both sides.

we isolate an imaginary horizontal cylinder there, it should not be pushed sideways. So we conclude

$$P_A + \rho_1 g h_1 + \rho_2 g \Delta = P_A + \rho_2 g (h_2 + \Delta). \tag{20.6}$$

Canceling P_A and $\rho_2 g \Delta$ from both sides, we find the desired relation between the densities

$$\frac{h_1}{h_2} = \frac{\rho_2}{\rho_1}. \tag{20.7}$$

If you know the density of one fluid, you can find the density of the other.

Notice that even though you cannot connect points lying on the light dotted line with a cylinder of water, you can equate pressures there by starting at the solid dotted line and working your way up the *same* fluid. But you cannot compare the pressures at two points on a horizontal line above the light dotted line, say a line that passes the top of the water column, because it does not pass through the same fluid.

20.2   The hydraulic press

Now for yet another application of the fact that the pressure in a fluid is equal at two points at the same height: the *hydraulic press*. Figure 20.5 depicts two pistons of different areas A_1 and A_2 connected as shown and filled with a body of some *incompressible* fluid, which means its volume

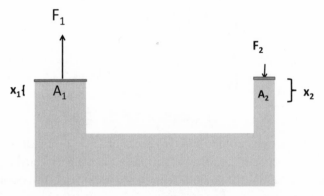

Figure 20.5   A hydraulic press that amplifies the force is made of two cylinders of different cross sectional areas A_1 and A_2, connected by a horizontal segment. Given $P_2 = \frac{F_2}{A_2} = \frac{F_1}{A_1} = P_1$, the equality of the pressures, and $A_1 x_1 = A_2 x_2$ (the equality of volumes displaced in the two sides), we find $\frac{F_2}{F_1} = \frac{A_2}{A_1} = \frac{x_1}{x_2}$. A small force F_2 acting on the right cylinder pushes the incompressible fluid by a distance x_2, and this leads to large force F_1 on the left that moves the piston over a smaller distance x_1. Energy is conserved because $F_1 x_1 = F_2 x_2$, as in a lever.

cannot be changed by changing pressure. Now, no liquid is really incompressible, but close approximations exist, like water. On the left piston will be some object I want to lift by applying a force on the right piston.

When I push down on the right piston with a force F_2, what happens at the other side? If these pistons are at the same height, we know that $P_1 = P_2$. However, this equality is essentially true even if the pistons move a bit at the two ends, because most of the pressure is due to the forces on the pistons and not the $\rho g h$ contribution of the fluid. On equating the pressures

$$\frac{F_2}{A_2} = \frac{F_1}{A_1} \quad \text{which means} \quad F_1 = F_2 \frac{A_1}{A_2}. \tag{20.8}$$

Suppose we have an elephant at the left of weight F_1, and I want to lift it by applying a force F_2 at the right. If $\frac{A_1}{A_2} = 1000$, if I apply one Newton at the right, I'll lift a 1000–N elephant on the other side.

But this is not the oldest trick in the book. An even older one, invented by cave people, is that if you have a lever you can amplify the force. But you must know from that example that you don't get something for nothing. In other words, when I lift the elephant, the fact that $F_1 \neq F_2$

is perfectly okay. But the work I do at the right must be the work delivered at the other side; that is the law of conservation of energy. The work I do is the force multiplied by the distance, and the force in turn is the pressure times area. So we want to confirm that

$$F_1 x_1 = F_2 x_2 \quad \text{or} \quad P_1 A_1 x_1 = P_2 A_2 x_2. \tag{20.9}$$

Because $P_1 = P_2$, we require that $A_1 x_1 = A_2 x_2$, which merely equates the volume displaced at the two sides. This means the total volume of the fluid is unchanged, which in turn is assured by incompressibility. I cannot get more joules out than in, using any device. But the press is still useful, because I may be willing to move a whole meter at the right to lift the elephant by one millimeter.

This is also how the brake in your car works. You press the brake pedal, which pushes on a narrow cylinder filled with brake fluid. You push it quite a bit, several centimeters at your end; at the other end is a broad cylinder whose piston pushes on the drum, retarding the rotating wheel. It moves a very tiny amount but exerts an enormous amount of force. The brake fluid has the same pressure at both places, but the force you apply with your foot is much, much smaller than the force that the drum will exert on the rotating wheel. (In practice, this effect is compounded by an intervening lever that amplifies the force exerted by your foot.)

20.3 Archimedes' principle

When Archimedes was taking a bath, he noticed that something immersed in a fluid seems to weigh less. Imagine that you have attached an object to some kind of a spring balance and weighed it so that the $-kx$ of the spring was the mg of the object. If you do the same thing again with the object now immersed in a fluid, you will find it seems to weigh less. And the question is, "How much less?" Archimedes' answer is very simple. *The weight loss equals the weight of fluid displaced.* Now, how do you show that? One way, which I like, is as follows. Look at Figure 20.6, which shows an irregular object, say a stone, of weight Mg suspended in a fluid. Now, if the thing hanging here is itself a chunk of the same fluid shaped like that stone, you will not have to do anything, because that chunk of fluid can float at that height for free. But if you now take that chunk of fluid out and put a stone there of the same shape, the rest of the fluid doesn't know what you're doing. It applies the same force that it would to support its own

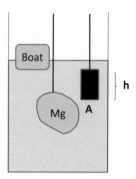

Figure 20.6   The irregular object of mass M displaces a volume of liquid equal to its own. The loss in its weight is the weight of that body of liquid. This is more readily seen for a regular object like the cylinder shown to the right. At the top left is a boat whose weight equals the weight of the liquid displaced by the fraction immersed.

kin. Namely, the rest of the fluid is in a configuration ready to support that amount of fluid. So, if you took that fluid out and put in something else with the same shape, the fluid will apply the same amount of force. The rest of the weight is of course your problem. You provide the remaining force, the reduced weight.

Now, one formal way to prove Archimedes' principle is to consider a cylindrical object of area A and height h as shown in Figure 20.6. What is the net force of buoyancy? It's P_B, the pressure at the bottom times the area, minus P_A, the pressure at the top times the area. We have already seen the difference in pressure is $P_B - P_A = \rho_l g h$ where ρ_l is the density of the liquid. The net force of buoyancy is then

$$F_{\text{buoyancy}} = \rho_l \cdot hA \cdot g, \qquad\qquad (20.10)$$

which is the weight of the liquid displaced. You can extend this result to irregular shapes by imagining them to be made up of many arbitrarily thin cylinders of various lengths glued together.

So, basically the body weighs less in water because the lower part of the body is being pushed up harder than the upper part of the body is being pushed down, which happens because the pressure increases with depth.

Figure 20.6 assumes the objects immersed are made of materials like iron. The weight of that chunk of iron will be more than the weight of the water displaced, and you will have to provide the difference to hold it up. But suppose the object was not made of iron, but made of cork. If it's cork, it won't want to be there, under water. Right? Because then the force applied by the water is more than the weight it takes to support it. So, the cork will bob up to the surface. The same goes for the human body, as the Mafia has figured out: if you want the body to stay under water you will have to add weights to it in the form of concrete shoes. Or, closer to home, it's like a rubber ducky in your bath. If you want to keep the rubber ducky under the water level, you have to hold it down. You relax for a moment, and it will bob right up to the surface.

Suppose you do not hold it down. We know some fraction f of the volume V of the object O (for example, the rubber ducky) is going to be inside the water and the rest outside. You can already guess what f is, but let's prove it. Equating the weight of the water displaced to the weight of the object gives us

$$\rho_w(fV)g = \rho_o Vg \quad \text{or} \quad f = \frac{\rho_o}{\rho_w}. \tag{20.11}$$

In other words, if the floating material has 90% the density of water, 90% of it will be submerged. That's exactly what happens with ice, which has a smaller density than water. Normally, when you cool something it decreases in volume and the density goes up, but water actually expands slightly when you cool it below 4°C. That's why ice floats on water and why icebergs are mostly under water, and why we have movies like *Titanic*.

Archimedes' principle has many applications. Say we want to build a boat of steel. Now, please don't say, "How do you make a steel boat float in water?" It's not a solid steel boat. If you are thinking about a solid steel boat, you should get into another line of work. The boat is made out of steel, but it's completely hollow. Look at the floating boat in Figure 20.6. You can easily calculate how deep this one should sink to balance its weight. If you give me the weight Mg of the boat and the area A of the base, I'll tell you it will sink to depth H such that

$$\rho_w(HA)g = Mg. \tag{20.12}$$

Then, of course, you can load some cargo, and the boat will sink even more to displace enough water to equal its weight plus that of the cargo. How much cargo can the boat take? With the maximum weight of the cargo, the boat is infinitesimally away from being fully immersed, that is, the volume of water displaced equals that of the hollow boat.

20.4 Bernoulli's equation

This is the first time we will consider fluids in motion. Once again I will only invoke $F = ma$, but we have to be clever, as when we mentally isolated a chunk of water and demanded that it be in equilibrium to find out how the pressure varied with depth.

Figure 20.7 is the standard picture in all the textbooks. This is the best we have been able to come up with after three hundred years. The incompressible fluid of density ρ is now flowing in a pipe. We pick two points along the flow labeled 1 and 2, and we focus on the fluid in between. The cross section of the pipe is changing, with values A_1 and A_2; the velocity has values v_1 and v_2; the altitude (from some reference height) has values h_1 and h_2; and finally the changing pressure has values P_1 and P_2. Remember that in the following discussion h_1 and h_2 are *heights* measured upward from some reference level, in contrast to the h used earlier (in formulas like $P = P_A + \rho g h$), which stood for the *depth* measured downward from the surface.

We are going to find a relation between the above-mentioned quantities at two locations 1 and 2 using the law of conservation of energy.

20.4.1 Continuity equation

First a purely kinematical result. If the fluid is incompressible it obeys the *continuity equation*, which says that the rate of volume flow in at 1 equals the rate of flow out at 2. What comes in has to go out of this fixed volume (between points 1 and 2) because the density cannot change. How much water do you think comes in through the pipe at point 1 in a time dt? Can you visualize in your mind that, in time dt, the fluid moves a distance $dx_1 = v_1 dt$ at the left end and therefore a volume $A_1 v_1 dt$ enters at the left end, and likewise at 2 a volume $A_2 v_2 dt$ exits? Equating the input and output rates and canceling dt we find

$$A_1 v_1 = A_2 v_2, \tag{20.13}$$

which is called the *continuity equation*.

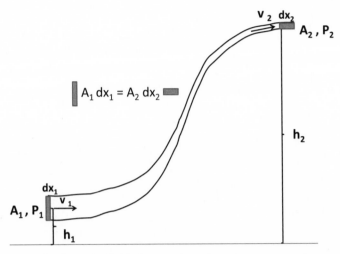

Figure 20.7   The flow of a fluid along a pipe of varying cross section, pressure, and height. We focus on the part initially between cross-sectional areas A_1 and A_2. In a time dt, it is left unchanged but for a sliver of volume $A_1 dx_1 = A_1 v_1 dt$, which has been removed from the left and appended to the right end as a sliver of equal volume $A_2 dx_2 = A_2 v_2 dt$.

Think of cars going down a freeway as the freeway gets narrow. Unlike in real life, let us not allow the cars to pile up. We want the density of cars to be the same. That means if there's a narrow section of the road, the cars have to go faster to maintain the traffic flow. It follows that if I go to one checkpoint and see how many cars pass me per second there, the same number will cross anywhere else. So the speeds will be in inverse proportion to the width of the road (or area in the case of the pipe) so that the product remains the same.

Now, we are going to find a constraint between the variables at 1 and 2. Think about what is going to happen before you derive any formula. Does it make sense to you that, left to itself, the fluid will slow down on the way to the top, because it's got to work against gravity? Therefore, there's going to be some connection between the height and the velocity of the fluid, just from the law of conservation of energy. But you must remember that if there are external forces on a system, $E = K + U$ is not fixed but changes by an amount equal to the work done by external force, which in our case is exerted by the rest of the fluid on either side of the chosen volume:

$$E_2 - E_1 = \text{Work done by external forces.} \tag{20.14}$$

We are going to apply this to the fluid between these two cross sections A_1 and A_2. Imagine it is colored differently from the rest. At every point in the region between 1 and 2 the fluid has a certain velocity and height and thus a certain potential and kinetic energy at time $t = 0$. A little later this body of colored fluid that we follow mentally has moved a bit to the right. If we compare the "before" to the "after," most of the fluid is doing the same thing except that a sliver of width $dx_1 = v_1 dt$ and mass $A_1 v_1 dt \rho$ is missing at the left and has been appended to the right end as a sliver of width $dx_2 = v_2 dt$ and mass $\rho A_2 v_2 dt$. The change in kinetic energy due to this transfer is

$$dK = A_2 v_2 dt \frac{1}{2}\rho v_2^2 - A_1 v_1 dt \frac{1}{2}\rho v_1^2 \tag{20.15}$$

and the change in potential energy is

$$dU = A_2 v_2 dt \rho g h_2 - A_1 v_1 dt \rho g h_1. \tag{20.16}$$

The work done by the pressure at 1 is $F_1 dx_1 = +P_1 A_1 dx_1 = P_1 A_1 v_1 dt$ and the work at the other end is $-P_2 A_2 v_2 dt$, which is negative because the displacement is opposite to the applied force. The work-energy theorem of Eqn. 20.14 now says

$$A_2 v_2 dt \frac{1}{2}\rho v_2^2 + A_2 v_2 dt \rho g h_2 - A_1 v_1 dt \frac{1}{2}\rho v_1^2 - A_1 v_1 dt \rho g h_1$$

$$= A_1 v_1 dt P_1 - A_2 v_2 dt P_2. \tag{20.17}$$

Canceling $A_1 v_1 dt = A_2 v_2 dt$ (continuity equation) from both sides we arrive at

$$\frac{1}{2}\rho v_2^2 + \rho g h_2 - \frac{1}{2}\rho v_1^2 - \rho g h_1 = P_1 - P_2, \tag{20.18}$$

which is rearranged to arrive at *Bernoulli's equation*

$$P_1 + \frac{1}{2}\rho v_1^2 + \rho g h_1 = P_2 + \frac{1}{2}\rho v_2^2 + \rho g h_2. \tag{20.19}$$

It is true there's a pump somewhere pushing the fluid in the beginning, but you don't have to go all the way to the pump. In the end you only ask what is in contact with the chosen segment of the fluid. It is the fluid to the left that is doing work on the segment and fluid to the right on which the segment is doing work.

In a real fluid, the walls will in fact exert a force parallel to themselves. There will be a drag on the fluid because the fluid really doesn't like to move right up against the walls. It will move more easily in the middle of the tube. Different parts of the fluid will be going at different speeds, and there will be a lot of dissipation due to this effect called *viscosity*. We're ignoring viscosity and all other losses in deriving Bernoulli's equation.

20.5 Applications of Bernoulli's equation

Figure 20.8 shows a tank of water filled to a height H. Your goal is to punch a hole on the side at a height h so that the water that sprays out lands on the dog's bowl. This is a two-part problem: (i) find the velocity v_2 of the water as it emerges horizontally from the hole; (ii) equate the distance it travels before hitting the ground to d.

For point 1 of Bernoulli's equation choose a point on the surface of the water in the tank. Assuming the height drops very slowly due to the leak, you may set $v_1 = 0$. The pressure $P_1 = P_A$, the atmospheric pressure. For point 2, choose a point just outside the hole. Clearly $h_2 = h$, and $P_2 = P_A$. We find the velocity v_2 using

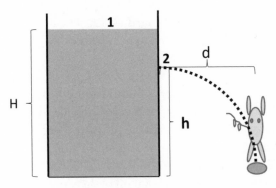

Figure 20.8   Choose the height h such that the dog gets a drink.

$$P_A + \rho_w gH + \frac{1}{2}\rho_w 0^2 = P_A + \rho_w gh + \frac{1}{2}\rho_w v_2^2, \text{ which means} \quad (20.20)$$

$$v_2^2 = 2g(H - h). \quad (20.21)$$

The water has the same speed it would have had if it had freely fallen vertically a distance $H - h$. However, the body of water coming out of the hole is not the water from the top. If you put in some coloring at the top and punch a hole, the water immediately coming out of the hole would not be colored.

If the jet of water travels for a time t before hitting the ground, then from familiar kinematics,

$$h = \frac{1}{2}gt^2 \quad (20.22)$$

$$d = v_2 t. \quad (20.23)$$

Eliminating t and using Eqn. 20.21 we find

$$h = \frac{d^2}{4(H - h)}, \quad (20.24)$$

which has two solutions spaced equally above and below the midpoint $\frac{H}{2}$:

$$h = \frac{H \pm \sqrt{H^2 - d^2}}{2}. \quad (20.25)$$

If you choose the higher solution for h, the water will come out slowly but have a large time-of-flight; if you choose the smaller one, the situation will be reversed. Observe that the farthest the jet can go is $d = H$ (beyond which the square root become imaginary), and for this you must punch the hole at the half-way point $h = \frac{H}{2}$.

The next example is the atomizer in the left half of Figure 20.9. There is some perfume in the container at the atmospheric pressure P_A. You now squeeze the bulb that sends a jet of fast-moving air over the nozzle at the top at some velocity v_2. This lowers the pressure there below P_A and sucks the perfume out through the nozzle and blows it on your face.

The last application of Bernoulli is the Venturi meter shown in the right half of Figure 20.9. Suppose that oil of density ρ_o is flowing in a pipe, and you want to know the flow rate in m^3/s. It is of course $A_1 v_1$, but you only know A_1 but not v_1. So you create a constriction of area A_2 in the flow

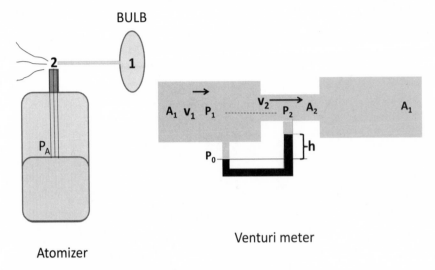

Figure 20.9   The atomizer and Venturi meter.

and then hook up a U-tube as shown in the figure. The U-tube has some other fluid of density $\rho_f > \rho_o$ that does not mix with oil. Because of the continuity equation $A_1 v_1 = A_2 v_2$. The oil will be moving faster in the constriction and the pressure there will be less. This difference is easily related to the height h shown in the figure as follows. Start at the solid horizontal line where the pressures must be equal in the two legs to some P_0 because they are in a static fluid and at the same height. The pressures P_1 and P_2 at the middle of the pipe (dotted line) are then P_0 plus the corresponding $\rho h g$ terms. Canceling a common term on both sides due to the oil above the height h, we get

$$P_1 - P_2 = (\rho_f - \rho_o)gh. \tag{20.26}$$

Thus the equations we have are

$$P_1 + \frac{1}{2}\rho_o v_1^2 = P_2 + \frac{1}{2}\rho_o v_2^2 \tag{20.27}$$

$$A_1 v_1 = A_2 v_2 \tag{20.28}$$

$$P_1 - P_2 = (\rho_f - \rho_o)gh \quad \text{which may be solved to give} \tag{20.29}$$

$$v_1 = \sqrt{\frac{2(\rho_f - \rho_o)gh}{\rho_o\left[(A_1^2/A_2^2) - 1\right]}}. \tag{20.30}$$

CHAPTER 21

Heat

21.1 Equilibrium and the zeroth law: temperature

This chapter—devoted to the study of heat, temperature, and heat transfer—sets the stage for our study of thermodynamics.

You already have an intuitive notion of temperature. Let us begin here with what may be new: the notion of *thermal equilibrium*. Systems are said to be in thermal equilibrium when their *macroscopic* properties, properties discernible by macroscopic probes like the naked eye or a thermometer, have stopped changing.

Take a cup of hot black coffee and take another cup of cold pink soda, and keep them both thermally isolated from each other and the outside world. No matter how long you wait, the coffee will be hot and uniformly black and the soda cold and uniformly pink. The coffee and soda are in their respective states of thermal equilibrium.

If you now pour the contents of one cup into the other, there will be a period when the system is not in equilibrium in the sense that it doesn't have a well-defined temperature or composition. For example, if you just poured the coffee from the top, the hot black coffee is initially on the top and the cold pink soda at the bottom. There will be a period of transition when you really cannot even say what the temperature of the mixture is. Some parts are hot, and some parts are cold; the system doesn't have a global temperature, and it is changing at a macroscopic scale. It also does not have a uniform composition; some parts are black, some are pink, and some are in between. But if you wait long enough, until the two parts have gotten to know each other, they will turn into some undrinkable lukewarm mess, but the nice thing about the mess is that it will have a well-defined temperature and uniform composition. Further waiting will not change anything. That's again a system in equilibrium.

Here's another example. Suppose you take a gas and you put it inside a cylinder with a massless piston of area A on top, as in the left half of Figure 21.1. You put on some weights mg, and they exert a pressure $P = mg/A$. (We assume there is no atmosphere outside pushing down on the cylinder. We also ignore the pressure $\rho g h$ due to the mass of the gas because the gas is so light.) We say that it's in equilibrium because the macroscopic things, things you can see with your naked eye, are not changing. It's just going to sit there. But if you suddenly remove a third of the weights, the piston is going to shoot up, shake around a little bit, and then settle down in a new location in a new state of equilibrium after some time. In between these two states of equilibrium you will see the piston moving, the gas turbulent with eddies and vortices, and the pressure high in some regions but low in other regions. These are non-equilibrium states of the system.

Here is a summary of the difference between the microscopic and macroscopic. At the microscopic level the atoms and molecules that form the liquid or the gas are always in well-defined states at every instant,

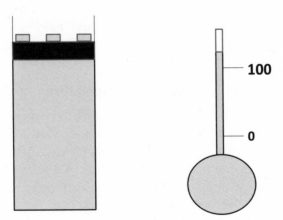

Figure 21.1   Left: The gas is inside the cylinder capped by a piston of area A. The weights on the piston exert a force F and pressure $\frac{F}{A}$. Right: A thermometer with a liquid that expands from the bulb into the thin calibrated stem.

even if the system is not in equilibrium. Each molecule has a definite location and velocity, which determine its future in accordance with Newton's laws. But, at a macroscopic level, when you don't look into the fine details, global attributes like temperature and pressure don't have a well-defined value away from equilibrium. Conversely, in equilibrium, although macroscopic attributes like temperature and pressure appear fixed and uniform, at a molecular level there is a lot of motion and activity.

Now, whenever a system is in equilibrium, we can assign to it a temperature that we call T. Right now we don't know anything about this temperature, other than your intuitive feeling for it. So we're going to build it up from scratch.

Let us begin with the *zeroth law of thermodynamics*—zeroth law because after the first two laws were codified, it was realized that there was a notion even more basic. The zeroth law says, "If A and B are at the same temperature, and B and C are at the same temperature, then A and C are at the same temperature."

You are probably asking: "You call this a law?" Yes, it is the key to our being able to speak about temperature globally. It is the assumption that if I use a thermometer to measure something at one place and then dip the thermometer into something that is elsewhere, and it reads the same number, then I may conclude these two entities, which never met each other directly, are also the same temperature. That means if I bring them into thermal contact they will continue to be in equilibrium. That seems pretty obvious to you, but the whole notion of temperature is predicated on the fact that you can define an attribute that you can globally compare between two systems that never met directly, but met a third common system.

21.2 Calibrating temperature

Once we have some idea of hot and cold, we wish to be more quantitative. Describing somebody as tall or short is a good start but we usually want to know how tall—how many feet, how many inches. The need for more quantitative information led to the idea, "Let's find things in the world that seem to vary with temperature and use that variation to quantify temperature, and build thermometers."

Here is an example. You take this metallic meter stick at the National Bureau of Standards, kept in some glass case, at some temperature. You make a duplicate of it and you lay it outside. What you find is that if the room is hotter than inside the glass case, the stick outside the case expands to a new length. It contracts when the room is colder. Nothing has been done to the stick in the air-conditioned glass case, so this one outside must be expanding and contracting. One way to define temperature is to correlate the length of the stick with temperature in any unique fashion. For example, the temperature could be the length of the stick in centimeters, measured by comparison with the one in the glass case. However, if this metallic stick is to be a portable thermometer, we need to compare it to something portable that does not expand or contract when heated or cooled. We could take a meter stick made of wood and notice that its length matches the one in the glass case no matter how hot or cold the room is. We could then carry this wooden meter stick and a metallic one and use the difference in lengths as a measure of temperature.

The very first thing we can do is to use the zeroth law to say when two things are at the same temperature. You successively dip the metallic stick in two buckets containing two fluids, giving it enough time to equilibrate. If it ends up having the same length (as measured by the wooden stick), you say the fluids are at the same temperature. You could go a step further and declare that if the lengths do not match, the greater length corresponds to the hotter fluid. To specify *how much hotter*, we may say the length of the metallic stick in centimeters (as measured by the wooden stick) is the temperature. You may not be used to measuring temperature in centimeters, but that unit of measurement is just as good as degrees, just as 760 millimeters (of mercury) is an accepted measure of atmospheric pressure.

In practice one picks something a little easier than this metal stick–wooden stick combination. We know liquids expand when you heat them. If you fill your gas tank on a hot day, you have to leave some room at the top for the expansion. One way to measure temperature is to take some liquid, put it in a jar, mark the level, and watch the liquid expand or contract to a new height and mark that height. (Here we neglect the expansion of the jar itself and need to verify that is valid by other means.) You can associate each marking with a certain temperature. This mark can be zero, that can be 5, that can be 19; you have to make sure that it's monotonic, so that 21 is hotter than 19 in some objective sense, say by noting that some substance melts at 21 but not 19.

A more practical design for the *thermometer* is shown in the right half of Figure 21.1. You have a lot of fluid in a big reservoir or bulb, connected to a very thin stem evacuated at the top. What's clever about this is that even if the liquid expands by one percent in volume, it can climb up the narrow stem quite a bit, that is, a little bit goes a long way, because the stem is so narrow. In fact, the stem is so narrow, it is embedded inside a prism that magnifies the mercury or alcohol column so you can see it with the naked eye.

Next we want to design thermometers so that people in different parts of the world, different countries, different labs, can all agree. So, we will make it possible for everyone to make his or her own thermometer by the following recipe. We will dip this thermometer in a bucket that contains some ice and water in equilibrium. Coexistence of ice and water seems to occur only at one temperature. That's called the melting point of ice or the freezing point of water. We go to the lake where ice and water coexist, we dip the thermometer there, and we find that the reading does not change as long as ice and water coexist. It changes only when all the water has turned into ice or vice versa. Whatever reading we get we will postulate to be 0 degrees centigrade or $0°C$. That is just a definition. In the thermometer in Figure 21.1, we mark the point on the stem as $0°C$. But right now you can only tell if another object is at, above, or below $0°C$ using this thermometer with just one marking.

So we need to find another universally accessible thing, which, as you all know, is the boiling point of water. If you take water in a pressure cooker and put it on a stove, it heats up and heats up and then begins to boil and evaporate. As long as water and steam coexist, the temperature remains fixed, as you can tell because the mercury in the thermometer does not rise. That coexistence temperature is going to be called 100 degrees centigrade or $100°C$. You stick your thermometer in when this happens and mark the height of the liquid as $100°C$. Then, you take the interval on the stem between the 0 and 100 markings (Figure 21.1), and you divide it into 100 equal parts. Because the markings can continue outside this interval, this unit defines the temperature outside as well. (This is like saying that a meter stick can be used to measure lengths more or less than a meter.) If the liquid has gone 79% of the way from the 0 mark and the 100 mark, the temperature is 79 degrees. That is the *centigrade* scale. You know there are different scales. You can have the Fahrenheit scale, or any other scale in which what you want to call the freezing point is a different number. Somebody thinks it's zero; another person thinks it's 32. You can again assign to the boiling point a different number: 100 or 212, and you can divide this interval into 100 parts, 180 parts, whatever you like. But the philosophy has the same three steps. You find two points that are reproducible conveniently, divide the region between them into some number of equal steps, and name the scale after yourself.

Now, there are some problems with this approach. One is that the boiling point of water does not seem to be a very reliable standard. If I boil water at high elevations, like in Aspen, for example, it doesn't seem to be as hot as it is at sea level. I know that because when I cook rice, I find it doesn't cook fully when the water is boiling in Aspen, but it does when the water is boiling in New Haven. I know it's boiling earlier in the mountains than in the plains, based on a physical phenomenon, the cooking of rice. A thermometer calibrated at sea level will not agree with one on the mountains. So, who's going to decide what the real boiling temperature is: the person on the plains or on the mountain? You have to be careful when you say boiling point and freezing point, because these are not independent of altitude, pressure, et cetera.

Nowadays, people have much fancier definitions for calibration, and I will tell you a little bit about that later. But, for now, don't worry about the fact that water boils differently at different altitudes; we could all agree to calibrate at sea level, and the relevant conditions at sea level are pretty much constant all over the world.

Now we will turn to a far deeper problem that exists even after you have figured out reliable and reproducible boiling and freezing points. If you make a thermometer with your favorite fluid, say mercury, and I make one with alcohol, they will agree at 0 and they will agree at 100 because that's how we fixed it. We rigged it so at 0 everyone says 0, at 100 everyone says 100. But how about 75 degrees? I say it's 75, if my fluid has climbed three-fourths of the way to the top. At that point, yours may not have climbed three-fourths of the way. In other words, we have the two graphs shown in Figure 21.2. Along the x-axis, we measure temperature according to, say, the mercury thermometer. Along the y-axis we plot the temperatures according to another thermometer, say, alcohol. The straight line $y = x$ is provided for comparison and simply equals the reading on the mercury thermometer. The two agree at 0 and 100 by definition. The curved one does not agree with the reference in between. When the mercury thermometer says it is 75, the other may say it is 55. You will have to pick one liquid and say, "We swear by that liquid, and when that liquid's gone halfway toward 100, we'll say it's 50 degrees." To pick a liquid you'll have to have an international convention, and before you know it, there will be an argument between the alcohol lobby and the mercury lobby.

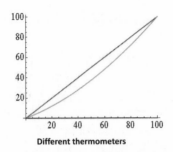

Different thermometers

Figure 21.2   The x-axis is labeled by the reading of a reference thermometer, say with mercury. Along the y-axis are plotted the readings as per another—say, an alcohol—thermometer (curved line) along with a straight line $y = x$, which is just the reading of the same mercury thermometer plotted for easy comparison.

What if I can show you hundreds of thermometers that agree with each other all the way? Clearly we should go with the hundreds that agree. Who are these guys? *They are the gas thermometers.* Here is the explanation of how they work. It takes some effort, but it's worth it.

First I take a small quantity of gas inside a cylinder at some pressure P. The volume V is just the volume up to the piston. This is my reference thermometer. I put it inside boiling water until it reaches equilibrium and then write down the value of the product $[PV]_0$, where the subscript 0 tells us it is the reference gas. I repeat with the tub of coexisting water and ice, and I keep track of the product again. Now I label points on the x-axis by the product (see Figure 21.3). To the place on the x-axis corresponding to freezing, I assign the value $0°C$ and I assign the value $100°C$ to the boiling point. I divide the interval between the ice value and boiling value into 100 equal parts. *This defines the centigrade scale according to this gas thermometer.* To find the temperature of a body, I place my gas thermometer against it, wait for equilibrium, and measure the product $[PV]_0$. Whatever number on the centigrade markings this product falls on, that is the temperature in centigrade. Thus the point $[PV]_0 = 350$ in the figure corresponds to the temperature $75°C$. I can continue this scale to the left of the ice point and to the right of the boiling point.

I now consider two more gas thermometers with different amounts of gases of different kinds, and I plot their PV as a function of my centigrade scale (based on $[PV]_0$) measured along the x-axis. Two such lines are shown in the figure. *The main point is that they are both straight lines.* This guarantees that the thermometers will agree not only at the end points

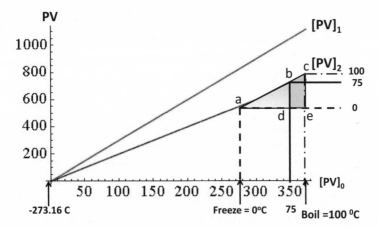

Figure 21.3   Comparison of different gas thermometers labeled 0, 1, and 2. The x-axis is labeled by one and the other two correspond to two other gas thermometers, possibly filled with different dilute gases. The linearity of the graphs guarantees that the thermometers will agree not only at 0 and 100, where they have to, but also everywhere in between given the similarity of triangles *abd* and *ace*. When $[PV]_0 = 350$, the temperature is $75°C$.

where they have to, but also in between and beyond. Let us understand why. Look at the lower one labeled 2 and the shaded right triangle that extends from 0 to 100 in the x-direction and the corresponding values of $[PV]_2$ in the y-direction. The value of $[PV]_2$ grows linearly between the ice and boiling points, as shown by the hypotenuse of the shaded triangle. Of course, no one wants to know what $[PV]_2$ is; we want the corresponding temperature in centigrade. This means we must associate with each value of $[PV]_2$ a certain temperature. By convention we label the ice point as $0°C$ and the boiling point as $100°C$. Next we divide the difference in $[PV]_2$ between freezing and boiling into 100 equal parts, and each one is one degree centigrade, as shown in the vertical side of the triangle. This way if $[PV]_2$ rises by 75% of the difference in $[PV]_2$ between the ice and boiling points, the temperature reading according to thermometer 2 will be $75°C$. But this will also be the temperature of the corresponding point in the x-axis, for it too would be 75% of the way between 0 and 100 by similar triangles *abd* and *ace*. By the same logic, thermometer 1, which uses $[PV]_1$ as the measure of temperature, will also agree with the other two everywhere. Indeed, any gas thermometer whose measure of temperature is via PV will be represented by a linear plot, and it will agree with its fellow gas thermometers everywhere. The slope of the plot will vary with the kind and amount of gas (for example, doubling the volume of gas at the same pressure will double the PV and the slope), but that will only change the spacing in PV per degree centigrade.

The basic point is that a straight line is completely determined by any two points on it.

It is essential that the x-axis itself be calibrated by a gas thermometer. If there were a thermometer that disagreed with the gas thermometer at intermediate points, and we used *that* to label the x-axis, the PV plots of all the gas thermometers would be curved.

There appears to be just one requirement for a good gas thermometer: the gas has to be very dilute. Given this requirement, everybody can build and use the gas thermometer.

21.3   Absolute zero and the Kelvin scale

Comparison of different gas thermometers (Figure 21.3) reveals not only that each gas thermometer is represented by a straight line, but that all these lines cross the x-axis at $-273.16°C$. (In reality, the PV plots stop short of 0, and one extrapolates linearly to see where they would hit zero.) There is something very special about that temperature because it is shared by all gases. It is called the *absolute zero of temperature.* It's called absolute zero for many reasons. One is that, unlike the 0^0 centigrade, which is by no means the absolute lowest possible temperature, the absolute 0 is the lowest possible temperature. Why? Because the gas pressure can't be reduced below zero. That's it. It cannot go below having no pressure. Later, we'll find reasons why no further cooling is even conceptually possible. That will require you to understand what hot and cold mean at the atomic level. But right now, Figure 21.3 shows that all gas thermometers point to this temperature. So people decided that calling the freezing point of water 0 is artificial. That's based on human preoccupation with water. But if you think laws of physics describe the whole universe, what about planets where there is no water? Suppose you're talking to a different civilization: Planet of the Apes, say. You tell them, "Apes, we're going to sync our temperatures; 0 is when water freezes," and they say, "What is this thing called water?" You cannot identify water as the stuff they drink, because you don't know what these apes are drinking. Maybe they drink methane or liquid hydrogen. On the other hand, if you say, "Take any gas and wait until the product of the pressure and volume go to zero, and then call that zero," that will be a universal standard. It's not tied to something called water. The absolute zero is referred to as $0K$, where K stands for (Lord) Kelvin.

Having picked the zero, we need one other temperature to form a scale. It was decided the triple point of water would be assigned a value $273.16K$. What's the triple point of water? Water and ice can coexist along a line in the $P - T$ plane, and water and steam can coexist along another line. There is a certain magical point at which the lines meet, and there ice, water, and steam can coexist. The system cannot choose among those three options. We will assign to that point a temperature $+273.16\,K$ in the Kelvin scale. (For our purposes the triple point is essentially the freezing point, which we called "Freeze" in the figure, at a particular pressure.) The centigrade and Kelvin scales differ only in the location of the zero; a *change* of $1°$ in the Kelvin scale is also a change of $1°$ in the centigrade scale. The *difference* between absolute zero and the triple point of water is 273.16 degrees in both scales.

Now there is a rule. (I do not say "unwritten" for it is written.) You may say, "50 degrees centigrade," but you're not supposed to say, "50 degrees Kelvin." You have to say "50 Kelvin." I keep forgetting this but so far nothing terrible has happened to me. But you should remember this rule when you take the GRE or go for a job interview. Once you get tenure, you too can say "50 degrees Kelvin," whenever you want, with no dire consequences.

21.4   Heat and specific heat

Heat is denoted by the symbol Q. What are we talking about when we talk about heat? Again, let's use your intuition. Say we have a bucket of water and we want to heat it up. We put the bucket on top of a stove that we think is hotter than the water, and when the two are brought together, somehow the water begins to feel hotter and hotter. We say we've heated the water, and we say we have transferred heat. Now, scientists were not always sure what really was being transferred. What goes from the stove to the water? Why is it that the stove, if it's not plugged in, gets cooler as the water gets hotter? Some theorists imagined there was a certain *caloric fluid* that is abundant in hot things, and not so abundant in cold things. When we put hot and cold together, this fluid flows from hot to cold, and in the process heats the cold thing. It was decided to measure heat transferred in *calories*. So next we have to define a calorie. You want to ask, "How much heat does it take to raise the temperature of this bucket of water by ten degrees?" Here is the rule: The number of calories you need is equal to the mass of water in grams times the change in temperature

$$\Delta Q = m\Delta T. \tag{21.1}$$

In other words, if you had 4 grams of water, and you did something to it and the temperature went up by 7 degrees, you have, by definition, pumped in 28 calories. If this was a kilogram of water, this would be 28 kilo-calories. Sometimes we use grams and calories; sometimes we use kilograms and kilo-calories. But the definitions are consistent: if you put a kilo in the gram, put a kilo in the calories.

Now, suppose you want to talk about heating something else, say a gram of copper. Then you write down the following rule. The amount of heat it takes to heat up anything must be proportional to the amount of stuff you're trying to heat up. That's our intuitive notion. If you have one chunk of gold that takes some number of calories, and you have a second identical chunk, that should take the same number of calories. If you put them together, it is clear that whatever this caloric fluid is, you will need to double what was required by one chunk. So, ΔQ has to be proportional to the mass of the substance. And it's got to be proportional to what you're aiming for, namely, increase in temperature. But this is true for any substance, whether you're heating copper, wood, or gold. No matter what you are heating, the heat needed is proportional to mass and to the change in temperature. So, what is it that distinguishes one material from another? We introduce the number c here, called the *specific heat*, and write

$$\Delta Q = mc\Delta T. \tag{21.2}$$

The specific heat c is the property of that material, and we will soon discuss how it is to be measured. Equation 21.1 tells us $c = 1$ cal/g for water:

$$c_w = 1 kcal/kg = 1 cal/g. \tag{21.3}$$

You have to understand that formulas will depend on certain parameters in a generic way, and on other things in a material-specific way. In this example the dependence on m and ΔT is generic, while that on c is specific to some material.

Here is another example of this logic. I can ask by how much a rod will expand if I heat it by ΔT. The increase ΔL has to be proportional to the original length. To see this, take a meter stick that expands by some amount and put another identical meter stick next to it. The two-meter stick will clearly expand by twice as much as the one-meter stick. So, we put the length L in the right-hand side. There is, of course, the ΔT, which causes the expansion. No matter what you are heating—a block of wood or a block of steel—it is true that

$$\Delta L \propto L \Delta T. \tag{21.4}$$

But then the fact that heat has different effects on copper versus wood is indicated by introducing a number α, called the *coefficient of linear expansion*, which depends on the material, to obtain

$$\Delta L = \alpha L \Delta T. \tag{21.5}$$

So copper will have a certain α, iron will have a different α, and so on. Suppose you say, "Well, I had some material and when I heated it up by one degree, its length increased by nine inches; another one increased by two inches." Is it clear that the first one expands more readily? No, because the first one could have been a mile long, and the second one could have been ten feet long. So, you have to take out certain factors that are universal, and the rest of it you put into a property of the material.

Given that water has (by definition) a certain specific heat $c_w = 1$ $kcal/kg$, I can measure the specific heat of other materials as follows. I take a container with some water in it. Let's assume the container has no mass, so I don't have to worry about it. The water of mass m_2 is at some initial temperature T_2. I have some new material, lead, and I want to find its specific heat. So, I take the lead of mass m_1 in the form of pellets, I heat the pellets to some temperature T_1, and I drop these pellets into this water. When I put the pellets into the water, there will be a period when the temperature is not defined. Soon the water and pellets will settle down to some common, final, equilibrium temperature called T_f and will have a mass $m_1 + m_2$.

We will now postulate that the total change in Q is zero. In other words, if heat Q is lost by one body and gained by another body, the loss and the gain must equal. It's a new law, called *the conservation of heat*. You can make up all the new laws you want. You don't know if they're right,

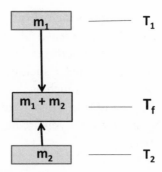

Figure 21.4   Calorimetric problem: What will be T_f?

but this is the law you first make up. What can you say in this particular problem? I urge you to draw the picture in Figure 21.4 for such calorimetric problems. It shows the masses m_1 (lead) and m_2 (water) at temperatures T_1 and T_2 before, and their final state at T_f after they are combined. The sum of all the ΔQ's is zero:

$$\Delta Q_1 + \Delta Q_2 = m_1 c_1 (T_f - T_1) + m_2 c_2 (T_f - T_2) = 0. \qquad (21.6)$$

Note that the ΔQ for the water will be positive because $T_f > T_2$ (water gains heat) while the lead pellets, which lose heat, will have a negative ΔQ because $T_f < T_1$.

You know the mass of the water is m_2, the mass of the pellets is m_1, and $c_2 = 1 kcal/kg$. You can measure T_1, T_2, and T_f and hence solve for c_1.

It turns out that the specific heat of materials is not really a constant; it can vary with temperature and generally vanishes as $T \to 0K$. There is a big industry calculating the specific heats of materials as a function of T, starting from atoms and quantum mechanics. We can treat them as constants over some limited range, say, near room temperature, which is roughly $300\ K$.

None of the things treated as constants is ever truly constant. The previous description with constant specific heats was espoused before we even knew about atoms. Physicists were doing the best they could. They found empirically that using the specific heat of lead, determined as above, and the specific heat of gold, determined the same way, in a third experiment performed with lead and gold using *those* values of specific heat, the ΔQ's indeed added up to zero, that Q was conserved. But this was based on the best possible measurements of that epoch. In a later epoch, when more accurate measurements were performed, over a bigger range of temperatures, it was found that the specific heat could not be taken as temperature-independent, if calories were to be conserved in calorimetric experiments.

21.5 Phase change

Now for a twist. I take some ice at $-30°C$. I've gone to centigrade now so we can better relate to ice. I put the ice on some source of heat, a device that will pump in a fixed number of calories every second. As a function of time, I'm expecting the temperature of the ice to go up. Every second, I get some number of calories, and those calories are going to produce for me some $mc_i\Delta T$ where c_i is the specific heat of ice. Remember the specific heat of ice is not the same as the specific heat of water. Even though ice is also made up of water molecules, the calories needed to heat one gram of ice by one degree is roughly half what it takes to heat one gram of water. Because m and c_i are constants, ΔQ is proportional to ΔT. The rate of temperature rise will be proportional to the rate at which the heat flows into the system. The temperature of the ice goes initially from $-30°C$ to

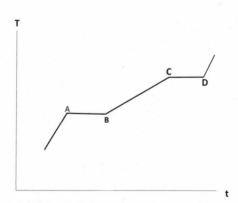

Figure 21.5   Different phases of water from ice below $0°C$ to water above $100°C$. As time t increases, a steady flow of calories first heats the ice below $0°C$ to A, where it melts at $0°C$ between A and B. The water then heats up between B and C, and then it evaporates at $100°C$ between C and D, before becoming superheated steam.

$-20°C$ to $-10°C$ and so on as in Figure 21.5. But once it hits 0, point A, it gets stuck. Although I know heat is coming in, the ice is not getting hotter. Then I notice that the ice is beginning to melt. There will be a period when, as I pump in calories, I don't get any increase in temperature, but I get conversion of ice into water. This is called a *phase change*. A phase change occurs when a substance changes its atomic arrangement, in this case, from a regular array that forms a solid into a liquid whose atoms are free to run around.

An atom in a solid is like a kid in a classroom: you have an assigned seat but you can fidget in that seat. A liquid in a container is more like a playground with a fence; the atoms get to run around within a certain contained space.

Return to the period when there is some water with chunks of ice floating on it, between A and B. Until all the ice is converted to water, the whole system is stuck at that temperature, $0°C$. It is called "No ice left behind." That's a very interesting property. Now, if you really put a pot on a stove and drop a chunk of ice in it, you know what will happen. The bottom of the ice will melt; it may even boil and evaporate, yet the top may still be solid. That's not what I'm talking about, because that's not a system where there's a globally defined temperature. I want you to heat the ice ever so slowly; the minute you put in a little bit of calories, you give it enough time for the whole system to share that heat, so that the whole system has one single common temperature. When you do this, you will find the temperature stays fixed until all the ice has melted.

The caloric price you have to pay to bring about the melting is called the *latent heat of fusion*, L. It has a value of $80 kcal/kg$ for water, and it varies from substance to substance.

Once everything has become water, and the point B is reached, the uniform system of water starts increasing in T. I guess you know the next stopping point. When you come to $100°C$, point C, again it gets stuck until everything vaporizes, and beyond point D it is all steam. Then, you can have superheated steam, which is even higher than $100°C$. The caloric cost involved per kilogram in this phase change is the *latent heat of vaporization*, L_v. It is around $500 Kcal/kg$. That's information I don't carry in my head.

So, if I give you some ice at $-30°C$ and I give you a budget of $5,000$ calories, where will it end up? Look at in Figure 21.5. You have to first spend a few calories going from below A to A at $0°C$, and, if you have some calories left, you can start melting the ice; maybe you'll run out of stuff there and get stuck between A and B, with a mixture of ice and water. If you have still more calories at your disposal, you can melt it all (point B) and start heating the water and keep going till it is all steam, and heat that steam and so on.

The kind of problems you can get are fairly simple most of the time. The only time when you can really get in trouble is in the following situation. Suppose I mix some water at $+20°C$ and some ice at $-40°C$. What will be the end product? Now, this is a subtle problem. If you had water at $+20°C$ and you added more water at $+60°C$, you can easily guess that it would end up as water at an in-between temperature, which you can easily calculate. Now it's more subtle. The answer will depend on how much ice and how much water you have. If by water at $+20°C$ you mean the Atlantic Ocean and by ice you mean a couple of ice cubes, we know what's going to happen. These ice cubes are going to get clobbered; they're going to melt. You will end up with all water. Then, you can easily calculate the final temperature T_f of the water by writing down (all in centigrade) and solving

$$0 = M_a c_w (T_f - 20) + M_i c_i (0 - (-40)) + M_i L + M_i c_w (T_f - 0) \tag{21.7}$$

$$= M_a c_w (T_f - 20) + M_i c_i 40 + M_i L + M_i c_w T_f \tag{21.8}$$

where M_a and M_i are masses of the Atlantic and the ice cubes, c_i and c_w the specific heats of ice and water, and L the latent heat of melting. The first term is the ΔQ for the ocean as it drops from $20°C$ to T_f. The next is the ΔQ for ice as it goes from $-40°C$ to $0°C$, the next is the heat of melting for the ice, and finally the heat gained by the melted ice to become water at T_f.

In general, given some ice below zero and some water above, you will not know in advance whether you will end up with all ice, a mixture at $0°C$, or all water. You can first make the optimistic assumption that you will end up as all water at an unknown temperature T_f. Write your equations, including the heat it takes to melt the ice, and solve for T_f. If you get a positive answer you can use it, because the assumption that you heated up the ice, you melted the ice into water, and then heated up the water from 0 to the final water at T_f is correct. But if you did the calculation and obtained a negative value of T_f, that result is inconsistent with the assumptions in the equation. Then, you can try something else; you can assume it is all ice at the end. If you think T_f is down there, below A, then you've simply heated the ice from very cold ice to not-so-cold ice. You brought this water down from $T > 0$ to 0, sucking out the $mc\Delta T$ to do that, then you've taken out the latent heat of melting to turn the water at 0 into ice at 0, and then you've cooled that ice down to the final T_f below A. Then, all those losses of the original water equal the gain of the initial ice. If, when you solve for T_f, you get a negative number, then you're okay. That will certainly be the case if I sprinkle two drops of water on a big iceberg; we know it's going to end up as ice.

But if I give you numbers that are wishy-washy, when it is not so clear whether ice will win or water will win, you may be forced to consider a third possibility if these two fail: you end up with some amount of water and some amount of ice at 0 degrees, between A and B. Then the question is not what the final temperature is—it is 0—but how much is ice and how much is water.

You solve that as follows. Start with the water at some point between B and C, say at 25°C. You extract heat $\Delta Q = m_w c_w 25$ from that water to bring it to water at 0°C, which is the final equilibrium temperature. You start giving that heat to the ice, first to move it from below A to A. It does that by absorbing that $m_i c_i \Delta T$, the mass of the ice times specific heat of ice times ΔT. If the ice was at −40°C, then $\Delta T = +40$. You certainly have enough heat to do that because you know that, in the end, any ice we have is at 0°C. Whatever remains of the heat you extracted from the water, you now use to melt the ice at the rate of L kcal/kg. The unmelted ice is the amount of ice at the end, and the rest of the mass is water.

21.6 Radiation, convection, and conduction

We now discuss different ways of transferring heat.

Radiation is heat energy transfer from some body to another without the benefit of any intervening medium, like heat from the sun, which is really electromagnetic radiation. Electromagnetic radiation doesn't need air; it doesn't need any medium to propagate. In fact, if it needed a medium like air, we would not get any heat from the sun because between the earth and the sun is mostly vacuum. Thus, if you were in front of one of these space heaters, with glowing red coils, and you were feeling warm and someone started pumping the air out of that room, your last thought would be, "Yeah, but I am still warm."

The second way of heat transfer is called *convection*, explained by the following example. You have water in a pan; you put it on a hot plate. Then, in the lower part of it, the water gets hot. When it gets hot it expands, and when it expands the density goes down; therefore, by buoyancy it will start raising up. Remember, a chunk of water belongs in water. A chunk of something else with lower density will float to the top. But the point is, water doesn't have a fixed density. If you heat it up, the density goes down, so the water downstairs has a lower density—like a piece of cork, it will rise to the top. When it rises to the top, the cold water with the higher density will fall down. So, you set up a current. Hot rises to the top and cold comes down. And this also happens in the atmosphere. On a hot day, the air next to the ground gets really heated up and it rises, and the cold air comes down, and this process creates thermal currents. So, here you're trying to *equalize the temperature between a region that is cold and a region that is hot by the actual motion of some material.* In radiation, you don't have the medium transferring heat because a medium is not even present. In convection, the medium actually moves and by that process heat is transferred.

The final form of heat transfer, the one I want to focus on more quantitatively, is *conduction*, which is something you've all experienced. Why does a skillet have a wooden handle? There is a simple reason: given that your body is at 98 degrees and the skillet is, say, 200 degrees, a steel handle would have heat flowing from the skillet to your hand. We want to understand the rate at which heat flows from the hot end to the cold end.

I'm now going to introduce a new term called *heat reservoir.* A reservoir is another body like you and me, except it's not at all like you and me. It's enormous. It is so big that its temperature cannot be changed. If you can sit on it, you will fry and you'll evaporate, but its temperature will not change. No body is really a reservoir. If you drop an ice cube in the Atlantic, you'll lower the temperature of the Atlantic but only by a negligible amount. So, take the limit Atlantic goes to infinity, and you have a reservoir. Reservoirs have one label, namely, their temperature. The room in which you are sitting is a good approximation to a reservoir. If you put a cup of hot coffee here, we say it will come to room temperature. Actually,

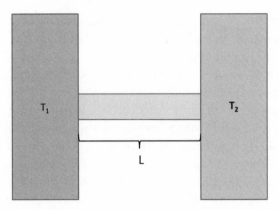

Figure 21.6   Heat flow from a hot reservoir to a cold one through a conducting rod of length L and cross-sectional area A.

the room temperature rises a bit to meet that of the coffee, but this increase is negligible in practice. We can attribute to the room a temperature quite independent of bodies that go in and out of it.

Look at Figure 21.6. A rod of length L and cross section A is connected at the right to a reservoir at T_2 (maybe a tank of water-ice mixture at 0 degrees) and at the left to a reservoir at T_1 (say a water-steam mixture at $100°C$). We know heat is going to flow from the hot end to the cold end. We want to write a formula for how much heat flows per second. What will it be proportional to? Cross section is one correct answer. We can understand this result as follows. You take one rod, and for convenience let's just take it to be of rectangular cross section. Take another identical rod, and it will also transfer the same amount of heat for a given amount of time. Just glue them together to form a rod of twice the cross section. It's going to transmit twice the amount of heat, and it has twice the area. So, the rate of heat flow is going to be proportional to the cross-sectional area A. And why is the heat flowing? It's flowing because of a temperature difference ΔT. So, that's always there; that's the underlying cause behind heat transfer. That's the dynamics in thermodynamics; that's what makes the heat flow. But then, we find as an empirical fact that the rate of heat flow decreases as the distance between the reservoirs, the length of the rod, increases. Heat flow seems to depend not just on the temperature difference but on its gradient in space. So, you want to divide by L, the length of the rod separating the hot and cold ends. These features of heat transfer happen to be truly independent of what material we are talking about. So far we have

$$\frac{dQ}{dt} \propto A \frac{(T_1 - T_2)}{L}. \tag{21.9}$$

Again, having put all these universal factors, you have to ask, "How does the formula differentiate between a copper rod and a wooden rod?" The answer again is that you have to put one more number κ, the *thermal conductivity*, which depends on the material:

$$\frac{dQ}{dt} = -\kappa A \frac{(T_1 - T_2)}{L}. \tag{21.10}$$

The minus sign in front says that if T decreases with increasing x (that is, it gets colder as we go to the right), the *heat current* flows along the positive x-direction.

Let us raise the same question once more with feeling: "I have two reservoirs, hot and cold. I connected them with two different rods. This rod carried twice the amount of heat per second as the other rod. Is it necessarily a better conductor?" No. Maybe it had $10,000$ times the cross section. So, what you want to do is to make the playing field level: compare rods of the same cross section, same temperature difference, and same length, and then ask which conducts more heat. That depends on the material and that's the role of κ.

21.7 Heat as molecular kinetic energy

In the old days, people just said that heat was a caloric fluid, and they postulated the conservation law for the fluid. You can postulate what you want, but you have to make sure it works The conservation of calories did seem to work, in the sense that all the ΔQs in any reaction did seem to add up to zero. But then people started getting hints that this fluid was not conserved and that this thing we call heat is not entirely independent of other things we have studied.

So, where do you get the clue? When we studied mechanics, we talked about two cars in a totally inelastic collision—they slam into one big static lump; in the end we have no kinetic energy and no potential energy. We just gave up and said, "Look, conservation of energy does not apply to this inelastic collision." On the other hand, we find that in such a collision the bodies become hot. So calories are not conserved either: they are created in a collision. There are more clues. You take cannon balls and drop them from a big tower. When they hit the sand, they start heating up. Or you drill a hole in a cannon to guide the ball, which is what Count Rumford did in 1790, and you find that you need to constantly pour water to keep the drill bit from heating up. You find all the time that, when mechanical energy is lost, things heat up, that is, calories appear from nowhere. So, you get a suspicion that maybe there's a rule connecting the mechanical energy that you cannot account for, that seems lost, to a corresponding gain in calories. Maybe the conservation of mechanical energy and of heat could both be salvaged into a single law of conservation of energy *if joules measure energy you can see, and calories energy you cannot see.* That is the premise. If that is the case, we must first determine the exchange rate between calories and joules.

James Prescott Joule performed the experiment shown in Figure 21.7. You have an insulated container in which there is water and a shaft attached to fins. The shaft can rotate. At its top is a pulley with a weight hanging as shown. There is a rope wrapped around the shaft, and when you let this weight go down, it's going to spin the shaft. The fins churn up the water. Now, you can keep track of how much mechanical energy is lost, right? This mass was at rest at first and it lost potential energy Mgh as it came down. Let's say it's got some kinetic energy at the end that is less than

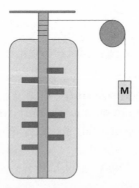

Figure 21.7   Joule's experiment to find the mechanical equivalent of heat, $4.2kJ/kcal$. As the weight descends, the fins turn and heat up the water. The missing mechanical energy in the weight becomes heat energy. (It is assumed that the rest of the apparatus does not absorb any appreciable part of the generated heat.)

Mgh. Some number of joules are gone but the water heats up. When the water gets hot, you can ask how many calories were supplied to the water by looking at the mass and specific heat of the water and the increase in temperature. Then you ask if there is a proportionality between the joules and calories. You find it is 4.2 joules to a calorie, called the *mechanical equivalent of heat*. In other words, if you can expend 4.2 joules of mechanical energy, you get one calorie to be used for heating. In the example of the colliding cars, one car had some kinetic energy, and the other had some kinetic energy, all measured in joules; they slammed together, and they came to rest. That means you can take all those joules, divide by 4.2, and get some number of calories ΔQ. Then those calories will produce an increase in temperature of the cars, such that $\Delta Q = mc\Delta T$, where c is specific heat of whatever material the cars were made of.

In practice, there will be other losses. You heard the crash: that's some sound energy gone; you won't get it back. Some sparks were flying; that's light energy that's gone. You subtract all that out, and you find that in the end, the calories explain the missing joules.

Counting heat as another form of energy, we can save the law of conservation of energy. It is not violated, even during inelastic collisions, if you include heat as a form of energy using 4.2 joules per calorie. What right do we have to call heat energy? When we say some kid is energetic, we mean the kid is always running around mindlessly, back and forth. Energy is associated with motion. These two cars were moving, and we have every right to say they have kinetic energy. How about potential energy then? Well, if the car starts climbing up a hill and slows down, we think it has potential. If you let it go, it will come back and give you the kinetic energy. So, most people think of energy as just kinetic energy. That is what seems to have been lost in the collision. And yet, you get calories in return, so you ask yourself, "How is the calorie related to kinetic energy?"

The correct answer to that came only when we understood that everything is made of atoms. It turns out that the kinetic energy of atoms is what we call heat. But you have to be very careful. I take a cylinder full of very cold gas. I throw it at you. That whole cylinder is moving. That's not what I call heat; the cylinder is not hotter due to its overall motion. That is motion you can see. I'm talking about a cylinder of gas that doesn't seem to be going anywhere; yet it has motional energy because the little guys inside are going back and forth. If you kept track of the kinetic energy of every single molecule in this car, every single molecule in that car, and you added them up, you would get exactly the same number, before and after.

The only difference will be that, originally, each car also had some global common velocity, macroscopic velocity you could see. On top of it, it had random motion of the molecules that made up the car. So did the other car. When they slammed together, the macroscopic motion was traded for thermal motion.

CHAPTER 22

Thermodynamics I

22.1 Recap

In the last chapter we took the notion of temperature, for which we have an intuitive feeling, and turned it into something more quantitative, so you can not only say this is hotter than that, you can say by how much, by how many degrees. We agreed to use the absolute Kelvin scale for temperature and to use the product of PV of a gas thermometer as a measure of temperature T. The Kelvin scale has its origin ($T = 0$) at $-273.16°C$, which is where PV vanished for any dilute gas. In other words, it appears that pressure times volume is some constant times this temperature T. That's the scale chosen by nature, and it doesn't seem to depend on the gas that you use. I can use one; you can use another one. People on another planet can use a different gas. There is really nothing below this $T = 0$.

The next thing I mentioned was that initially people believed in the caloric fluid. Hot things have a lot of it, and cold things have less of it, and the caloric flows from hot to cold. Caloric is conserved, which allows you to do some problems in calorimetry. That promoted heat to a new and independent entity, different from all other things.

Then it turned out that heat was not so unrelated to other things in mechanics. You can heat up water by putting it on the stove, and the caloric fluid flows from the stove into the water. But there's a different way to heat the water. This is not the most economical way to make your coffee but I'm just telling you as a matter of principle. Buy two Ferraris, slam them into each other, take a pot of water, and put it on top of the cars. It will heat up because the Ferraris will be hot. What happened to the kinetic energy of the two cars? It is really gone. In the old days, we would have said that we cannot apply the law of conservation of energy to this inelastic collision. That was our way out of the whole issue. But then we had another problem: caloric fluid was not conserved, because slamming the two cars produced this extra heat. So both the law of conservation of energy and the law of conservation of the caloric fluid seemed to be violated. *Luckily both could be saved if heat was counted as a form of energy.* When mechanical energy disappears, a definite amount of heat energy appears. How many calories can you get if you sacrifice one joule of mechanical energy? Joule did the experiment with his gadget, which had a shaft with some fins that turned in some water as some weight went down and heated the water. Equating the loss of mechanical energy to the calories needed for the heating, he found the *mechanical equivalent of heat*: 4.2 joules = 1 calorie.

22.2 Boltzmann's constant and Avogadro's number

When we say something is hotter, what do we mean on a microscopic level? The answer is based on the profound fact that everything is made up of atoms.

Take the simple example where temperature enters: the gas thermometers from the last chapter. The linearity of the PV plots as a function of T means that PV is proportional to the absolute temperature T. What parameters do you think enter the right-hand side of the proportionality relation

$$PV \propto T? \tag{22.1}$$

The amount of gas is correct. The amount of atoms is even more correct. But suppose you were not aware of atoms. Then what would you mean by "amount of gas"? It would have to be the mass. What's the reasoning? We know that if you have some amount of gas producing the pressure in a box, and you put in twice as much stuff, it should produce twice as much pressure. That's actually correct. We can write for one particular gas sample of mass m:

$$PV \propto mT. \tag{22.2}$$

As usual, after these generic factors are incorporated, you still need a constant of proportionality α that is specific to the gas. (It will be renamed before you get a chance to confuse it with the coefficient of linear expansion.) So now we write

$$PV = \alpha m T. \tag{22.3}$$

Experimentally we find that α depends on the gas, not unlike the specific heats of various substances. But there is an intriguing relation between the α's for various species and the α_H for hydrogen. *For this discussion, and to grasp the main point, let us pretend that there are only three atomic species, hydrogen, helium, and carbon, and that as gases they always occur in atomic and not molecular form.* It is found that for helium $\alpha_{He} = \frac{1}{4}\alpha_H$ and for carbon $\alpha_C = \frac{1}{12}\alpha_H$. In other words, one gram of helium produces the same pressure of $\frac{1}{4}$ grams of hydrogen, and one gram of carbon produces the same pressure as $\frac{1}{12}$ grams of hydrogen (at some fixed V).

If m_H, m_{He}, and m_C denote the masses of H, He, and C, the preceding results may be summarized as follows:

$$PV = \alpha_H m_H T = a_H \left[\frac{m_H}{1}\right] T \tag{22.4}$$

$$= \alpha_{He} m_{He} T = \frac{\alpha_H}{4} m_{He} T = \alpha_H \left[\frac{m_{He}}{4}\right] T \tag{22.5}$$

$$= \alpha_C m_C T = \frac{\alpha_H}{12} m_C T = \alpha_H \left[\frac{m_C}{12}\right] T \quad \text{and so on.} \tag{22.6}$$

So it looks like the mass of the gas has to be divided by numbers like 4 or 12 to find its true effectiveness in contributing to pressure and that *if these rescaled masses are used, the same proportionality constant α_H may be used for all gases.* What do these rescaled masses mean and why are the rescaling factors nice round numbers? Here is the answer, with no details of how it was arrived at, because that would take way too long.

- All matter, including gases, is made of atoms (and also molecules, but I will refer to them all as atoms).
- The masses of hydrogen, helium, and carbon atoms are in the ratio

$$H : He : C = 1 : 4 : 12. \tag{22.7}$$

Thus in Eqns. 22.4–22.6, the mass of each gas is really being divided by a number proportional to the mass of the constituent atoms. The result is clearly proportional to the number of atoms in each sample. In other words, PV is proportional to the product of T and N, the number of atoms in the sample *with a constant of proportionality that is independent of the gas.*

Thus we may write

$$PV = kNT \qquad\qquad (22.8)$$

where this new constant $k = 1.38 \cdot 10^{-23} J/K$ *is independent of the gas* and is called *Boltzmann's constant.*

Bear in mind that we are discussing an *ideal gas*, one whose atoms move independently of each other with no interatomic force. Gases tend to become ideal at low densities, when the typical separation between atoms is large and the interatomic forces negligible.

The number N of gas atoms in a typical sample is huge. In particular, one gram of the lightest atom, hydrogen, has $N_A = 6 \cdot 10^{23}$ atoms, where N_A is called Avogadro's number. (So, N_A is the reciprocal of the mass of a hydrogen atom in grams.) It's simply a number, like a dozen, and it is also called a *mole*. Sometimes the word *mole* is also used to stand for the amount of any substance that has a mole of atoms in it. Thus a mole of carbon has a mass of 12 grams. It is a natural unit to use for counting atoms, the way a dozen is a natural unit for eggs, a light year is a natural unit for cosmological distances, and a kilogram is a natural unit for the mass of humans.

In terms of n, the *number of moles* defined by

$$n = \frac{N}{N_A}, \qquad\qquad (22.9)$$

the *equation of state*, the relation between P, V, and T for the ideal gas becomes

$$PV = NkT = nN_A kT \qquad\qquad (22.10)$$

$$= nRT \quad \text{where} \qquad\qquad (22.11)$$

$$R = N_A k = 8.31 \frac{J}{K \cdot \text{mole}}, \qquad\qquad (22.12)$$

where $R = N_A k$ is the *universal gas constant*, equal to 2 *cals/°C/mole*, easy to remember and the same for all gases. Initially R was what was measured and people worked with the moles n. Then they looked under the hood and found that the gas is made of individual atoms, and they switched from the macroscopic parameters n and R to the microscopic N and k.

22.3 Microscopic definition of absolute temperature

Look at $PV = NkT = nRT$. Is there a microscopic basis for this equation? In other words, once we believe in atoms, do we understand why there is a pressure at all in a gas? For this purpose, we will take an $L \times L \times L$ cube of gas as shown in Figure 22.1. Inside this is some gas and it has some pressure, and I want to know the value of the pressure. Consider the shaded face of the cube. It has to be nailed down to the other faces; otherwise, it'll just come flying out because the gas is pushing it out. The pressure is the force on this face divided by area. The atoms are constantly bouncing off the wall, and every time one bounces on a wall, its momentum changes. So, who's changing the momentum? Well, the wall is changing the momentum. It's reversing it, for example, if the atom bounces head-on and goes back, as in the figure. That means the atom pushes the wall with some force, and the wall pushes back with the opposite force. It's the force

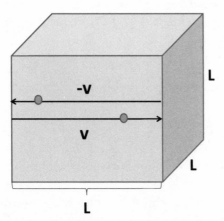

Figure 22.1 A cubic box with N gas atoms inside. The figure shows an atom of velocity **v** hitting the shaded wall and reversing its velocity. The corresponding momentum change is due to the wall, which exerts a force on the atom. The opposite force the atom exerts on the wall is its contribution to the pressure.

that the atom exerts on the wall that I'm interested in. I want to find the force on this particular face and divide by its area to find the pressure. You can find the pressure on any face; it is going to be the same in equilibrium.

I have N atoms, randomly moving inside the box, each suffering collisions with the walls, bouncing off like a billiard ball would at the walls of the pool table, and going to another wall and doing it again. Now, that's a very complicated problem, so we're going to simplify it. We are going to assume that one-third of the atoms are moving from left to right or right to left (perpendicular to the shaded face), one-third are moving up and down, and one-third are moving in and out of the page. Of course, you will have to assign equal numbers to these directions, for nothing in the gas favors any one. So $\frac{1}{3}N$ atoms are going back and forth between this shaded wall and the one opposite to it. The figure shows you a side view.

The force on an atom is the rate of change of momentum:

$$F = ma = m\frac{dv}{dt} = \frac{d(mv)}{dt} = \frac{dp}{dt}. \tag{22.13}$$

Next we assume all the atoms have the same speed, which I'm going to call v. Take one particular atom that hits the shaded wall and bounces back. Its momentum changes from mv to $-mv$; therefore, the change in momentum is $2mv$. How often does that collision take place? Once it hits that wall, it has to go to the other wall and come back. It has to go a distance $2L$, at a speed v. The time it takes is $\frac{2L}{v}$ and the frequency of collisions with the shaded wall is $\frac{v}{2L}$. So

$$\frac{dp}{dt} = \text{momentum change per collision} \times \text{collisions per second}$$

$$= 2mv \cdot \frac{v}{2L}. \tag{22.14}$$

That is the force due to one atom. It's not a continuous force. The atom hits the wall, there's a little force exchange between the two, then there's nothing, and then you wait until it comes back and hits the wall again. If that were the only thing going on, most of the time there would be no pressure, and suddenly there would be a lot of pressure, and then nothing. But, fortunately, this is not the only atom. There are roughly 10^{23} guys pounding against the wall. In a short time, even 10^{-5} seconds, there will be a large number of atoms hitting the wall. The force will appear to be steady rather than staccato. The average force due to all the atoms hitting this wall is

$$\bar{F} = \frac{mv^2}{L} \cdot \frac{1}{3}N.$$ (22.15)

The other $\frac{2}{3}N$ atoms are moving parallel to this wall and apply no force on it.

We're almost done. What about the average pressure? It is the average force divided by the area of that face

$$P = \frac{1}{3}N\frac{mv^2}{L \cdot L^2}.$$ (22.16)

This is very nice because L^3 is just the volume of my box. I send it to the other side and find

$$PV = \frac{1}{3}Nmv^2.$$ (22.17)

This is what the microscopic theory tells you: if your atoms all have a single speed, if they're moving randomly in space so that a third of them are moving back and forth against that shaded wall and the opposite wall, then $PV = \frac{1}{3}Nmv^2$. Experimentally, you find $PV = NkT$. So, you compare the two expressions and out comes one of the most beautiful results:

$$\frac{1}{2}mv^2 = \frac{3}{2}kT.$$ (22.18)

This profound formula not only confirms the *kinetic theory*, that there are atoms, but it also gives for the first time the microscopic meaning of temperature. What you and I call the absolute temperature of a gas is, up to the factor $\frac{3}{2}k$, simply the kinetic energy of the atoms. If you put your hand into a container with some gas and it feels hot, the temperature you're measuring is the kinetic energy of the atoms. We see why absolute zero is absolute. As you cool your gas, the kinetic energy of atoms steadily decreases, but you cannot go below not moving at all, right? That's the lowest possible kinetic energy. That's why it's absolute zero. At that point, everybody stops moving. That's why you have no pressure. Now, these results are modified by the laws of quantum mechanics, but we don't have to worry about that now. In classical physics, it's correct to say that when the temperature goes to absolute zero, all motion ceases.

So, bear in mind that absolute temperature is a measure of atomic kinetic energy for a gas. Every ideal gas, whatever it's made of, has the same kinetic energy per atom at a given temperature. The kinetic energy will be the same, but not the velocity. As the carbon atom is heavier than hydrogen, it will be moving slower at that temperature in order to have the same kinetic energy.

22.4   Statistical properties of matter and radiation

In a gas, the atoms are moving anywhere they want in the box. In a solid, every atom has a place. If you take a two-dimensional solid, the atoms form a nice array; for example, a square grid or lattice. They are not absolutely locked to the points on the grid; instead, they execute simple harmonic motion about the lattice points. They experience a potential that looks like an egg carton. The minima of the potential are at the lattice points. If an atom moves off the minimum in any direction, there is a restoring force to bring it back to the minimum. At $T = 0$ all the atoms will sit at these minima. If there is an atom here, at this minimum, I know that if I go 100 times the lattice spacing in the x or y direction (for a square lattice), there will be another atom sitting there. That's called long-range order. If you heat up that solid, the atoms start vibrating. If you put the solid on top of a hot plate, the atoms in the hot plate will bump into these atoms and start them vibrating. Their average locations still exhibit long-range order. But, in a hot solid, the atoms are making violent oscillations around their assigned positions. If you energize them more and more, there is nothing to prevent them from rolling over to the next minimum. Once that happens, all hell breaks loose, because they don't have any reason to stop there. They start going everywhere. That's what we call melting.

A liquid is more subtle. If you look at a liquid locally, the interatomic spacing is very tightly constrained. Locally, the environment around an atom is known, but if you go a short distance, I cannot give you a precise location where you may find another atom. So, we say a liquid has short-range positional order, but not long-range order. An ideal gas has no order at all. If I tell you there's a gas atom here, I cannot tell you where anybody else is because nobody has any assigned location in relation to others.

Going back to the gas, of course, a third of the atoms are not all moving in one direction. They're moving in random directions. They are also not moving at a fixed speed v either. If I give you a gas at $300K$ and you take this formula Eqn. 22.18 literally and calculate from it a certain v, you will not find every atom at that velocity. Not only are the atoms moving in random directions, they are also moving with essentially all possible velocities. The velocity you are getting from this formula is some kind of average. If you have the ability to see the gas atoms and measure each velocity, what is the probability $P(v)$ that you get a magnitude v? The answer for a typical case is shown in Figure 22.2 and has the form

$$P(v) \propto v^2 e^{-\frac{mv^2}{2kT}}. \tag{22.19}$$

It has a certain peak, a most probable velocity (actually speed). The average kinetic energy will obey $\frac{1}{2}mv^2 = \frac{3}{2}kT$. This is called the *Maxwell-Boltzmann velocity distribution*, though it should be called the speed distribution. This is the detailed description of what's happening in a gas. A given temperature does not pick a unique velocity, but it picks a unique graph for $P(v)$ parametrized by T. You will understand this distribution better upon reading the end of Chapter 24.

Here's a digression. Take a box containing just radiation; in other words, go inside a pizza oven. Take out all the air; the oven is still hot,

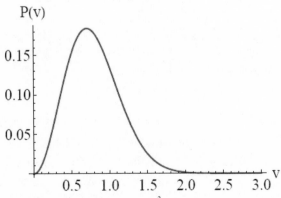

Figure 22.2 The probability $P(v) \propto v^2 e^{-\frac{mv^2}{2kT}}$ that an atom in a gas will have a speed v.

and the walls of the oven are emitting electromagnetic radiation. Electromagnetic radiation comes in different frequencies, and you can ask how much energy is contained in every possible frequency range. You know each frequency is a color if it is in the visible range. So, how much energy is in the red and how much is in the blue? That graph $P(\omega)$ as a function of ω also looks like the Maxwell-Boltzmann distribution and is called the *Planck distribution*. Whereas for atoms, $P(v)$ is determined by temperature, the Boltzmann constant k, and the mass of the molecules, $P(\omega)$ is determined by temperature, the Boltzmann constant, the velocity of light, and Planck's constant $\hbar \simeq 10^{-34} J.s.$ You give me a temperature, and I will draw you another one of these roughly bell-shaped curves. As you heat up the furnace, the shape will change, shifting to the right, toward higher frequencies.

In summary, a temperature for electromagnetic radiation implies a particular distribution of energy at each frequency, while for a gas it means a distribution of atomic velocities.

One prediction of the Big Bang theory is that the universe was formed roughly 13.8 billion years ago. In the earliest stages the temperature of the universe was incredibly high, and then as the universe expanded it cooled. Today, at the current size, it has a certain average temperature, which is a remnant of the Big Bang. And that temperature means that we are sitting in the furnace of the Big Bang. But the furnace has cooled a lot over the billions of years. The temperature of the universe is around $3K$. You determine that by pointing your telescope at the sky. Of course, you're going to get light from this star and that star. Ignore all the pointy things and look at the smooth background, and it should be the same in all directions. Plot that radiation as a function of ω, and you'll get a perfect fit to the Planck distribution with $T \simeq 3K$. If you go to intergalactic space, that is your temperature. We're all living in that heat bath at 3 degrees, and it is getting colder as the universe expands.

22.5 Thermodynamic processes

Now we are going to study thermodynamics in detail. There is only one system we will study, which is an ideal gas sitting inside a cylinder as shown in Figure 22.3. It has a pressure P_1 and a volume V_1. I'm going to put a dot in the $P - V$ plane at (P_1, V_1), and that's my gas. The state of my gas is summarized by where I put the dot. Every dot here is a possible state of equilibrium for the gas. Remember, if you look under the hood,

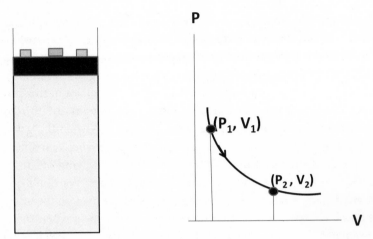

Figure 22.3   Left: A piston and cylinder combination. The weights balance the pressure P of the gas inside. (The atmospheric pressure is assumed to be zero or added to that of the weights.) The volume V is just the volume below the piston. Right: A point (P_1, V_1) representing a state of the gas and a possible quasi-static path in the $P - V$ plane to state (P_2, V_2). To move quasi-statically, the weights must be replaced by fine grains of sand, which may be added or removed.

the gas is made up of roughly 10^{23} atoms. The microscopic state of the gas is obtained by giving 10^{23} locations and 10^{23} velocities. According to Newton, that's the maximum information you can give me about the gas right now. With that and Newton's laws, I can predict the future state. But when you study thermodynamics, you don't really want to look into the details. You want to look at gross macroscopic properties, and there are two that you need: pressure and volume. Now, you might say, "What about temperature?" Why don't I have a third axis for temperature? Why is that not a property? Because $PV = NkT$. I don't have to give you T, if I give you P and V. By the way, $PV = NkT$ only applies to ideal gases, whose atoms and molecules are so far apart that they don't feel any forces between each other unless they collide. In general $PV = NkT$ will be replaced by an *equation of state*, which is in general a complicated relation between P, V, and T. But we are going to study only dilute ideal gases.

Back to the gas in the cylinder. It had three weights on top of the piston. I suddenly pull out one weight. What do you think will happen? The piston will now shoot up and bob up and down a few times. Then, after a fraction of a second, it will settle down at a new location (P_2, V_2). By "settle down," I mean after a while I will not see any macroscopic motion. Then the gas has a new pressure and a new volume. It's gone from being at 1 to being at 2. What happened in between the starting and finishing points? You might say, "Look, if it was here in the beginning, and if it was there later, it must've followed some path." Not really. Not in this process, because if you do it very abruptly, suddenly throwing out one-third of the weights, there's a period when the piston rushes up, when the gas is not in equilibrium. There is no single pressure you can associate with the gas. The bottom of the gas doesn't even know the top is flying off. It's at the old pressure. At the top the gas is at a lower pressure. We don't call that equilibrium. So, the dot, representing this system, moves off the graph. It's off the radar, and only when it has finally settled down, when the entire gas has made up its mind on what its pressure is, can you represent it once again as a point (P_2, V_2).

22.6 Quasi-static processes

Now we have a little problem. We have these equilibrium states, but when you try to go from one to another the system flies off the $P - V$ plane. So, you want to find the means by which you can stay on the PV plane as you change the state of the gas. That brings us to the notion of a *quasi-static process*. A quasi-static process is trying to have it both ways, in which you want to change the state of the gas, and you don't want it to leave the PV diagram. You want it to be always infinitesimally close to equilibrium. So, what you really want is not three big fat weights on the piston, but many, many tiny grains of sand that produce the pressure. Now, remove one grain of sand. The piston moves a tiny bit and very quickly settles down. It is again true that during the tiny bit of settling down you didn't know what it was doing, but you certainly nailed it at the second location very close to the first. You remove one grain at a time and make the grains smaller and smaller, and you wait longer and longer between these changes to permit equilibration. Then, in a mathematical sense, you will be able to connect the dots representing equilibrium states to form a continuous line. That is a quasi-static process. Our discussions are not totally academic because many processes at real speeds are approximately quasi-static. For example, the internal combustion engine in your car completes thousands of cycles per minute, and yet at each instant the gas inside is close enough to equilibrium to be represented in the $P - V$ plane.

The above quasi-static process is also *reversible*, which means the following. If, when I took off a grain of sand, the representative point moved from one dot in the $P - V$ plane to a nearby dot, then, when I put the grain back, it'll go back to where it was. So you can go back and forth. But now, that's an idealized process. If you have friction, and you take out a grain and it goes up, then when you put the grain back, it will not come back to quite where it was. You cannot put Humpty Dumpty back. Thus most processes are irreversible, even if you do them slowly, due to effects like friction. We will assume in our discussion that idealized reversible processes exist.

In the old days, when we studied a single particle in the $x - y$ plane, I just said the particle goes from here to there. There was no restriction on how fast it moved. Particles had trajectories no matter how quickly they moved. As for thermodynamic systems, you cannot move them too fast. They are extended and you have a huge number of atoms described by a few macroscopic numbers like pressure. You cannot change one part of the gas without waiting for the rest of it to respond, readjust, and achieve a global value for the new pressure and other thermodynamic variables.

22.7 The first law of thermodynamics

Every dot in the $P - V$ plane denotes an equilibrium state. In every state of the system, I'm going to define a new variable, U, called the *internal energy of the gas*. For the ideal gas, to which we will restrict most of our discussions, it is simply the kinetic energy of the gas molecules. (For non-ideal gases, solids, and liquids, U is the total energy including potential energy.) Thus

$$U = \frac{3N}{2}kT = \frac{3}{2}nRT = \frac{3}{2}PV \qquad (22.20)$$

using $PV = NkT = nRT$. That means that, at any given point in the $P - V$ diagram, you have a certain internal energy. Notice that *the internal energy of an ideal gas depends only on the temperature.* That's something very, very important. If the temperature has not changed, the internal energy has not changed.

Note the change in notation: in mechanics E stood for the total energy and U for the potential energy. Now U *stands for the total energy.* What is worse, in the case of the ideal gas the thermodynamic U is all kinetic! You just have to learn to live with such conflicting notations. When different parts of the subject are invented by different people, this will happen: k is used to denote the spring constant, the momentum of the photon, and now Boltzmann's constant! The best we can hope for is that two different definitions of the same symbol do not appear in the same discussion.

I'm ready to write down what's called the *first law of thermodynamics* that talks about what happens if you make a move in the $P - V$ plane from point 1 to point 2. The internal energy will change by $dU = U_2 - U_1$. We want to ask what causes the internal energy of the gas to change. There are two ways you can do it: you can do work on the gas by moving the piston, or you can put the gas on a hot plate. If you put it on a hot plate, we know it's going to get hotter. If it gets hotter, the temperature goes up. If the temperature goes up, the kinetic and hence internal energy goes up. Of course, you can put it on a cold plate and take out some heat. So in general

$$dU = \Delta Q - \Delta W. \tag{22.21}$$

Here ΔW is positive if work is done *by* the gas, and it is negative if work is done *on* the gas. As for ΔQ, it is positive (negative) if heat is put in (taken out).

Note that the infinitesimal change in U is denoted by dU while small quantities of heat and work are denoted by ΔQ and ΔW. The reason will follow later.

What's the formula for work done *by* the gas? If it expands against the applied pressure

$$\Delta W = F dx = PA \, dx = P dV. \tag{22.22}$$

If the gas is compressed so that $dx < 0$, then the work done *by* the gas is negative. That leads to this great first law of thermodynamics, as applied to the gas:

$$dU = -P dV + \Delta Q \quad \text{(first law)}. \tag{22.23}$$

It expresses the law of conservation of energy. It says the energy of the gas changes either because you pushed the piston or the piston pushed you; or because you put it on a hot plate or cold plate to add or take out heat. We are now equating putting it on a hot plate or cold plate as also equivalent to giving or taking out energy, because we recognize heat as energy.

If you fix the piston so it cannot move, and you put it on a hot plate, the PdV part will vanish because there is no chance of a dV. On a hot plate there are fast-moving molecules. When they collide with the slow-moving gas molecules, typically the slow ones become a little faster and the fast ones a little slower, and therefore there will be a transfer of kinetic energy to the gas. Another thing you can do is thermally isolate your gas so no heat can flow in or out of it, and then you can either have the volume increase or decrease. If the gas expands, dV is positive and the $-PdV$ is negative, and so is dU. That's because the molecules are beating up on the piston and moving the piston. Remember, applying a force doesn't cost you anything. But if the point of application moves, you do work. And who's going to pay for it but the gas? It'll pay for it through its loss of internal energy. Conversely, if you push down on the gas, dV will be negative and $dU = -PdV$ will be positive, and the energy of the gas will go up.

Let us now calculate the work done in a process where a gas goes from a point V to a nearby point $V + dV$ when the pressure is at $P(V)$, as in Figure 22.4. The infinitesimal work done, $dW = P(V)dV$, is the shaded

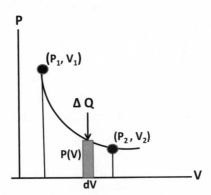

Figure 22.4   During an infinitesimal change by dV, the work done by the gas is the shaded area $dW = PdV$. If this were an isothermal process, some heat ΔQ would flow in from the reservoir to keep T and U constant. For a finite process the work done is the area between the V axis, the graph $P(V)$, and the vertical lines at $V = V_1$ and $V = V_2$.

area, and the work done over a macroscopic path is

$$W_{1\to 2} = \int_{V_1}^{V_2} P(V)dV. \tag{22.24}$$

The work is the area enclosed by the function $P(V)$, the V-axis and the vertical lines $V = V_1$ and $V = V_2$. To get an analytical expression for the work you need to know P as a function of V. We are going to consider a special case of an *isotherm*, a graph of a gas at a given temperature. Because $PV = nRT$, if T is constant, PV is a constant. The graph is a rectangular hyperbola: the product of P and V constant, so when the P increases, V decreases, and vice versa.

We want to take the gas for a slow quasi-static ride from 1 to 2 along the curve $P(V) = \frac{NkT}{V}$. The work done by the gas is

$$W_{1\to 2} = \int_{V_1}^{V_2} P(V)dV = NkT \int_{V_1}^{V_2} \frac{dV}{V} = NkT \ln \frac{V_2}{V_1}. \tag{22.25}$$

If I make it go backward from 2 to 1, the work done by the gas is given by the same area, but with a minus sign. If you go to the right, the area is considered positive. If you go to the left, the area is considered negative. The integral Eqn. 22.25 of course gives the right answer for both cases because the ln will change sign under $V_1 \leftrightarrow V_2$.

If that is $W_{1\to 2}$, the work done by the gas, what is the heat input? Recall the first law

$$dU = \Delta Q - PdV. \tag{22.26}$$

This gas did not change its temperature, and $U = \frac{3}{2}NkT$ implies U didn't change. So

$$\Delta Q = PdV \tag{22.27}$$

every step of the way and the gas has had a heat input of

$$Q = NkT \ln \frac{V_2}{V_1}. \tag{22.28}$$

What do you think is happening to the gas here? Think of the piston and weight combination as the gas expands a little bit isothermally. The expansion should cool the gas if it were isolated because expansion costs energy. However, the gas is held at a fixed T by placing it on some reservoir, like a hot plate. Thus when it tries to cool down, heat ΔQ flows from the reservoir, maintaining the temperature, as indicated by the vertical arrow in Figure 22.4. So, what the gas is doing in this case is taking heat energy from below (assuming that is where the reservoir is) and working against the atmosphere and weights above. It takes in with one hand and gives out to the other, converting heat into work, without changing its own energy.

22.8 Specific heats: c_v and c_p

You have to be careful when you talk about specific heats of gases, and here is why. For liquids and solids, given $dQ = mcdT$, we define $c = \frac{dQ}{dT}$ per unit mass of the substance. For a gas, you've already seen that what you want to count is not the actual mass, but the moles. Every molecule gets a certain amount of energy, namely $\frac{3}{2}kT$, and so you just want to count the number of molecules, or the number of moles. Now, there are many ways in which you can pump heat into a gas, but let's agree that we will always take one mole from now on and not one kilogram. Take a mole of some gas and call the *molar specific heat* as the energy needed to raise the temperature of one mole by one degree. If you take one mole, it has energy $U = \frac{3}{2}N_A kT = \frac{3}{2}RT$. From the first law

$$\Delta Q = dU + PdV \tag{22.29}$$

$$\frac{dQ}{dT} = \frac{dU}{dT} + P\frac{dV}{dT}. \tag{22.30}$$

There's a problem here. Did you or did you not allow the volume to change while adding heat? That's going to determine what the specific heat is. In other words, when a solid is heated, it expands such a tiny amount that we don't worry about the work done by the expanding solid against the atmosphere. But, when you heat a gas, the volume changes so much that the work it does against the external world is non-negligible. Therefore, the specific heat is dependent on what you allow the volume term to do.

Consider c_v, *the specific heat at constant volume* where you don't let the volume change; you clamp the piston. You pump in heat by putting it on a hot plate. All the heat goes directly to internal energy. None of that is lost to expansion. Setting $dV = 0$ in Eqn. 22.30 and $n = 1$ in $U = \frac{3}{2}nRT$,

$$c_v = \left.\frac{dQ}{dT}\right|_V = \frac{dU}{dT} = \frac{3}{2}R. \tag{22.31}$$

Then, there's c_p, the *specific heat at constant pressure* defined as follows. You have some gas at some pressure. You pump in some heat but you don't clamp the piston. You let the gas expand freely at the same pressure. For example, if it's being pushed down by the atmosphere, you let the piston move up if it wants to, maintaining the same pressure. If it moves up a little bit, some of the heat that you put in goes to changing its internal energy and some into doing the work PdV. So now

$$c_p = \left.\frac{dQ}{dT}\right|_P = \frac{dU}{dT} + P\frac{dV}{dT} = \frac{3}{2}R + \frac{d(PV)}{dT} = \frac{3}{2}R + \frac{d(RT)}{dT}$$

$$= \frac{3}{2}R + R = \frac{5}{2}R = c_v + R \tag{22.32}$$

where we have taken P, which is being held fixed, into the T derivative and used $PV = RT$ for one mole.

You should have expected $c_p > c_v$ because some of the heat goes into expanding the gas and only the rest into raising its T. This is confirmed and quantified by the result

$$c_p = c_v + R. \tag{22.33}$$

Finally, consider the ratio that will come in later:

$$\gamma = \frac{c_p}{c_v} = \frac{5}{3}. \tag{22.34}$$

Notice that neither c_p nor c_v depends on what particular monoatomic gas is involved. All will have the same specific heat per mole or *molar specific heat*. They won't have the same specific heat per gram, because one gram of two different gases will have a different number of moles or atoms.

One final caveat: all this is for a mono-atomic gas, whose constituents are essentially points, with just the kinetic energy of translation. Compare this to, say, a diatomic gas, whose molecules are two atoms joined together like a dumbbell. The kinetic energy of the dumbbell has two parts, as you learned long ago. It can rotate around some axis and its CM can also move in space. So, the internal energy has also got two parts: due to the motion of the center of mass and due to rotation. Some molecules also vibrate. The value of γ is not $\frac{5}{3}$ for these.

CHAPTER 23

Thermodynamics II

23.1 Cycles and state variables

Let us begin with the first law

$$dU = \Delta Q - \Delta W = \Delta Q - PdV, \qquad (23.1)$$

which says that dU, the change in the energy of the gas, equals the heat input ΔQ minus the work done *by* the system, $\Delta W = PdV$. Why do we refer to some infinitesimals with a Δ and some with a d?

This has to do with whether these refer to simply small quantities or to small quantities that correspond to a change in a *state variable*, which is some function that depends on the state of the gas, specified by P and V. Consider the internal energy U. It is a state variable. That means that at each point in the $P - V$ plane it has a definite value. If we take the gas for a spin and come back to where we started, U will return to its original value. It will not matter what path we took: if you are back where you started, U is back to its value. Thus in Figure 23.1, if we start at some (P, V) and go around the shaded area in the clockwise sense and return, so does U to $U(P, V)$. For example, for an ideal gas, $U = \frac{3}{2}PV$, and dU is the change in this function due to a slight change in state. At any point (P, V) we may speak of the internal energy resident in the gas. If you peek in and add the kinetic energies of all the molecules, you will get this number.

This is not so for the work W. We cannot speak of the work in the gas—it does not correspond to anything resident in the system. Consider

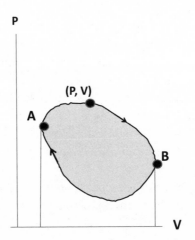

Figure 23.1   The work done by a gas as it goes over a closed cycle equals the area enclosed by the cycle in the $P - V$ plane. During AB the gas expands and does work, and during BA it is compressed and has work done on it. The shaded area is the algebraic sum of the two.

what happens when we go around the closed loop shown in Figure 23.1. What is the work done? We have seen in the last chapter that, for an open segment, it is the area under the graph if we move in the direction of increasing V and minus that if we move to the left. For the closed region shown, it must be clear that as we move along the upper part, from A to B in the direction shown, the work done is the area under that part of the curve. Then as we move back along the lower part, from B to A, we must subtract the area under that segment. The algebraic sum of these is the area of the shaded region. This is written as follows

$$W_{\text{cycle}} = \oint P(V)dV. \tag{23.2}$$

If there were a definite value of W attached to each state, there would be a problem: after adding W_{cycle} to the system, you have come back to the same state! So there isn't a definite value of W attached to each state; W is not a function $W(P, V)$, unlike U, which is. On the other hand, we can always say some tiny amount of work ΔW was done by the system; it is just a number.

Just like W, there is no function $Q(P, V)$ associated with each state. You cannot peek into the gas and say, "I see this amount of heat there." Let me prove it. Say there is some Q that you claim is resident in the system when it is at (P, V). I then send you away and take the gas through the cycle shown. Because $U(P, V)$ is back to its old value, and work W_{cycle} was done by the system, the Q in the system must have increased by W_{cycle}. Now you come back, and I ask you how much heat there is in the system. You will give the old answer because the system is in exactly the same state, but I know Q has increased by the area of the closed loop. So Q is not a state variable, though we can always say some number of calories ΔQ were added to the system.

In summary, only if $F(P, V)$ is a state variable will we use dF to represent the change in that function as we move to a neighboring point.

23.2 Adiabatic processes

We have already considered the work done under an isothermal process: $W^T_{1\to 2} = NkT \ln \frac{V_2}{V_1}$. In the isothermal case, you keep the gas on a hot plate at a given temperature T, and as the volume changes, heat comes in or goes out to maintain T. A process at constant pressure P is called *isobaric*. The work done is trivial: $W^P_{1\to 2} = P(V_2 - V_1)$, which is just the area of a rectangle of height P and width $V_2 - V_1$.

Now I'm going to consider the *adiabatic process* in which the gas is thermally isolated, so $\Delta Q \equiv 0$. You wrap this guy in a blanket and you do things to it. Consider point A in Figure 23.2, which under isothermal expansion ends up at B, on the same isotherm $T = T_1$. Likewise, from A' the system can expand isothermally at $T = T_2$ to C. Which is higher, T_1 or T_2? Take a point on A' on T_2 directly below A. It has the same volume as A but lower pressure, so $PV = NkT$ is less. So $T_2 < T_1$.

But suppose I start at A and let the gas expand adiabatically. The gas expands against the external pressure but no heat is allowed to come in. It'll pay for the work through its own internal energy, which will go down, and that means T will go down. The gas will be cascading down from one isotherm to another, until you stop somewhere at the lower temperature, such as the point D at T_2. Another way to say this is that the drop in pressure for a given increase in volume would be more precipitous for the adiabatic process. The evolution will not be $P \propto 1/V$, but something steeper. What is the equation for an adiabatic process? What is P as the function of V?

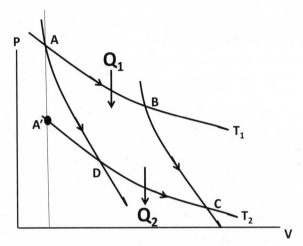

Figure 23.2   The figure shows two isothermal curves at T_1 and T_2 and two adiabatic curves crossing them. In the latter case P falls more rapidly with increasing V because it gets no help from outside.

This will be determined by the first law of thermodynamics. First we combine the definition of adiabatic, $\Delta Q = 0$, with the first law to obtain

$$\Delta Q = dU + PdV = 0. \tag{23.3}$$

This equation relates the change in U to the change in V in an adiabatic process. Given $U = \frac{3}{2}nRT$, it follows that $dU = \frac{3}{2}nRdT = nc_v dT$. Using $PV = nRT$ we find

$$nc_v dT + \frac{nRT}{V}dV = 0. \tag{23.4}$$

Canceling the n and rearranging

$$\frac{c_v}{R}\frac{dT}{T} + \frac{dV}{V} = 0. \tag{23.5}$$

Note that because n drops out, we could have analyzed just one mole of the gas from the outset.

Not only does this equation tell us what we expect intuitively—that if the volume increases ($dV > 0$), the temperature decreases ($dT < 0$)—it quantifies the dT, drop in temperature under adiabatic expansion (or its opposite under adiabatic compression). However, what we wanted to do was relate dV to dP, to see how the adiabatic evolution takes place in the $P - V$ plane. Let us postpone that and first get a result connecting V and T, because it will be very useful later. We can then use $PV = RT$ to eliminate T from that result.

Integrating Eqn. 23.5 between states 1 and 2 yields

$$\frac{c_v}{R} \ln \frac{T_2}{T_1} + \ln \frac{V_2}{V_1} = 0. \tag{23.6}$$

We may rewrite this as follows:

$$\ln \left(\left[\frac{T_2}{T_1} \right]^{c_v/R} \cdot \frac{V_2}{V_1} \right) = 0 \tag{23.7}$$

$$\left[\frac{T_2}{T_1} \right]^{c_v/R} \cdot \frac{V_2}{V_1} = 1 \quad \text{using } \ln 1 = 0 \tag{23.8}$$

$$T_2^{c_v/R} V_2 = T_1^{c_v/R} V_1, \tag{23.9}$$

which may be rewritten as

$$T^{c_v/R} V = C, \tag{23.10}$$

where C is a constant along the adiabatic path connecting state 1 to state 2. For an ideal mono-atomic gas, c_v/R would be just $\frac{3}{2}$, but let us keep it in this form for now.

To find pressure as the function of volume on an adiabatic curve, we eliminate T in

$$T^{c_v/R} V = C \tag{23.11}$$

$$\left[\frac{PV}{R} \right]^{c_v/R} V = C \tag{23.12}$$

$$P^{c_v/R} V^{(1+c_v/R)} = C', \text{ another constant} \tag{23.13}$$

$$PV^{\frac{c_v+R}{c_v}} = C'' \text{ on raising both sides above} \tag{23.14}$$

$$\text{to the } R/c_v \text{ power.}$$

Finally, we have the familiar version

$$P_1 V_1^\gamma = P_2 V_2^\gamma = PV^\gamma = C \quad \text{where} \tag{23.15}$$

$$\gamma = \frac{c_v + R}{c_v} = \frac{c_p}{c_v}. \tag{23.16}$$

Here P and V refer to a generic point on the adiabatic curve, and we have renamed the final constant C'' in Eqn. 23.14 as C. It's not a constant you can look up in a book like the velocity of light. It depends on this particular sample of gas. (When we say the energy is constant for a particle, it is not a universal constant like Boltzmann's constant k; instead, it is constant for this particle on this particular trajectory, say, as it gains height and loses speed. Here, as the gas loses P and gains V in an adiabatic process, PV^γ remains fixed at a value C that depends on the particular sample.)

Now we calculate the work done in an adiabatic process, say from B to C in Figure 23.2.

$$W_{B \to C}^{Q=0} = \int_{V_B}^{V_C} P(V) dV \tag{23.17}$$

$$= \int_{V_B}^{V_C} CV^{-\gamma} dV \tag{23.18}$$

$$= \left[C \frac{V_C^{1-\gamma}}{1-\gamma} - C \frac{V_B^{1-\gamma}}{1-\gamma} \right] \tag{23.19}$$

$$= \left[P_C V_C^\gamma \frac{V_C^{1-\gamma}}{1-\gamma} - P_B V_B^\gamma \frac{V_B^{1-\gamma}}{1-\gamma} \right] \tag{23.20}$$

$$= \frac{P_B V_B - P_C V_C}{\gamma - 1} \tag{23.21}$$

where we have used $C = P_C V_C^\gamma = P_B V_B^\gamma$ in the penultimate step.

23.3 The second law of thermodynamics

We come to the last part of thermodynamics, which to me is the most beautiful. It starts out with the following consideration. There are certain things in this world that seem perfectly allowed but don't seem to happen. Take the Joule experiment. You take water in a cylinder with a shaft with fins that can spin. You hang some weight over a pulley, and, as the weight goes down, the shaft spins; the water heats up. You film a movie of that, and the movie is so detailed you can even see individual molecules. Now you play the movie backward. What do you find? Suddenly, the weight starts moving up, the shaft spins the other way, and the water cools down. This does not violate any of the laws you have learned, including the first law of thermodynamics. When the weight went down and the water heated up, some amount of work was done on the water, and the energy of the water went up. In the reverse process, the energy of the water went down and the weight went up, and an equal amount of work was done by the water. But that reverse process doesn't seem to happen, even though microscopically there is nothing funny about it. If you looked at the collisions between the water molecules and the fins, you would not know at the microscopic level whether the movie was going forward or backward; every collision would obey Newton's laws, and energy and momentum would be conserved. Why is the reverse process, allowed by all the laws of mechanics, somehow forbidden from happening in real life?

Here is another example. Consider an eraser on a table. I give it a push and it moves a bit and stops. The table and eraser heat up. Take a movie of that. Play it backward. You will find the table and eraser cool down, and the eraser moves backward, picking up speed. You might laugh, but you have no reason to do so right now, because the movie doesn't contradict any physics you have learned—just your expectations based on your daily experience. In the reverse process, every atom is made to stop, turn around, and move in the reverse direction at the reversed velocity. And every collision between atoms on the desk and atoms on the eraser will obey all the laws of mechanics in the forward and backward movies.

In the cases we have considered, the common feature seems to be that kinetic energy can turn into heat easily, but not so the reverse.

Now take a different kind of example. I put some gas molecules in one-half of a box, with a partition holding them in that side. Then I remove this partition and I wait a little bit. The gas will fill up the box. That's like perfume leaking out of the bottle. Now I take a movie of these molecules and play the movie backward. In the backward movie the gas which had spread over the box, will spontaneously gather in the left half. The movie does not violate any of the laws of mechanics. But what it portrays does not happen.

I take a chunk of some hot copper and a chunk of some cold copper. I place one on top of the other, thermally isolate them from the outside world, come back in an hour, and they have both become lukewarm. That's fine. But now maybe, if I wait long enough, the lukewarm chunks will spontaneously separate into hot and cold. That doesn't seem to happen. The heat seems to flow spontaneously from hot to cold but never back from cold to hot. But flowing from cold to hot would not violate the law of conservation of energy. As long as the same number of calories go from cold to hot, it doesn't violate anything. And yet that doesn't seem to happen.

This list of things allowed by all known laws and yet forbidden in nature can go on and on. Because we cannot explain this interdiction with any known laws of physics, we elevate it to a new law. The new law could say all of these things are forbidden. That's not a good enough law. The list of forbidden processes is a mile long.

Amazingly, there is one law, a single law, that not only is qualitative, but is quantitative, and it tells you exactly which things can happen and which cannot. That is the second law of thermodynamics. What form does it take? We will introduce a certain quantity called *entropy*. The second law will say that the *entropy of the universe will never decrease*. If the paddle spun the other way, the weight went up, and the water cooled down, you can show the entropy of the universe would have actually gone down. That's why that's not allowed. If you drop an egg and it splatters all over the floor, it cannot rejoin and rise back to your hand, for this too would cause a decrease of entropy of the universe. So will the unmixing of lukewarm into hot and cold and all the other forbidden things.

This great law was discovered following the investigations of an engineer called Sadi Carnot. One does not generally wake up and say, "I'm going to discover a great law." You just go about your business, but you need to recognize it when you've stumbled on something big, as Archimedes did in his bathtub. Carnot had a very practical question about engines, like the steam engine. You take some coal, you burn it to boil some water, it turns into steam, which in turn pushes the pistons that turn the wheels that make the train go forward. What happens in a steam engine is described schematically in the left half of Figure 23.3. Heat Q_1 from a reservoir at T_1 goes into the engine. Out comes some mechanical work W and some exhaust emission Q_2 at a lower temperature T_2. In the case of the steam engine T_1 is the temperature of the furnace, Q_1 the heat generated by burning the coal; T_2 is generally room temperature. If you ever have the opportunity to see a steam engine, you will notice that there's a lot of hot steam coming out of the side. That is Q_2. *Note that Q_2 is defined to be positive coming out of the engine.*

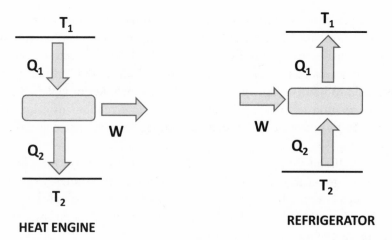

HEAT ENGINE   **REFRIGERATOR**

Figure 23.3   A heat engine (left) takes in heat Q_1 from a reservoir at T_1, does some work W, and rejects heat Q_2 to a colder reservoir at T_2. Things go the other way in a refrigerator (right): work W is pumped in to transfer heat Q_2 from T_2 to a higher temperature T_1.

The engine operates in a cycle, that is, it can do all of this (absorb Q_1, do work W, and reject Q_2) over and over again.

You can fill in the corresponding items for a gasoline engine.

On the right half of the figure is a device called the refrigerator: it takes in Q_2 from a cold place (freezer), has some work input W (from the electrical socket that powers the compressor), and emits $Q_1 = Q_2 + W$ into the environment (your kitchen). The refrigerator will play an important part later, but let us return to the engine for now.

By the first law, the work done by the engine (which manifests itself as the kinetic energy of the locomotive or automobile) is

$$W = Q_1 - Q_2. \tag{23.22}$$

We define a quantity called *efficiency* η. It's the work you want (W), divided by the heat you pay for (Q_1):

$$\eta = \frac{W}{Q_1} = \frac{Q_1 - Q_2}{Q_1} = 1 - \frac{Q_2}{Q_1}. \tag{23.23}$$

Every engine takes in heat. Some of it is converted to work and some of it is rejected. To the extent heat is rejected, the efficiency is less than 1.

Why not build an engine that doesn't reject any heat at the lower temperature T_2? Why not take all the heat and convert it to work? Carnot gave a great argument for the most efficient engine you can build operating between these two temperatures T_1 (at which heat is produced) and T_2 (at which heat is rejected). *He showed that there is an upper limit to η and the limit is less than* 1.

What is this limiting η and how did he find it? He needed a postulate. Carnot's postulate is the old version of the second law of thermodynamics, and it is fully equivalent to the modern version that makes reference to entropy. Here is Carnot's law: *You cannot build an engine whose* sole *effect is to transfer some heat from a cold body to a hot body.*

But we already have noticed that heat cannot flow from cold to hot spontaneously; that is why two lukewarm blocks of copper cannot separate into hot and cold blocks. All Carnot seems to say is that this cannot happen, because he postulates it cannot. In fact, from this one postulate, one may develop the notion of entropy, which in turn will be used to outlaw *all* the forbidden processes with a *single* restriction on entropy.

It doesn't take much effort to have heat flow from hot bodies to cold bodies: just connect them with a metal rod and wait. The sole effect of that is the transfer of heat from hot to cold. That's the natural order of things. Carnot is saying that you can never build a contraption whose *sole* effect is the transfer of heat from a cold body to a hot one. That is going to be taken as a reasonable postulate, and we want to see what we can get out of that postulate. It turns out that this postulate is going to put a bound on the efficiency of heat engines, by demanding that you must necessarily reject some heat Q_2. From this postulate evolved the second law, assuming more and more sophisticated forms, and culminating in the language of entropy.

23.4 The Carnot engine

The engine that Carnot conceived is depicted in Figure 23.4. Take an isotherm at T_1 with points A and B on it. Draw an isotherm at $T_2 < T_1$. You remember the adiabatic curves are much steeper. So draw two adiabatic curves leaving A and B at T_1 to join D and C on the colder isothermal at T_2. Then, take an ideal gas from A to B to C to D and back to A. That is the *Carnot cycle*. In the process AB, take the cylinder with the gas in it, sitting on top of a reservoir at T_1, and slowly lift the grains of sand on the piston so it expands to a volume B at the same temperature. Then, having

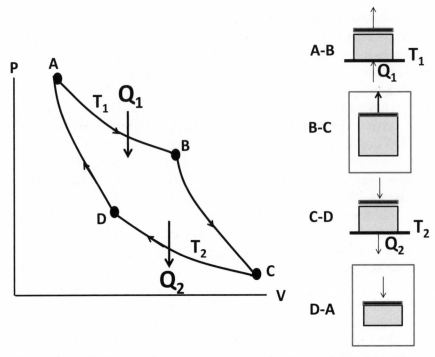

Figure 23.4   The Carnot cycle, running as an engine. It can also be reversed to run as a refrigerator. The process $A \rightarrow B$ is isothermal expansion at T_1 during which heat Q_1 is drawn in from the hot reservoir. The part $B \rightarrow C$ is adiabatic expansion, and the system cools down to T_2. (The box surrounding the system signifies thermal isolation.) The part $C \rightarrow D$ is isothermal compression at T_2 when Q_2 is rejected to the cold reservoir. Finally the adiabatic compression $D \rightarrow A$ closes the cycle.

reached B, you thermally isolate the gas (shown by a box surrounding it on the figures at the right). Take out even more grains of sand. Now, it's expanding without any energy coming in. It cools down to point C at T_2 along the adiabatic curve BC. Now you put it on a cold reservoir at T_2 and slowly start putting the sand grains back until you get to C. Finally, you isolate the gas and put more grains of sand back until you come back to A. Obviously, such a path exists because of the way isothermal and adiabatic lines crisscross the PV plane.

The important feature of the Carnot cycle is that it's *reversible*. It's reversible because at every stage, you are almost in equilibrium. If you take out a grain of sand and the piston moves up, you can put it back and move the piston back. If it absorbs some heat going one way, it will reject it going the other way. So, it can also run backward as a refrigerator (shown in Figure 23.4) following $A \to D \to C \to B \to A$. Some heat Q_2 is taken from the lower reservoir and the heat Q_1 is delivered to the upper reservoir because some work W is done on the refrigerator. You might say, "You violated the second law. You have transferred heat from a cold to a hot body!" But it was not the only thing that happened, because some compressor somewhere did some work. Some water in a dam flowed downhill to generate that electrical power. You paid your electric bill. Things are not how they used to be. What is not allowed is to have heat flow from cold to hot with nothing else changed.

Notice that after doing its cycle, the gas comes back to the starting point A. That means you can do it over and over again, and Q_1, Q_2, and W refer to one cycle. That's why it's a useful engine. (You can alway make a disposable, one-shot engine that converts heat fully to work: take a hot gas in a cylinder, and release the piston so it can rise up and lift some weights and cool down in the process. You have converted some heat fully to work, but things are not back where they were. We are only interested in engines that operate in cycles.) Carnot's question was, "What's the efficiency of this engine?" You might wonder why someone would be interested in the η of such a primitive engine. Before I explain that, let us first calculate η.

To find $\eta = 1 - \frac{Q_2}{Q_1}$ we need Q_1 and Q_2. Because AB is isothermal, the heat input Q_1 is the work done $W_{A \to B}^{T_1}$:

$$Q_1 = W_{A \to B}^{T_1} = nT_1 \ln \frac{V_B}{V_A}. \tag{23.24}$$

Likewise

$$Q_2 = W_{D \to C}^{T_2} = nT_2 \ln \frac{V_C}{V_D} \tag{23.25}$$

where in Q_2 the ratio of initial to final volumes $\frac{V_C}{V_D}$ appears rather than $\frac{V_D}{V_C}$ because Q_2 has been defined as the heat *rejected*. So

$$\eta = 1 - \frac{Q_2}{Q_1} = 1 - \frac{nT_2 \ln \frac{V_C}{V_D}}{nT_1 \ln \frac{V_B}{V_A}}. \tag{23.26}$$

First notice n drops out as it should, because the efficiency should not depend on how much gas we take. If we double the gas, we simply double Q_1, Q_2, and W. It turns out we can even cancel the logarithms using the following result I will prove in a moment:

$$\frac{V_B}{V_A} = \frac{V_C}{V_D}. \tag{23.27}$$

The final result is simplicity itself:

$$\eta = 1 - \frac{T_2}{T_1}. \tag{23.28}$$

In what follows, I will use η to always denote this maximum efficiency.

Now for Eqn. 23.27. We have seen in Eqn. 23.11 that in an adiabatic process $VT^{c_v/R}$ is a constant, that is, for two points x and y on the adiabatic curve,

$$V_x T_x^{c_v/R} = V_y T_y^{c_v/R}. \tag{23.29}$$

Apply this to the pair $x = B$, $y = C$ and $x = A$, $y = D$. Because A and B are at T_1 and C and D are at T_2, we find

$$V_B T_1^{c_v/R} = V_C T_2^{c_v/R} \tag{23.30}$$

$$V_A T_1^{c_v/R} = V_D T_2^{c_v/R}. \tag{23.31}$$

Dividing the left side by the left and the right side by the right we find Eqn. 23.27.

I've taken an ideal gas over a cycle and found out that it does function like a very primitive heat engine because it takes some heat Q_1, rejects some heat Q_2, and does some work $W = Q_1 - Q_2$ given by the area enclosed by the cycle. The ratio of work done to heat absorbed happens to depend only on the upper and the lower temperatures. It does not depend on the gas.

I will show that this efficiency is a theoretical maximum. *No engine can beat the Carnot engine in efficiency, not even one built in 2013.* This analysis is reminiscent of relativity. To show you why time slows down in a moving frame, I took a very simple clock, where a light beam went up and down between two mirrors. That's not your idea of a clock, but you can see why it slows down. But then you know that every clock has to slow down the same way, because all clocks in a moving rocket must run at the same rate. Otherwise, you can compare the two clocks and find out you are moving. Likewise, if you take a very primitive engine, but you can show that it is the most efficient engine, you are done finding the upper limit on the efficiency of all engines.

Here is Carnot's argument, based on the postulate that it's impossible to find a process the *sole* result of which is to transfer some heat from a cold body to a hot body. You have to grant Carnot that postulate, which we accept to be phenomenologically valid. Given that postulate, Carnot will now show you that no engine can beat his engine.

The key to the Carnot engine is that it is a reversible engine. That means the Carnot engine, starting at the point A, can go backward to D then C then to B and then back to A. If the Carnot engine were run backward, it would look like a refrigerator: it would take in heat Q_2 at T_2, somebody would do work W on it, and it would reject Q_1 at T_1. That's what we're going to use to show that you cannot beat the Carnot engine.

Here is a concrete demonstration that's good enough for our present purpose. Let us say our illustrative Carnot engine takes 100 calories, delivers 20 calories in work, and rejects 80 calories operating between some two temperatures T_1 and T_2. I am taking a particular example where the efficiency is $\eta = \frac{W}{Q_1} = \frac{20}{100} = .2$. Now, if you say you have a better engine, what you really mean is that your engine, also operating between the same two temperatures, can take 100 calories and deliver more than 20 calories of work, let us say 40 calories, and reject only 60 calories. This is shown at the left in the box in Figure 23.5. To disprove your claim, I am going to get a Carnot engine that is twice as big as the original one and then I am going to run it backward. Mind you, it is not more efficient; it just has twice as much gas. What will that Carnot engine, called $2 \times$ Carnot* in the right half of the box, do? It will take 160 calories from the colder reservoir, it will want 40 calories of work input, and it will dump 200 calories at the upper reservoir, as shown in the right half of the box. I want the reverse Carnot (refrigerator) to be twice as big for the following simple reason: Your engine is delivering 40 calories of work per cycle; my refrigerator needs 40 to run. *We can directly take the work output from your engine and feed it to my refrigerator.* Your heat engine produces work, my refrigerator needs work, and I've scaled its size so that its appetite matches your engine's output.

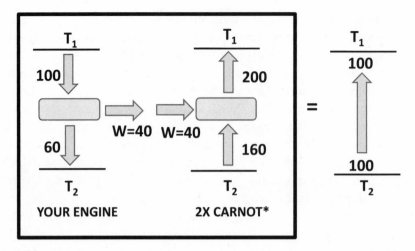

Figure 23.5   If you have an engine that is more efficient than Carnot's it can be combined with his (reversed and rescaled) to form a machine whose sole effect is to transfer heat from a cold to a hot reservoir. In this case the Carnot took in 100 calories, did 20 calories of work, and rejected 80. The supposed rival, which took in 100, did 40 of work, and rejected 60, is shown on the left half of the box. The machine called 2 × Carnot* in the right half of the box is the original Carnot reversed and doubled in size. The gadgets taken together (and enclosed within the box) equal a third one, shown in the far right, which transfers 100 calories from cold to hot with no other effects.

Now, let's draw a box around these two guys and not look under the hood to see what we've got. At the end of a full cycle, when everything is done, we find that all the gases, all the pistons, have come back to where they were at the outset. No need to plug this gadget into the wall because the refrigerator at the right is getting the power from this heat engine at the left. I look at the lower reservoir; I see 100 calories leaving: 60 coming down and 160 going up. I look at the upper reservoir; I see 200 calories in and 100 out. So basically, the combined gadgets are equal to a single gadget that transfers 100 calories per cycle from the cold reservoir at T_2 to a hotter one at T_1 *with no other changes anywhere in the universe*, and that is not allowed. Therefore, your claim that you have an engine more efficient than the Carnot engine has to be false.

The numbers I picked above are simply representative, but you can take any set of numbers as long as your engine does better than mine:

instead of 40 calories of work output, it could have been 30 calories. If it produced 30, I'll get an engine that is $\frac{3}{2}$ times as big as the standard Carnot engine and run it backward. Your 30 calories will feed my refrigerator, and you will find 50 calories of heat flowing from a cold body to a hot body per cycle, with no other changes anywhere in the universe. This is how Carnot's engine, even though very primitive, is the standard for all engines. The key to the result is that the Carnot engine is reversible.

The Carnot engine was based on the ideal gas. Given its reversibility, we were able to show that no engine could be more efficient. Consider now any other reversible engine, running on any substance. We already know it cannot be more efficient than Carnot's. It also cannot be less efficient. To show this, repeat our previous argument, but with this reversible one running backward, coupled to the Carnot running forward, to produce a combination that transfers heat from cold to hot with no other effects. For example, if the less efficient but reversible engine took 100 calories and converted only 10 to work, you can run it backward and power it using a Carnot engine half as big as our illustrative one, which takes in 50 calories and delivers 10 calories of work. You should verify that the combination will transfer 50 calories from cold to hot in every cycle, with no other effects, which is not allowed. The only acceptable result is that all reversible engines are equally efficient, Carnot's being the one that we have studied in depth. The efficiency of actual engines will always be less than this limit because of losses.

23.4.1 Defining T using Carnot engines

We have shown that

$$\eta = 1 - \frac{T_2}{T_1} \tag{23.32}$$

where the temperature T is defined by the ideal gas equation $PV = NkT$. We can introduce an absolute temperature T, with no reference to any specific substance, using reversible engines as follows.

We *define* the *ratio* of absolute temperatures T_1 and T_2 of any two objects by the relation

$$\eta = 1 - \frac{T_2}{T_1} \tag{23.33}$$

where η is the efficiency of a reversible heat engine operating between T_1 and T_2, known to be independent of any details of the engine, the substance it runs on, and so forth. Next we fix the absolute values by *defining* T at the triple point of water to be 273.16. We are done.

Suppose you get me a bucket of some fluid. To find its absolute temperature T_B, I will run any reversible engine between the bucket of fluid (the hot reservoir) and a bucket containing water at its triple point (the cold reservoir), measure its η, and obtain T_B by solving

$$\eta = 1 - \frac{273.16}{T_B}. \tag{23.34}$$

Here I assume $T_B > 273.16$. If not, I will switch reservoirs. The engine can be made arbitrarily small so as not to affect the temperature of either bucket.

Let us leave the practical domain and go back to more theoretical issues that were raised earlier, namely, how Carnot's result can be used to define entropy and how the single law of increasing entropy can succeed in forbidding all the processes we said were forbidden, like the unmixing of hot and cold.

CHAPTER 24

Entropy and Irreversibility

24.1 Entropy

As promised at the end of the last chapter, we will now go from Carnot's practical considerations on the efficiency of heat engines to the notion of entropy.

In computing the efficiency of the Carnot engine, we found that the heat absorbed from the reservoir at T_1 and the heat rejected into the reservoir at T_2 are in the ratio (consult Figure 23.4)

$$\frac{Q_2}{Q_1} = \frac{T_2}{T_1}, \tag{24.1}$$

which we can rewrite as

$$\frac{Q_1}{T_1} - \frac{Q_2}{T_2} = 0. \tag{24.2}$$

Note that in this convention, Q_1 was the heat *absorbed* from the hot reservoir during *AB*, while Q_2 was the heat *rejected in CD* into the cold reservoir.

Let us add two harmless 0's on the left and write

$$\frac{Q_1}{T_1} + 0 + \left[-\frac{Q_2}{T_2} \right] + 0 = 0 \tag{24.3}$$

so that the four terms correspond to the four parts in the Carnot cycle. The part AB contributes $\frac{Q_1}{T_1}$ because Q_1 was *absorbed* from the hot reservoir at T_1. The adiabatic segment BC contributes the first zero. The segment CD contributes the $-\frac{Q_2}{T_2}$ because Q_2 was *rejected* into the cold reservoir at T_2. The last adiabatic segment DA gives the final zero.

Let us change conventions and uniformly define ΔQ_i to be the heat *absorbed* by the system and rewrite Eqn. 24.3 as follows:

$$\sum_i \frac{\Delta Q_i}{T_i} = 0 \qquad (24.4)$$

where the sum is over the different segments in the closed loop.

Eqn. 24.4 is the heart of the whole entropy concept. It tells us there is another state variable lurking around besides the energy U. Remember a state variable is one that returns to its initial value when we go around a closed loop. The internal energy U is a state variable because if you go around a cycle and come back to the same point (P, V), U (which is given by $U = \frac{3}{2}PV$ for an ideal gas) returns to the old value. If the loop is made up of some discrete segments labeled by i (as in the Carnot cycle) and dU_i is the change in segment i we write

$$\sum_i dU_i = 0. \qquad (24.5)$$

If the cycle is some arbitrary continuous loop, we write

$$\oint dU = 0. \qquad (24.6)$$

By contrast, if you add the ΔQ_i's around a loop, you will get the area enclosed by the closed loop, which is the work done by the system in the cycle. So Q is not a state variable.

Now look at Eqn. 24.4. It says that even though the ΔQ_i's do not add up to zero in a cycle, if we divide each ΔQ_i by the value of T_i, the sum vanishes. This suggests that we define a new variable, the *entropy* S of the system, as follows.

The entropy S is a state variable that changes by

$$dS = \left.\frac{\Delta Q}{T}\right|_{rev} \tag{24.7}$$

when heat ΔQ is added reversibly, that is, with the system arbitrarily close to equilibrium.

By this definition S returns to its initial value after any Carnot cycle:

$$\sum dS_i = 0. \tag{24.8}$$

However, to merit the title of state variable Eqn. 24.8 should be valid for *any closed loop*, not just the Carnot cycle bounded by adiabatic and isothermal curves. In other words we want

$$\oint dS = 0 \tag{24.9}$$

for all loops. *This happens to be true.* The proof (which you may skip if you wish) relies on the fact that *you cannot convert heat fully into work with no other effects* because this means $\eta = 1$, in violation of Carnot's result.

Now for the proof that over any closed cycle $\oint dS = 0$. Imagine the system going on a loop as shown in Figure 24.1 and an auxiliary reservoir at T_0. Take one loop segment labeled i. If during this it needed heat input ΔQ_i, let it be delivered by a Carnot refrigerator acting between the reservoir at T_0 and the system temperature T_i. Let ΔQ_{0i} be the heat extracted in this step from the reservoir at T_0, and let ΔW_i be the work needed by this refrigerator to pump the heat. Being a Carnot refrigerator it obeys

$$\frac{\Delta Q_{0i}}{T_0} = \frac{\Delta Q_i}{T_i}. \tag{24.10}$$

Summing this over the closed loop (in parts of which ΔQ_i and ΔQ_0 may be negative, as in segment j in the figure) we find

$$\frac{Q_0}{T_0} = \oint \frac{dQ}{T} \tag{24.11}$$

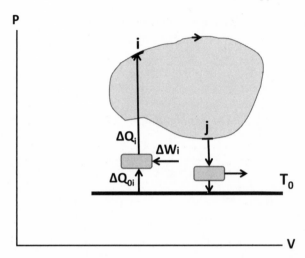

Figure 24.1   A system is taken quasi-statically through a cycle. In the segment labeled i it needs some heat ΔQ_i, which it gets from a Carnot refrigerator operating between the system's current temperature and a reservoir at T_0. The refrigerator needs work ΔW_i. If it happens that $\Delta Q < 0$, as is the case at point j, the heat will be sucked out of the system by a Carnot engine and dumped into the reservoir at T_0.

where Q_0 is the total heat extracted from the reservoir. If Q_0 is zero, we have the desired result

$$\oint \frac{dQ}{T} = \oint dS = 0. \tag{24.12}$$

We will see that this has to be so. (In the expression for dS we use dQ/T rather than $\Delta Q/T$ because this ratio stands for the differential dS.)

If $Q_0 > 0$, it means some heat has been lost by the reservoir. Because the system and all the auxiliary Carnot engines have come back to where they started, by conservation of energy this heat must have been converted *fully* to an equivalent amount of work delivered by all the Carnot engines and our system, that is, with $\eta = 1$, which is impossible. If $Q_0 < 0$, we can run the whole thing backward (because all steps are reversible in all the Carnot engines and our system) and change the sign of Q_0 to positive and get the previous contradiction.

Even if you did not follow this argument you must at least note it is important that ΔQ has to be added reversibly if we want to relate it to dS by $\frac{\Delta Q}{T} = dS$.

Thus there are now two state variables U and S. Because we have only defined the change in S, its absolute value is indeterminate up to a constant, just like the potential. Using this freedom we may arbitrarily assign a value S_0 to the entropy at some point (P_0, V_0) and find its value at any arbitrary point (P, V) experimentally by adding the changes $dS = \frac{dQ}{T}$ in going from (P_0, V_0) to (P, V). We can take any path because each point has a well-defined value of S and the difference in S between two points is independent of the path joining them.

To summarize, we know S is a state variable and that its change is given by Eqn. 24.7. But we have no idea what this quantity S stands for, in contrast to ΔQ, dV, or dU, which appeal to our intuition directly. But even in the early days people like Rudolf Clausius realized, "Here's another variable we have unearthed. We may not fully know what it means, but it is a state variable, so we better take it very seriously."

Let us bring S into the picture by rewriting the first law

$$dU = \Delta Q - PdV \qquad (24.13)$$

as follows. Because U is a state variable, the change dU is independent of how heat is added to the system. So let us assume it is added reversibly, allowing us to write

$$dU = TdS - PdV. \qquad (24.14)$$

Mathematically this tells us U is a function of S and V and that

$$T = \frac{\partial U}{\partial S} \qquad (24.15)$$

$$P = -\frac{\partial U}{\partial V}. \qquad (24.16)$$

Because we are now focused on S, let us rewrite the first law as

$$dS = \frac{1}{T}dU + \frac{P}{T}dV. \qquad (24.17)$$

This equation tells us that S (for a fixed amount of gas, say 197 moles) is a function of its macroscopic properties: volume V, energy U, and that

$$\left.\frac{\partial S}{\partial U}\right|_V = \frac{1}{T} \qquad \left.\frac{\partial S}{\partial V}\right|_U = \frac{P}{T}. \tag{24.18}$$

These two equations are worth committing to memory.

Before we develop a feeling for what S means, I want you to get some practice calculating the entropy change for a couple of processes armed with just

$$dS = \left.\frac{dQ}{T}\right|_{rev}. \tag{24.19}$$

First consider m grams of ice that are going to be melted at $0°$. We have to add the latent heat $L = 80cal/g$ reversibly, that is, add a little heat from a reservoir at $(0 + \varepsilon)°C$ (with $\varepsilon \to 0$) and let it be fully absorbed by the entire specimen of ice and water till it reaches a new equilibrium state with a little more water. And then keep doing this till all the ice has melted, at which point

$$S_2 - S_1 = \int_1^2 \frac{dQ}{T} = \frac{mL}{T_{ice}} \; cal/K \text{ where } T_{ice} = 273.16K. \tag{24.20}$$

In this simple case where T is fixed at T_{ice}, it can be pulled out of the integral, which gives mL.

Next we consider a more complicated problem: heating m grams of water, from T_1 to T_2. The entropy change is

$$S_2 - S_1 = \int_1^2 mc_w \frac{dT}{T} = mc_w \ln \frac{T_2}{T_1}. \tag{24.21}$$

Once again, remember that the heat should be added reversibly: you don't just dump the water at T_1 on a saucepan at T_2: instead you bring it in contact with a succession of reservoirs, each infinitesimally above the previous one, to take the water from T_1 to T_2, giving it enough time to equilibrate with each reservoir.

If you cooled it from a higher to lower temperature, you can use the same formula and the result will be negative because $T_2 < T_1$.

Finally we turn to an entropy calculation involving a gas, whose results are going to be very instructive. Take a gas and let it expand isothermally at temperature T from volume V_1 to V_2 as shown in Figure 24.2.

As when the ice was melted, T is constant and can be pulled out of the integral

$$S_2 - S_1 = \int_1^2 \frac{dQ}{T} = \frac{1}{T} \int_1^2 dQ = \frac{Q}{T} \tag{24.22}$$

where Q is the total heat absorbed by the gas. Since U is fixed in this isothermal process, Q is the work done by the gas. Thus

$$S_2 - S_1 = \int_1^2 \frac{dQ}{T} = \int_1^2 \frac{PdV}{T} = \int_1^2 \frac{nRTdV}{VT} = nR \ln \frac{V_2}{V_1}. \tag{24.23}$$

If at every point there is a unique entropy, then the entropy difference between 2 and 1 should be independent of how we go from 1 to 2. So, instead of following the isotherm let us go straight down the P axis to the point 0, which has the same volume as 1 and the same pressure as 2, that is, $V_0 = V_1$ and $P_0 = P_2$. The entropy changes are

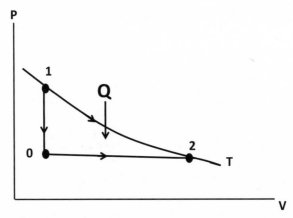

Figure 24.2   The entropy change between 1 and 2 at the same T can be computed by integrating $\frac{dQ}{T}$ along the isotherm or any other path, such as the one that goes via point 0.

$$S_0 - S_1 = nc_v \int_{T_1}^{T_0} \frac{dT}{T} = n\frac{3R}{2} \ln \frac{T_0}{T_1} = n\frac{3R}{2} \ln \frac{T_0}{T} \quad (\text{because } T_1 = T)$$

$$(24.24)$$

$$S_2 - S_0 = nc_p \ln \frac{T_2}{T_0} = n\frac{5R}{2} \ln \frac{T}{T_0} \quad (\text{because } T_2 = T) \qquad (24.25)$$

$$S_2 - S_1 = n\frac{3R}{2} \ln \frac{T_0}{T} + n\frac{5R}{2} \ln \frac{T}{T_0}$$

$$= n\frac{3R}{2} \ln \frac{T_0}{T_0} + n\frac{2R}{2} \ln \frac{T}{T_0} = nR \ln \frac{T}{T_0} \qquad (24.26)$$

$$= nR \ln \frac{V_2}{V_1} \quad (\text{upon using } \tfrac{T}{T_0} = \tfrac{T_1}{T_0} = \tfrac{P_1 V_1}{P_0 V_0} = \tfrac{P_1}{P_2} = \tfrac{V_2}{V_1}),$$

$$(24.27)$$

using many equalities implied in Figure 24.2: $T = T_1 = T_2$, $V_1 = V_0$, and $P_0 = P_2$. So, we can find the entropy change anyway we like. We usually pick the easiest path.

Remember I told you there are many, many phenomena that seem forbidden in our world. A lot of things can happen one way but not the other way, and we asked what law might prevent all of them from happening. Now I'm ready to state the mega law that enforces all that.

24.2   The second law: law of increasing entropy

The second law of thermodynamics says

$$dS \geq 0 \text{ for the universe.} \qquad (24.28)$$

There you have it; that's the great law: *Any process that reduces the entropy of the universe is forbidden.*

Now we have to see how this law can forbid the processes that seem disallowed. First consider Carnot's version of the second law, that you cannot have a process in which some heat goes from a cold body to a hot body with no other changes anywhere. Let's first calculate the entropy change when the two bodies are reservoirs fixed at T_1 and T_2, briefly connected by a conducting rod of negligible mass (whose heat content is negligible). Because these are reservoirs they are always in equilibrium and have well-defined temperatures, allowing us to invoke

$$dS = \left. \frac{\Delta Q}{T} \right|_{rev}. \tag{24.29}$$

So if ΔQ flows from T_1 to T_2,

$$dS = -\frac{\Delta Q}{T_1} + \frac{\Delta Q}{T_2} = \Delta Q \left[\frac{1}{T_2} - \frac{1}{T_1} \right]. \tag{24.30}$$

If $T_1 > T_2$ (hot to cold), then $dS > 0$ and the process is allowed, while if $T_1 < T_2$ (cold to hot), it is not allowed. Notice that energy is conserved but entropy increases: even though the energy gains are equal and opposite, the changes in entropy are not.

Next consider some mass m of copper at some temperature T_1 and an equal mass of copper at T_2 that are isolated from each other. Each has a well-defined entropy. Let us now place them in contact, keeping them thermally isolated from the rest of the world. They will end up at a common temperature $T^* = \frac{1}{2}(T_1 + T_2)$ just by symmetry. Energy, of course, is conserved, and that's in fact how we determine T^*. But look at the entropy.

The naive answer may be that for the combined system $dS = 0$ follows from Eqn. 24.29 because $\Delta Q = 0$, which in turn follows from the fact that the system was thermally isolated. This is wrong, because the process was not reversible; the system was nowhere near equilibrium when the samples at two different temperatures were brought together abruptly. There was a period of non-equilibrium when they went off the radar, without even a globally defined temperature.

However, after a long time, when they have settled down to T^*, they do have well-defined entropies, and we can ask what these final entropies are. To find them, we forget about the actual process and instead imagine taking each block from its initial to its final temperature in a sequence of steps never far from equilibrium, during which the blocks have well-defined entropies and for which dS could be computed using Eqn. 24.29.

Here is an analogy. Suppose you are on some mountain at some point 1 with height h_1. You hike to a new place 2. I want to know the height difference $h_2 - h_1$. It is simply the sum of the height differences for every step you took. Suppose now that instead of hiking from 1 to 2, you used your yogic skills to disappear at 1 and reappear at 2. Even though you were invisible during the transit, there is a well-defined height difference between your initial and final locations. However, it cannot be found by adding up the height changes for each step you took, because you did not get from here to there by taking steps; you went off the radar and reappeared. But I can find the height difference between the end points by simply walking myself from 1 to 2 and adding up all the height changes. I can take any route I want connecting the same two end points. My hike is like the reversible path used to compute the entropy change, while your yogic journey is like the actual irreversible process.

So let us imagine the cooler block being steadily heated by putting it in contact with a sequence of reservoirs, each a little hotter than the previous one, keeping the block never far from equilibrium and slowly bringing it to T^*. Similarly, the hotter block is slowly brought down to $T^* = \frac{1}{2}(T_1 + T_2)$. The entropy change for the two of them is, in terms of the specific heat c:

$$S_{\text{final}} - S_{\text{initial}} = mc \int_{T_1}^{T^*} \frac{dT}{T} + mc \int_{T_2}^{T^*} \frac{dT}{T} \qquad (24.31)$$

$$= mc \left[\ln \frac{T^*}{T_1} + \ln \frac{T^*}{T_2} \right] = mc \ln \frac{T^{*2}}{T_1 T_2}. \qquad (24.32)$$

Let us now confirm that the change in S is positive. At every stage, when we brought the hot one down a little by sucking out ΔQ and pushed the cold one up a little by pumping in ΔQ, the entropy change was positive *because we divide the heat gained by a smaller temperature than the heat lost*.

To absolutely establish the positivity of $S_{\text{final}} - S_{\text{initial}}$ we need to show that $T^{*2} > T_1 T_2$. Is $\frac{1}{4}(T_1 + T_2)^2 > T_1 T_2$? Since this can be rearranged to give "Is $(T_1 - T_2)^2 > 0$?" the answer is of course yes.

So, when hot and cold meet and create lukewarm, the entropy of the universe goes up. It follows, therefore, that if lukewarm spontaneously separated into hot and cold, the entropy would go down, and that's why that doesn't happen. That's why a jar of lukewarm water doesn't spontaneously separate into a cold part and a hot part. Such a separation would not violate the law of conservation of energy, but it would violate the law of increasing entropy. So evolution in one direction is allowed, because entropy goes up, but the opposite is not allowed, because that would correspond to a spontaneous decrease of entropy.

State 1   **State 2**

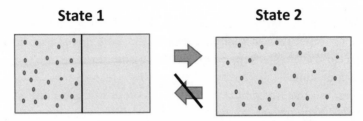

Figure 24.3   (Left) The gas is originally restricted to the left half of the thermally isolated box by a partition. (Right) When the partition is suddenly removed, the gas expands and fills the whole box. If a movie of this is made and played backward, the gas will go back to the left half. This is, however, forbidden by the second law.

Finally, consider the following process depicted in Figure 24.3.

State 1 refers to a gas confined by a partition to the left half of a thermally isolated box. The gas has reached equilibrium, and it has spread out uniformly over this volume. State 1 corresponds to a point in the $P - V$ plane. Now, I suddenly remove the partition. There is a period when the gas goes off the radar because it is not in a state of equilibrium. Just after the partition is removed, the gas doesn't have a well-defined pressure: it is non-zero in the left half of the box; it is zero in the right half, which is still a vacuum. That is why the gas has gone off the $P - V$ plane. After you wait long enough, the gas is again in equilibrium, in state 2.

What is the entropy change now? Once again here's the wrong way to do the calculation. You say, "Well, $dS = \frac{\Delta Q}{T}$, but this whole box is thermally isolated, and so $\Delta Q = 0$, and the entropy change is zero." This could not possibly be the right answer. If 2 had the same entropy as 1, there should be no obstacle to going back to 1 spontaneously. But we know that the process $2 \rightarrow 1$ will never happen spontaneously—you cannot put the genie back in the bottle.

So the entropy must have gone up under this "free expansion" against a vacuum. We should not find the change in entropy by blindly applying $dS = \frac{\Delta Q}{T}$ because this process was abrupt, and it was far from equilibrium in the intermediate stages, whereas the correct formula says the process must be reversible, never far from equilibrium:

$$dS = \left. \frac{\Delta Q}{T} \right|_{rev} . \tag{24.33}$$

What are we to do? The answer is the same as with the two copper blocks: to forget what actually happened and take the system through a sequence of reversible steps from the initial equilibrium state to the final equilibrium state.

Because the end points of the gas are very much present as points in the $P - V$ plane, we can find the entropy difference by adding up the changes in S along *any* reversible path connecting the end points.

While any path will do, there is a very simple path we can use *because the initial and final temperatures are the same.* This is because the gas does no work against the vacuum into which it is expanding, and it gets no heat because it is thermally isolated. So $U_2 = U_1$, and being an ideal gas, this implies $T_2 = T_1$. We may therefore join the end points by an isotherm. The entropy change under an isothermal expansion is already known from Eqn. 24.23:

$$S_2 - S_1 = nR \ln \frac{V_2}{V_1}. \tag{24.34}$$

So in this experiment, the way to calculate the entropy change is to forget about what actually happened and do the following instead. Take the initial gas, put a piston at the halfway point, apply a counter pressure to balance the internal pressure, and slowly let it expand, keeping it all the while on a reservoir at T. At every stage the dS for the reservoir cancels that for the system because the ΔQ's are equal and opposite, and their temperatures are essentially equal. (I say "essentially" because the reservoir had to be infinitesimally hotter than the gas for the gas to absorb heat.) In this process, the entropy of the universe does not change at all. The entropy gain of the gas balances the entropy loss of the reservoir. But it is the entropy gain of the gas that I am after. I used this reversible process as a device for finding the entropy gain of the gas alone, because that was the same gain it incurred in the irreversible free expansion.

This was the same story when two chunks of copper that were at T_1 and T_2 were brought together to form a chunk at $T^* = \frac{1}{2}(T_1 + T_2)$. In the actual irreversible process the blocks were thermally isolated and the entropy of the universe ΔS_U went up, because the universe in this case was just the two chunks of copper. Thus in this case, $\Delta S_U = \Delta S_{\text{chunks}}$. However, to find the ΔS_{chunks}, we cooked up a reversible process in which a whole sequence of reservoirs was called into play, to slowly bring the blocks to their final temperatures. Here the entropy gain for the blocks was the same as in the original process, but that of the universe was zero: at every stage, the entropy change of the reservoirs exactly negated that of the blocks, because heat transfer between either chunk and the reservoirs took place at essentially the same temperature.

24.3   Statistical mechanics and entropy

I am finally ready to explain what entropy stands for at a microscopic level according to *statistical mechanics*, whose key ideas I will now briefly introduce.

Consider the free expansion of a gas, and, in particular, Eqn. 24.34 for the simple case where the gas expands from some volume $V_1 = V$ to $V_2 = 2V$. We saw that

$$S_2 - S_1 = nR\ln\frac{V_2}{V_1} = nR\ln 2. \qquad (24.35)$$

Note that n, the number of moles, is a macroscopic quantity. For example, it is the mass of the hydrogen gas in grams or the mass of the carbon gas in grams divided by 12. The universal gas constant $R = 2 cals/mole$ is likewise relevant to the macroscopic description.

By the microscopic description, I mean the one where we are armed with the fact that the gas is made of atoms and in particular that we may rewrite Eqn. 24.35 as

$$S_2 - S_1 = nR\ln\frac{V_2}{V_1} = Nk\ln\frac{V_2}{V_1} = Nk\ln 2 = k\ln 2^N \qquad (24.36)$$

where we have used $R = N_A k$ and that the number of atoms is $N = nN_A$. The formula $S_2 - S_1 = nR\ln\frac{V_2}{V_1}$ was discovered long before atoms were proven to exist. You recall that the derivation never referred to atoms. Equation 24.36 is the first time the number of atoms has entered the formula for entropy. Rewriting it in terms of N requires the knowledge that there are atoms. From this, or by some other means, Boltzmann divined the formula for the entropy of a gas in terms of what the individual atoms are doing. That is at the heart of statistical mechanics, which gives you the microscopic basis of thermodynamics.

The truly microscopic theory would follow the evolution of every atom using Newton's laws, thereby tracking the state of the gas in maximum possible detail. In practice, this voluminous data for 10^{23} atoms is neither computable nor digestible, if computed. Here is where statistical mechanics comes in with a more modest and realizable goal.

Statistical mechanics focuses on only the following attributes of the gas: energy U, volume V, and number of particles N. Despite the microscopic chaos, with atoms darting about here and there and colliding with each other and the walls, these three numbers do not change. They are easily measured by macroscopic means, including N, which is the mass of the gas divided by the mass of the atoms, presumed known from independent microscopic measurements. By contrast, the individual atomic momenta and coordinates are perpetually changing.

Remarkably, the entropy of the gas in equilibrium is determined in statistical mechanics solely by these three unchanging macroscopic quantities. Boltzmann's formula for $S(U, V, N)$ is so important that it is carved on his gravestone. When we physicists go to Vienna, we skip the orchestras and first go to Boltzmann's grave and read this formula once more with reverence. It summarizes a lifetime of work. Here it is:

$$S(U, V, N) = k \ln \Omega(U, V, N), \tag{24.37}$$

where $\Omega(U, V, N)$ is the number of different microscopic states or microstates *of the system compatible with its macroscopic properties, namely U, V, and N.* We will verify that it gives the correct entropy change under free expansion from V_1 to V_2.

The first step is to count the microstates of the gas. For this we must know what they are. A microstate is a collection of data that fully specifies the complete and exhaustive state of the gas, using which its future may be determined by Newton's laws. Thus it is the collection of every atomic coordinate and every atomic momentum. We need to count the number of states in which every atomic coordinate lies inside the box of volume V and the momenta are such that the sum of the individual kinetic energies adds up to U.

To follow the upcoming arguments, you need to know this: If a system has N members (atoms in our example) and each can be in m states, independent of what the others are doing, the total number of allowed states for the system is $m \times m \times \ldots \times m = m^N$. For example, a coin can be only heads or tails ($m = 2$), and, if we toss 3 of them ($N = 3$), we can get $2^3 = 8$ results. If we throw 42 dice, we can get 6^{42} results. Make sure you get this.

For pedagogical reasons the complete dependence of S on N will not be computed here. But we will find that as long as N is fixed, the partial derivatives of S with respect to U and V can be evaluated with no error, and these will confirm beyond any doubt that Boltzmann's S indeed corresponds to the one in thermodynamics by reproducing $PV = NkT$ and $U = \frac{3}{2}NkT$.

First consider the spatial coordinates. Because each atom is point-like in our description, its position is a point. If we equate the number of possible positions to the number of points inside the volume V, the answer will be infinite no matter what V is! So what one does is divide the box mentally into tiny cells of volume a^3 where a is some tiny number determined by our desired accuracy in specifying atomic positions in practice. Let us say we choose $a = 10^{-6}m$. In a volume V, there will be V/a^3 cells indexed by $i = 1, 2, \ldots, V/a^3$. We label the atoms A, B, and so forth, and we say in which cell each one is. If A is in cell $i = 20$ and B in cell $i = 98000$, et cetera, that's one microscopic arrangement. We can assign them to other cells and in case we permute them, say with $A \to B \to C \to D \to A$, that is counted as another arrangement (except when two exchanged atoms are in the same cell). Thus when the gas is restricted by the partition to volume V_1, and each of the N atoms has V_1/a^3 possible cell locations,

$$\Omega_1 = \left[\frac{V_1}{a^3}\right]^N, \tag{24.38}$$

and

$$S_1 = k \ln \left[\frac{V_1}{a^3}\right]^N = Nk \ln \frac{V_1}{a^3}. \tag{24.39}$$

If, after free expansion, the volume is V_2,

$$S_2 = k \ln \left[\frac{V_2}{a^3}\right]^N = Nk \ln \frac{V_2}{a^3}. \tag{24.40}$$

Notice that S depends on the cell size a. If we change a, we will change S by a constant, because of the $\ln a^3$ term. This is unavoidable until quantum mechanics comes in to specify a unique cell size. However, *changes* in S will be unaffected by the varying a, just as adding a constant to the potential energy does not affect the application of the law of conservation of energy $K_1 + U_1 = K_2 + U_2$.

But there is another problem with this result. The state of the atom is not given by just its location, but also its momentum **p**. Thus every Ω above should be multiplied by a factor $\Omega_p(U)$ that counts the number of momentum states open to the gas at a given value of U. Again one divides the possible atomic momenta into cells of some size. Whereas the atoms could occupy any spatial cell in the box independently of the others, now they can only assume momentum configurations in which the total kinetic energy of the gas adds up to a given fixed U.

Thus, the formula to use is

$$\Omega = \left[\frac{V}{a^3}\right]^N \times \Omega_p(U) \tag{24.41}$$

$$S(U, V) = Nk \ln \frac{V}{a^3} + k \ln \Omega_p(U). \tag{24.42}$$

Luckily we do not need to evaluate $\Omega_p(U)$ in this example because U does not change during free expansion and so neither does $\Omega_p(U)$. It will drop out in the difference $S_2(U, V_2) - S_1(U, V_1)$. Later we will compute $\Omega_p(U)$ for the general processes in which U varies as well.

The *change* in entropy is

$$S_2 - S_1 = Nk \ln \frac{V_2}{a^3} - Nk \ln \frac{V_1}{a^3} = Nk \ln \frac{V_2}{V_1} = Nk \ln 2 \text{ in our example}$$

$$\tag{24.43}$$

independent of a and in agreement with what we got from thermodynamics, before we knew there were atoms! However, in earlier times $S_2 - S_1$ was written in terms of the macroscopic attribute, the number of moles n and the macroscopic parameter, the universal gas constant $R = 2$ *cal/mole*:

$$S_2 - S_1 = nR \ln \frac{V_2}{V_2}. \tag{24.44}$$

Because a drops out, in computing the change in S, we sometimes choose the cell size a^3 to be half the volume for pedagogical reasons. In this case each atom has only two positions: left or right. If initially all are in the left (with the partition in place) there is just one way, $\Omega_1 = 1$, $S_1 = 0$, while if both choices are open (with the partition removed), $\Omega_2 = 2^N$ and $S_2 = Nk \ln 2$ and $S_2 - S_1 = Nk \ln 2$ as before.

If every equilibrium state has a unique entropy (up to a constant) what do we mean when we say S is maximized in equilibrium?

The preceding case of free expansion illustrates the answer to this common question. Initially the gas occupies the left half and is in equilibrium. The entropy assigned is S_1. Nothing will change if we leave it alone, including S_1. Now we remove the partition or constraint and wait until a new equilibrium is reached. The law of increasing entropy says that the new S_2 will be greater than the old S_1. Thus the entropy increased, not when the system was in equilibrium, but when some external conditions keeping it there were changed to allow a new equilibrium state. In other words, the change in entropy is a result of our lifting some constraint (the partition) that made available a new equilibrium state to the system that was initially constrained to live in the old equilibrium state. Removing the constraint can only open up more options (or keep them the same), which is why S either rises or stays put. We will return to this theme shortly.

Here is another example. Two different gases A and B occupy the two halves of a partitioned box. The two halves settle down, reach equilibrium, and have some total well-defined entropy S_{initial}. This entropy does not change with time. Suppose we now remove the constraint, the partition. The separated gases do not represent an equilibrium state in the absence of the partition. There will be macroscopic changes as the gases begin to mix, and there will be a period when S is not defined. Finally the system will assume a new equilibrium state with both gases uniformly spread out over the entire box and an entropy $S_{\text{final}} > S_{\text{initial}}$.

So here is what statistical mechanics and in particular $S = k \ln \Omega$ have done for us.

- Given us a microscopic basis for entropy that is a lot more comprehensible than $dS = \frac{\Delta Q}{T}$ and that is also capable of reproducing all thermodynamic results, as illustrated in the case of free expansion.
- Made it clear why S will go up (or stay fixed) in a spontaneous process: when a constraint is removed more states Ω become available.
- Explained why $A \rightarrow B$ happens spontaneously but not $B \rightarrow A$: because the macroscopic state B can be realized in many more ways than state A.
- Given us a deeper and more accurate picture of equilibrium states. Consider the gas after the partition is removed and it fills the whole box

evenly. This uniform density corresponds to our image of the equilibrium state in thermodynamics. Statistical mechanics tells us that while this is *typically* what happens, small deviations from this will occur with a calculable probability determined by the following sole postulate of classical statistical mechanics.

Postulate of statistical mechanics: *In equilibrium, every allowed microscopic state of an isolated system has the same probability.*

It is this postulate that ensures S will go up in a spontaneous process when a constraint is removed: less constraints mean more microscopic states are allowed.

To illustrate the other implications of this postulate for an ideal gas, let us assume every atom has just two positional states: left or right half of the box. There are 2^N allowed microscopic states and all occur with the same probability. *But this does not mean every possible value of a macroscopic observable is equally probable.* The fraction $f = \frac{n}{N}$ of particles in the right half is a macroscopic variable. Not every f is equally likely. Consider the configuration with all the particles in the left half: $n = 0$ and $f = 0$. It can occur in only one way. It is like tossing 10^{23} coins and getting all heads—possible but highly improbable. If you now let one of the particles be in the right half, that is, $n = 1$, there are N ways for this to happen because there are N choices for who gets to be the odd atom. So this situation is N times more likely than the one with all in the left half. More generally, the situation with n atoms in the right and $N - n$ in the left, by elementary combinatorics, can occur in

$$\Omega(N, n) = \frac{N!}{n!(N - n)!} \tag{24.45}$$

ways. Therefore $f = \frac{n}{N}$ is $\Omega(N, n)$ times more likely than $f = 0$, the one with all atoms in the left half.

As n increases, so does $\Omega(N, n)$, which reaches a maximum at $n = \frac{N}{2}$:

$$\Omega = \Omega_{\text{max}} = \frac{N!}{\frac{N}{2}! \cdot \frac{N}{2}!}. \tag{24.46}$$

Beyond the half-way point, Ω falls and reaches the value 1 when $n = N$. Figure 24.4 shows $\Omega(30, n)$. This graph allows us to say, for this gas of 30 atoms, what the odds are for any partitioning of atoms between left

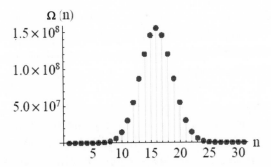

Figure 24.4 The number of microscopic configurations with $N - n$ particles on the left half of the box and n on the right, for $N = 30$.

and right. If every value of n is allowed, it means that in equilibrium the density will be inhomogeneous and not be uniquely determined by U and V. But this is not the situation to which statistical mechanics is applied. Rather it is intended for a typical gas for which $N \simeq 10^{23}$. Now the distribution will become very sharply peaked at $f = .5$ and the chances of getting anything really different from a 50/50 split will be negligible. It can be shown that typically f will not deviate from $\bar{f} = .5$ by more than a number of order $1/\sqrt{N}$. Thus the variable f will lie in the range $f = \bar{f} \pm \frac{1}{\sqrt{N}}$. Likewise, deviations from the average $\bar{P}$ will exist but will be relatively very small, and one may associate $\bar{P}$ with the P in $PV = NkT$ as $N \to \infty$.

In general the likelihood of obtaining a certain value for a macroscopic observable in equilibrium is proportional to the underlying number of microscopic arrangements.

We now return to our discussion of entropy. It is often said that entropy is a measure of how disordered your system is. To have all the atoms of the two species confined to two different halves of the box is perfectly natural *if there is a partition to enforce the separation*. However, once the partition is removed, this condition would be unnatural and extremely ordered in the face of the newly opened possibilities. This initial order, introduced by design, will quickly and spontaneously evolve into a fully mixed state. Thus opening the partition and allowing the gases to mix increases the disorder. The technical measure of disorder is the (log of the) number of microscopic arrangements that can lead to a given macroscopic situation. The spontaneous evolution from order to disorder under the removal of a constraint occurs because there are so many more ways to be disordered than to be ordered.

Here is another example. You take a container with a vertical partition, and pour clear water in one side and red water in the other. You have an ordered state. It can last forever as long as there is a partition. If you remove the partition, the red and clear water will soon mix, because there's no reason for the red particles to stay on one side forever. Eventually the mixture will become pink. This disordered configuration is inevitable, because there are many more ways to remain pink than to remain separated. Likewise, pink spontaneously separating into red and colorless, the emergence of order from chaos, is very, very improbable and points in the direction of lower entropy.

Although the unmixing seems disallowed by the law of increasing entropy, it is not strictly disallowed, just overwhelmingly unlikely. For example, if you are in one side of a room, there is a chance of one part in roughly $1/(2^{10^{23}})$ that all the air molecules spontaneously end up in the other half. But don't hold your breath; it is not likely to happen in the remaining age of the universe. The second law of thermodynamics may be a statistical law, but when it says something is "unlikely" it means "so overwhelmingly, ridiculously, unlikely you should forget about its happening in the entire history of the universe."

Be aware of one thing: the entropy of a part of the universe can go down. It's just the entropy of the whole universe that cannot go down. All of life is an example of lowering entropy: the creation of tomatoes out of mud and fertilizer is a highly organizing process where the entropy is really going down. But if you keep track of the rest of the world, you will find that there is a greater increase of entropy somewhere else. Or take your freezer: your refrigerator sucks heat out of your freezer, and the entropy of the freezer goes down, but elsewhere in the room there's a bigger increase due to the heat emitted from the back of the refrigerator. If you take account of everything, the entropy of the universe will go up or remain the same; it will never go down.

24.4 Entropy of an ideal gas: full microscopic analysis

We will now compute the entropy (up to a constant) of an ideal gas of N atoms, energy U, and restricted to a volume V using Boltzmann's formula. We will assume N never changes but that U and V could. So we want $S(U, V)$.

The energy of an ideal gas is entirely kinetic and independent of particle positions. We have to find the logarithm of the number of states $\Omega(U, V)$ compatible with the prescribed values of U and V.

We have already seen that

$$\Omega(U, V) = \left[\frac{V}{a^3}\right]^N \Omega_p(U) \tag{24.47}$$

where V/a^3 counts the possible positions for each atom and $\Omega_p(U)$ is the number of microscopic arrangements of momenta for the entire gas that respect the condition that the total kinetic energy be U. (In computing the *change* in S during free expansion we ignored $\Omega_p(U)$ because U was the same before and after free expansion. Here we want U to vary and so need to work a little harder and find $\Omega_p(U)$.)

The internal energy is (for the allowed configuration with every atom inside the box)

$$U = \sum_{i=1}^{N} \frac{1}{2} m |\mathbf{v}_i|^2 = \sum_{i=1}^{N} \frac{|\mathbf{p}_i|^2}{2m}$$

$$= \sum_{i=1}^{N} \frac{p_{ix}^2 + p_{iy}^2 + p_{iz}^2}{2m} \qquad \text{where } \mathbf{p} = m\mathbf{v} \text{ is the momentum.}$$

$$\tag{24.48}$$

Let us now form a vector $\mathbf{P}$ with $3N$ components

$$\mathbf{P} = (p_{1x}, p_{1y}, p_{1z}, p_{2x} \ldots\ldots, p_{Nz}), \tag{24.49}$$

which is simply the collection of the 3 components of the N momentum vectors $\mathbf{p}_i$. If we renumber the components of $\mathbf{P}$ with an index $j = 1, \ldots, 3N$,

$$\mathbf{P} = (P_1, P_2, \ldots P_{3N}), \tag{24.50}$$

that is to say,

$$P_1 = p_{1x}, P_2 = p_{1y}, P_3 = p_{1z}, P_4 = p_{2x}, \ldots, P_{3N} = p_{Nz}, \tag{24.51}$$

we may write

$$U = \sum_{j=1}^{3N} \frac{P_j^2}{2m}.$$ (24.52)

Regardless of their position, the atoms can have any momentum as long as the components satisfy Eqn. 24.52. So we must see how many possible momenta exist obeying this condition. The condition may be rewritten as

$$\sum_{j=1}^{3N} P_j^2 = 2mU.$$ (24.53)

This is the equation for a hyper-sphere of radius $\mathcal{R} = \sqrt{2mU}$ in $3N$ dimensions just as

$$x^2 + y^2 = \mathcal{R}^2 \quad \text{is a circle or 1-sphere and}$$ (24.54)

$$x^2 + y^2 + z^2 = \mathcal{R}^2 \quad \text{is the usual sphere or 2-sphere.}$$ (24.55)

In mathematical literature the circle in $d = 2$ and the usual sphere in $d = 3$ are both referred to as spheres, and the circumference and area are both referred to as areas. The areas are $2\pi\mathcal{R}$ and $4\pi\mathcal{R}^2$ respectively in the familiar examples. By dimensional analysis, a sphere of radius $\mathcal{R}$ in d dimensions has an area that goes as $\mathcal{R}^{d-1}$. In our problem $\mathcal{R} = \left[\sqrt{2mU}\right]$ and $d = 3N - 1 \simeq 3N$. If we divide the individual momenta into cells of size b^3, which like a^3 is small but arbitrary, the total number of states allowed to the gas behaves as

$$\Omega(V, U) = V^N U^{3N/2} F(m, N, a, b)$$ (24.56)

where we have focused on the dependence on U and V and lumped the rest of the dependence on m, a, b, and N in the unknown function $F(m, N, a, b)$. We do not need F because we just want to take

$$S = k \ln \Omega = k\left[N \ln V + \frac{3N}{2} \ln U\right] + k \ln F(m, N, a, b)$$ (24.57)

and find its V and U partial derivatives, to which F makes no contribution. These derivatives are

$$\frac{\partial S}{\partial V}\bigg|_U = \frac{kN}{V} \tag{24.58}$$

$$\frac{\partial S}{\partial U}\bigg|_V = \frac{3kN}{2U}. \tag{24.59}$$

If we identify the two derivatives above with $\frac{P}{T}$ and $\frac{1}{T}$ following Eqn. 24.18, that is, as they were in thermodynamics, we obtain

$$\frac{kN}{V} = \frac{P}{T} \quad \text{which is just } PV = NkT \tag{24.60}$$

$$\frac{3kN}{2U} = \frac{1}{T} \quad \text{which is just } U = \tfrac{3}{2}NkT. \tag{24.61}$$

Thus we are able to derive these *equations of state* of the ideal gas from the Boltzmann definition of entropy.

With more work we could get the full N-dependence of Ω as well. It has interesting consequences, but I will not go there, leaving it to you to pursue the topic on your own.

Two final remarks on Boltzmann's formula. First,

$$S = k \ln \Omega \tag{24.62}$$

is valid not only for the ideal gas but any thermodynamic system. However, computing Ω is generally impossible except for some idealized models.

Next, consider two systems that are independent. Then

$$\Omega = \Omega_1 \times \Omega_2 \tag{24.63}$$

that is, the number of options open to the two systems is the product of the numbers open to each. This ensures that the total entropy, S, is additive:

$$S = S_1 + S_2. \tag{24.64}$$

24.5 Maximum entropy principle illustrated

I have already explained briefly what it means to say that the entropy is defined only in equilibrium, and also that it is a maximum in equilibrium. I resume that discussion with an illustrative example.

Imagine a box of gas separated by an insulating partition, with energies U_1^0 and U_2^0 in the left and right halves respectively. The initial entropy of the system is

$$S_{1+2}^0 = k \ln \left[\Omega_1(U_1^0) \cdot \Omega_2(U_2^0) \right] \qquad (24.65)$$

in obvious notation. The two sides may not be at the same temperature, but there is nothing they can do about it, because of the insulating barrier. Suppose we now let the barrier conduct heat, so that energy can flow between the two sides.

Intuitively we expect that energy will flow till the two sides have reached the same T. On the other hand, statistical mechanics says the final state will be one of maximum entropy. We will now verify that the two conditions are equivalent.

Let us find out how statistical mechanics determines the final partition of the total energy

$$U_0 = U_1^0 + U_2^0. \qquad (24.66)$$

The new equilibrium entropy is

$$S_{1+2} = k \ln \left[\sum_{U_1=0}^{U_0} \Omega_1(U_1) \cdot \Omega_2(U_2 = U_0 - U_1) \right] \qquad (24.67)$$

where the sum runs over all possible assignments of U_1, from the lowest value (assumed to be 0) to the maximum, the entire energy U_0. The initial entropy comes from just one term in the sum with $U_1 = U_1^0$ and $U_2 = U_2^0$, a condition imposed by the barrier.

Without the barrier, all the partitions of the total energy are allowed. The final entropy is obviously larger, because all the extra terms in the sum are positive. But the crucial point is not that there are more terms, but that there are usually some new *individual terms that are astronomically larger than the initial one* $\Omega_1(U_1^0) \cdot \Omega_2(U_2^0)$. The largest term can be found by maximizing the product $\Omega_1(U_1) \cdot \Omega_2(U_2 = U_0 - U_1)$ with respect to U_1. Let U_1^* be the value of U_1 that maximizes the product and let $U_2^* = U_0 - U_1^*$. Let us maximize the logarithm of the product instead, to find

$$0 = \frac{d\ln\left[\Omega_1(U_1) \cdot \Omega_2(U_2 = U_0 - U_1)\right]}{dU_1} \qquad (24.68)$$

$$= \frac{d\ln\Omega_1(U_1)}{dU_1}\bigg|_{U_1^*} + \frac{d\ln\Omega_2(U_2)}{dU_2}\bigg|_{U_2^*}\frac{dU_2}{dU_1} \qquad (24.69)$$

$$= \frac{1}{kT_1} - \frac{1}{kT_2} \qquad \left(\text{because } \tfrac{dU_2}{dU_1} = \tfrac{d(U_0 - U_1)}{dU_1} = -1\right). \qquad (24.70)$$

In other words, the largest term corresponds to the partitioning of energy such that the temperatures are equal on the two sides. This confirms the assertion that the equality of final temperatures when the partition becomes conducting is synonymous with maximizing entropy.

The product $\Omega_1(U_1) \cdot \Omega_2(U_2 = U_0 - U_1)$ drops off very rapidly as we move away from the maximum at U_1^*. So Ω_{1+2}, the sum over partitions, or "the area under this graph," is then $\Omega_{max} \cdot W$, where W is some effective width, which will be some tiny fraction of U_0, the maximum allowed range for U_1. We will see that the details of W will not matter. Continuing,

$$\Omega_{1+2} = \Omega_1(U_1^*) \cdot \Omega_2(U_2^*) \cdot W \qquad (24.71)$$

$$S_{1+2} = k\ln\Omega_1(U_1^*) + k\ln\Omega_2(U_2^*) + k\ln W \qquad (24.72)$$

$$\simeq S_1(U_1^*) + S_2(U_2^*). \qquad (24.73)$$

We have dropped the $\ln W$ term compared to the first two terms, which are typically of order $N\ln U$ (see Eqn. 24.57) with $N \simeq 10^{23}$. It is often the case in statistical mechanics that the sum over terms is replaced by the largest term (or the area under the graph is replaced by the maximum height of the function), with negligible error in the logarithm, which is eventually and inevitably taken. Physically, it means in the example under consideration that, even though the total energy can be divided in all possible ways, you are not likely to find any division that differs macroscopically from the one at the maximum. Observe that only when we approximate the sum over partitions with the dominant term (which causes an utterly negligible error in the logarithm) does the total Ω_{1+2} become the product of the individual Ω's, and the total S_{1+2}, the sum of individual entropies.

Let us get a sense of the numbers involved. Imagine that initially we had ideal gases with N atoms each on both sides, with the temperature and energy on the right being three times that on the left:

$$U_1^0 = U \quad U_2^0 = 3U. \tag{24.74}$$

The initial entropy, with the insulating partition is, from Eqn. 24.57,

$$S_{\text{initial}} = \frac{3Nk}{2} \left[\ln \frac{U}{N} + \ln \frac{3U}{N} \right]$$

$$+ \text{ terms that are unaffected by the partition.} \tag{24.75}$$

The final state with equal T and hence energy $2U$ in each side has entropy

$$S_{\text{final}} = 2 \times \frac{3Nk}{2} \left[\ln \frac{2U}{N} \right] + \text{ the same unaffected terms.} \tag{24.76}$$

The change in entropy is

$$S_{\text{final}} - S_{\text{initial}} = \frac{3Nk}{2} \ln \frac{4}{3}, \tag{24.77}$$

which means the ratio of allowed configurations is

$$\frac{\Omega_{\text{final}}}{\Omega_{\text{initial}}} = \left[\frac{4}{3} \right]^{3N/2} \simeq 10^{1.87 \cdot 10^{22}} \tag{24.78}$$

for $N = 10^{23}$.

You are invited to show that if the conducting partition is also movable, the maximization of entropy will require that the total volume will be shared in such a manner that

$$\frac{P_1}{kT_1} = \frac{P_2}{kT_2}, \tag{24.79}$$

which reduces to $P_1 = P_2$ because $T_1 = T_2$.

24.6   The Gibbs formalism

Josiah Willard Gibbs was the greatest homegrown physicist produced by the United States. He spent his whole life at Yale, where his father was a professor of sacred languages. He obtained his bachelor's degree and doctorate there, joined the faculty, and taught until his death in 1903. He seems to have been an extraordinarily modest, generous, and cheerful person, in addition to being an exceptionally deep and original physicist. Among his many contributions to mathematics, physics, and chemistry, I will single out his alternative to Boltzmann's statistical mechanics.

Recall that Boltzmann considered an isolated system, with a conserved energy U. He postulated equal probability for every allowed microscopic state of this energy. The central function was the entropy,

$$S(U) = k \ln \Omega(U), \tag{24.80}$$

where $\Omega(U)$ was the number of states available to the system with energy U. (We hold V and N fixed for this discussion.) When internal constraints were removed, the systems would try to maximize entropy. The absolute temperature emerges as a partial derivative:

$$\frac{\partial S}{\partial U} = \frac{1}{T}, \quad \text{which can be rewritten as} \tag{24.81}$$

$$\frac{\partial \ln \Omega}{\partial U} = \frac{1}{kT}. \tag{24.82}$$

In contrast to Boltzmann, who gave a statistical description of isolated systems with a *definite energy* U, Gibbs wanted to describe systems at a *definite temperature* by virtue of being in thermal equilibrium with a heat reservoir at a fixed T. For example, the system could be a gas, confined to a heat-conducting box, placed inside a gigantic oven at that T.

Due to the coupling to the reservoir, the energy of the system can vary without limit, and the goal now is to find the probability $P(i)$ that the system would be found in a particular state i, of energy ε_i, and to find the appropriate version of the law of maximizing entropy.

There is no need for a new postulate. All results will follow from the fact that the *system plus reservoir is an isolated system*, with a fixed total energy U_0, to which Boltzmann's treatment applies.

Let $\Omega_R(U)$ be the number of states available to the reservoir when it has energy U. Let the lowest energy state of the system carry an index 0, and let us set the corresponding energy $\varepsilon_0 = 0$ for convenience. When the system is in the *one particular* state 0, the reservoir has the entire energy U_0, and it can be in any one of $\Omega_R(U_0)$ states. Thus the system plus reservoir can be in one of $1 \times \Omega_R(U_0)$ states. If the system is in *one particular* state i of energy $\varepsilon_i > 0$, the number of states available to the reservoir plus system is now reduced to $1 \times \Omega_R(U_0 - \varepsilon_i)$. Since every state of the joint system is equally probable, the ratio of probabilities for the system to be in state i versus state 0 is simply the ratio

$$\frac{P(i)}{P(0)} = \frac{1 \times \Omega_R(U_0 - \varepsilon_i)}{1 \times \Omega_R(U_0)}. \tag{24.83}$$

Let us take the logarithm of both sides and manipulate as follows:

$$\ln\left[\frac{P(i)}{P(0)}\right] = \ln\left[\Omega_R(U_0 - \varepsilon_i)\right] - \ln\left[\Omega_R(U_0)\right] \tag{24.84}$$

$$= -\left.\frac{\partial \ln \Omega_R(U)}{\partial U}\right|_{U_0} \varepsilon_i + \frac{1}{2}\left.\frac{\partial^2 \ln \Omega_R(U)}{\partial U^2}\right|_{U_0} (\varepsilon_i)^2 + \dots \tag{24.85}$$

$$= -\frac{\varepsilon_i}{kT} + \dots \tag{24.86}$$

In going from Eqn. 24.85 to Eqn. 24.86, I have recalled the definition of the temperature T of the reservoir (Eqn. 24.82),

$$\frac{\partial \ln \Omega_R(U)}{\partial U} = \frac{1}{kT}, \tag{24.87}$$

and dropped all but this first derivative in the Taylor series. The second and higher derivatives are successive derivatives of $1/kT$, and they vanish since the T of a reservoir, by definition, remains fixed no matter what energy the system has. (It is like saying that when you stick a thermometer in your mouth to measure the body temperature, the drop in body temperature is negligible.) Remember, we do not require the system itself to be small, only that the reservoir be overwhelmingly larger.

Taking antilogarithms of both sides of Eqn. 24.86, we find the ratio of probabilities to be

$$\frac{P(i)}{P(0)} = e^{-\varepsilon_i/kT}. \tag{24.88}$$

From the ratio of probabilities we may construct an *absolute probability* $P(i)$ (which gives 1 when summed over all values of i):

$$P(i) = \frac{e^{-\varepsilon_i/kT}}{Z} \quad \text{where} \tag{24.89}$$

$$Z = \sum_i e^{-\varepsilon_i/kT}. \tag{24.90}$$

We call $Z = Z(T)$ the *partition function* and $e^{-\varepsilon_i/kT}$ the *Boltzmann weight*.

Many interesting quantities can be deduced from $Z(T)$. For example, the average energy is

$$\bar{U} = \sum_i \varepsilon_i P(i) = \frac{\sum_i \varepsilon_i e^{-\varepsilon_i/kT}}{Z} \tag{24.91}$$

$$= \frac{kT^2 \frac{\partial}{\partial T}\left[\sum_i e^{-\varepsilon_i/kT}\right]}{Z} = kT^2 \cdot \frac{1}{Z} \cdot \frac{\partial Z}{\partial T} \tag{24.92}$$

$$= kT^2 \frac{\partial \ln Z(T)}{\partial T}. \tag{24.93}$$

If we know $Z(T)$ in closed form, we can extract the average energy by differentiation. It is a lot easier to compute $Z(T)$ than $\Omega(U)$, because the sum over i is unrestricted in energy.

As an illustration, consider an ideal gas in a box of volume V, in contact with a reservoir of temperature T. Choose as the system just one atom, and consider the rest of the gas and the reservoir as part of a new reservoir. (The Gibbs approach applies to systems of any size, as you may verify by going over the derivation above.) Its partition function is

$$Z_1(T) = \frac{1}{a^3} \int_{box} dx\,dy\,dz \cdot \frac{1}{b^3} \int_{-\infty}^{\infty} \int_{-\infty}^{\infty} \int_{-\infty}^{\infty} dp_x\,dp_y\,dp_z$$

$$\exp\left[-\frac{p_x^2 + p_y^2 + p_z^2}{2mkT} \right] \tag{24.94}$$

$$= VT^{3/2}f(m,k,a,b) \tag{24.95}$$

where $f(m,k,a,b)$ is a function that will not affect the mean energy since it is T-independent. In performing the p-integrals, I introduced rescaled variables $w_x = p_x/\sqrt{T}$ et cetera, and wrote each p-integral as $T^{1/2}$ times an integral over the corresponding w that did not depend on T. Now we see that for one atom,

$$\bar{U}_1 = kT^2 \frac{\partial \ln Z_1(T)}{\partial T} \tag{24.96}$$

$$= kT^2 \frac{\partial \left[\ln T^{3/2} + \ln V + \ln f(m,k,a,b) \right]}{\partial T} \tag{24.97}$$

$$= \frac{3}{2}kT. \tag{24.98}$$

For an ideal gas with N non-interacting atoms,

$$\bar{U}_N = \frac{3}{2}NkT. \tag{24.99}$$

In Gibbs's approach, T is fixed, U_N can fluctuate, and $\bar{U}_N$ is its average. It can be shown that as $N \to \infty$, deviations from $\bar{U}_N$ are negligible compared to $\bar{U}_N$. It is this average, with negligible (relative) fluctuations, that plays the role of the internal energy U in thermodynamics, for which statistical mechanics provides a microscopic foundation.

The *maximum entropy* principle applied to the system and the reservoir may be expressed in terms of the system alone as follows: When internal constraints are removed in a system in equilibrium with a reservoir at some T, the system evolves to *minimize its free energy* $F(T)$ defined as follows:

$$F(T) = -kT \ln Z. \tag{24.100}$$

I will not prove this here, and I hope that you are now motivated to pursue the subject on your own.

24.7 The third law of thermodynamics

We have no time to delve into the third law of thermodynamics, which declares that the entropy of all systems approaches 0 as $T \to 0$. You will realize that quantum mechanical considerations must have gone into this assertion, because entropy in classical statistical mechanics is defined only up to a constant. In quantum theory, the allowed states of a system are discrete and countable. As $T \to 0$, we see from Eqn. 24.88 that any state with energy higher than the ground state (chosen to have $\varepsilon_0 = 0$) has vanishing probability. The system has only one accessible quantum state, which ensures that $S = 0$. This law is definitely true for all systems of finite extent. Occasionally, infinite systems could have multiple ground states. Physical considerations show that if the system is found in one of these ground states, it will never evolve into any of the others.

Index